Tropical Forest Ecology and Management for the Anthropocene

Tropical Forest Ecology and Management for the Anthropocene

Special Issue Editors

Grizelle González
Ariel E. Lugo

MDPI • Basel • Beijing • Wuhan • Barcelona • Belgrade

MDPI

Special Issue Editors

Grizelle González
International Institute of Tropical Forestry, United States Department of Agriculture Forest Service
USA

Ariel E. Lugo
International Institute of Tropical Forestry, United States Department of Agriculture Forest Service
USA

Editorial Office
MDPI
St. Alban-Anlage 66
4052 Basel, Switzerland

This is a reprint of articles from the Special Issue published online in the open access journal *Forests* (ISSN 1999-4907) from 2017 to 2019 (available at: https://www.mdpi.com/journal/forests/special_issues/tropical_anthropocene).

For citation purposes, cite each article independently as indicated on the article page online and as indicated below:

LastName, A.A.; LastName, B.B.; LastName, C.C. Article Title. *Journal Name* **Year**, *Article Number*, Page Range.

ISBN 978-3-03921-964-3 (Pbk)
ISBN 978-3-03921-965-0 (PDF)

Cover image courtesy of Marinelis Talavera.

Contents

About the Special Issue Editors

Grizelle González, Ph.D., is Project Leader of the Research and Development Unit, and Assistant Director of Research and Development of the USDA Forest Service International Institute of Tropical Forestry, at Río Piedras, Puerto Rico. She is an adjunct faculty member of the Departments of Biology and Environmental Sciences at the University of Puerto Rico's Río Piedras Campus. González serves as co-principal investigator of the Luquillo Long-Term Ecological Research Program and the Critical Zone Observatory, and is a member of the editorial boards of the Caribbean Journal of Science, Caribbean Naturalist, Frontiers in Forests and Global Change - Forest Disturbance section, and MDPI's Forests. She has authored over 120 scientific publications and served as the editor of special synthesis volumes on diverse topics such as Earthworms of Latin America (2006), Ecological Gradient Analyses in a Tropical Landscape (2013), Tropical Forest Responses to a Simulated Hurricane Experiment (2014), Tropical Forest Responses to Large-Scale Experiments (2015), and Network Collaborations on Forest Dynamics and Regional Forestry Initiatives of the Caribbean Foresters (2016). González's broad research objectives include integrating new knowledge with concepts of ecosystem function and best management practices to ensure delivery of ecosystem services from tropical forests, and to explore art and science collaborations as a tool for science delivery to multiple audiences. Having experienced firsthand the impacts, aftermath and recovery dynamics of Hurricanes Hugo (1989), and Irma and María (2017) on the social–ecological systems in Puerto Rico, González believes science is central to the sustainable development and resiliency of tropical America.

Ariel E. Lugo, is the Director of the USDA Forest Service International Institute of Tropical Forestry, at Río Piedras, Puerto Rico. Lugo is a native Puerto Rican educated in public schools, the University of Puerto Rico (BS and MS), and the University of North Carolina at Chapel Hill (Ph.D.). Before his Forest Service career, Lugo was a professor of Botany and Plant Ecology at the University of Florida in Gainesville and a staff member of the Council on Environmental Quality in Washington, DC. For a time he also held the position of Assistant Secretary at the Puerto Rico Department of Natural Resources. Lugo is an ecologist whose primary focus is on the functioning of tropical forests including tree plantations, forested wetlands, and novel forests in dry, moist, and wet environments. In collaboration with a diverse group of colleagues he has edited books and special journal issues on forested wetlands of the world, tropical forests in the Caribbean, and social–ecological studies of tropical cities. During his tenure with the Forest Service Lugo has focused his attention on developing a conservation ethic for Puerto Rico and other tropical countries, providing education and research opportunities to under-represented communities and individuals, and seeking resilient solutions to the problems associated with the Anthropocene Epoch.

Preface to "Tropical Forest Ecology and Management for the Anthropocene"

This special issue takes both a historical and a forward outlook to the forest conservation challenges of the Caribbean, based on 75 years of research and applications by the United States Department of Agriculture, International Institute of Tropical Forestry (Institute) in Puerto Rico. It transforms Holocene-based scientific paradigms of the tropics into Anthropocene applications and outlooks of wilderness, managed forests, and urban environments. The content of this special issue furthers on the knowledge contained in the volume that commemorated the 50th anniversary of the Institute (Lugo and Lowe 1995) and together help transition tropical forest conservation from the Holocene to the Anthropocene epoch. Coincidentally, but yet instructive, both volumes were delayed in their production by extreme social-ecological events: hurricane Hugo and Desert Storm in Saudi Arabia for the 50th anniversary, and hurricanes Irma and María and government shutdowns for the 75th anniversary. Extreme social-ecological events and their effects are primary components of the uncertain environment of the Anthropocene and dealing with this uncertainty is a challenge that scientists, conservationists, land managers, and citizens cannot avoid. This volume showcases how Institute programs are evolving in their focus and content to support sustainable tropical forest conservation under conditions of uncertainty.

The manuscripts contained here highlight the importance of shared stewardship and an all-hands approach to conservation, long-term focus of research programs, and novelty in organizations to meet contemporary conservation challenges. Policies relevant to the Anthropocene as well as the use of experiments to learn about the future responses of tropical forests to global warming are reexamined in the following pages. Urban topics include how cities can co-produce new knowledge to spark sustainability and resilience transformations. Long-term results and applications of research in topics such as soil biota, migratory birds, tropical vegetation, substrate chemistry, and the tropical carbon cycle are also described in the volume. Moreover, the question of how to best use land while striking the right balance that provides for multiple uses in a tropical island is addressed.

This volume should be of interest to all actors involved in long-term sustainable forest management and research in light of the historical lessons and future directions that further our understanding of tropical cities and forests in the Anthropocene epoch. It should be of particular interest to:

- Undergraduate and graduate students of biology, ecology, and environmental sciences

- Tropical scientists, managers and enthusiasts

- General ecologists and conservation biologists

- Social scientists, students, and practitioners of urban systems

- Managers of natural protected areas

- Scientists engaged in international cooperation

- Individuals and organizations interested in social-ecological-technological systems

The advantages of this book over other publications in the field include: 1) the comprehensive, broad scope of topics, 2) the long-term nature of the studies and perspectives, and, 3) the integration of historical and new knowledge in the study of tropical forestry and social sciences in the Neotropics.

We thank the 27 authors to this special volume for contributing their insights to this collection of manuscripts; as well as to the many anonymous reviewers and editors at MDPI for providing timely, and thorough peer-review reports. Special thanks to the Institute's field and laboratory technicians, and supporting staff whom over 75 years have worked tirelessly at gathering and building long-term data sets to the highest quality and ethical research standards. Support for much of the research presented in this volume was provided by the USDA Forest Service, the University of Puerto Rico (UPR), and grants (DEB-0620910, DEB-0218039, DEB-0080538, DEB-9705814, DEB-1239764, DEB-1831952) from the National Science Foundation to the Institute of Tropical Ecosystem Studies, UPR, and the Institute as part of the Long-Term Ecological Research Program in the Luquillo Experimental Forest. González was additionally supported by the Luquillo Critical Zone Observatory (National Science Foundation grant EAR-1331841).

Literature Cited

Lugo, A.E.; Lowe, C. (Eds.) *Tropical Forests: Management and Ecology*; Springer: New York, NY, USA, 1995; 461p.

Grizelle González, Ariel E. Lugo
Special Issue Editors

forests

MDPI

Editorial

Sandra Brown (1944–2017): A Distinguished Tropical Ecologist

Ariel E. Lugo * and Grizelle González

United States Department of Agriculture, Forest Service, International Institute of Tropical Forestry, Jardín Botánico Sur, 1201 Ceiba St.-Río Piedras 00926, Puerto Rico; ggonzalez@fs.fed.us
* Correspondence: alugo@fs.fed.us; Tel.: +1-787-764-7800; Fax: +1-787-766-6302

Received: 20 June 2017; Accepted: 2 July 2017; Published: 8 July 2017

We dedicate this Special Issue commemorating the 75th Anniversary of the US Department of Agriculture, Forest Service—International Institute of Tropical Forestry to the late Dr. Sandra Brown.

Sandra Brown was a superb analytical scientist. Her notable contributions to the understanding of the global carbon cycle include the synthesis of ecological data from the tropics and the realization that those data were biased towards high biomass values. Her analysis of Food and Agriculture Organization (FAO) inventory data for the tropics led to a new biomass estimate for tropical forest that was considerably smaller than those used in global models that could not balance the carbon cycle. Dr. Brown also developed methods for estimating tropical forest biomass from inventory data, methods that were published by the FAO and are still used internationally by many researchers and government analysts. As a professor at the University of Illinois and later as Chief Scientist at Winrock International, she led and collaborated with staff to improve landscape-level visualization of carbon density data for the tropics, development of intensive inventory methods for more accurate estimates of carbon density, and use of remote sensing techniques to expand and plot carbon data to larger scales. Dr. Brown was also a wetlands scientist and contributed to the understanding of freshwater-forested wetlands functioning.

In 2014, the Institute bestowed on Dr. Brown its Conservation Award (Figure 1) in recognition of her scientific collaboration with Institute scientists and the results of such collaborations, which she summarized in her plenary talk entitled Trailblazing the Carbon Cycle of Tropical Forests from Puerto Rico, and as published in the article with the same title in this Special Issue entitled Tropical Forest Ecology and Management for the Anthropocene. The written article was her last publication after a distinguished career as a tropical ecologist. While we are saddened by her passing, we feel immensely fortunate to have had the opportunity to collaborate with her, a distinguished tropical ecologist, who shared her talents and time freely with us and with those who sought her advice.

Figure 1. Dr. Sandra Brown (center) at the 75th Anniversary of the USDA Forest Service International Institute of Tropical Forestry. She received the Conservation Award from Dr. Jim Reeves, Deputy Chief for Research and Development (left), and Mary Wagner, Associate Chief of the USDA Forest Service (right). Also in the picture are Grizelle González (Project Leader) and Ariel E. Lugo (Director), International Institute of Tropical Forestry.

Conflicts of Interest: The authors declare no conflict of interest.

![forests logo] *forests*

MDPI

Editorial

Introduction to the Special Issue on Tropical Forests: Management and Ecology in the Anthropocene

Ariel E. Lugo * and Grizelle González

United States Department of Agriculture, Forest Service, International Institute of Tropical Forestry,
Jardín Botánico Sur, 1201 Ceiba St.-Río Piedras, Puerto Rico 00926, USA; ggonzalez@fs.fed.us
* Correspondence: alugo@fs.fed.us; Tel.: +1-787-764-7743; Fax: +1-787-766-6302

Received: 17 December 2018; Accepted: 19 December 2018; Published: 10 January 2019

Abstract: This Special Issue of *Forests* is based on papers presented at the 75th anniversary of the United States Department of Agriculture (USDA) Forest Service International Institute of Tropical Forestry as well as other papers relevant to the topic of the Special Issue. The Institute is but one leg of a conservation relay among cultures and institutions that began in Puerto Rico a millennium ago. The Institute began operations in 1939 and celebrated its 75th anniversary on May, 2014. Over its 75 years of operation, the Institute has focused its research on tropical forests, with the scope of the research expanding over the years. An analysis of the lines of research of the Institute showed that over its history about 69 lines of research have been established and that of the original 17 lines of research between 1939 and 1949, all but one remained active in 2014. This history and continuity of the research program has allowed the Institute to observe ecological phenomena over decades, including the evolving forest structure and functioning on degraded land restoration experiments that began before the formal establishment of the Institute and are now over 80 years old.

Keywords: Tropical Forestry Research; Long-Term Ecological Research; Tropical Forest Management; Tropical Forest Conservation

---

1. Tropical Forestry Research in the Anthropocene

In spite of the continuity of research focus at the United States Department of Agriculture (USDA) Forest Service International Institute of Tropical Forestry [1,2], there have been historical moments when a particular research emphasis or paradigm shift has taken place. For example, the volume celebrating the Institute's 50th anniversary [3] summarized silvicultural and ecological research and their relevance to tropical forests in general, with little attention paid to the importance of disturbances to tropical forest functioning and species composition. The passage in September 1989 of hurricane Hugo over the Luquillo Mountains, where most of the research was focused, caused a paradigm shift not reflected at the time of the 50th anniversary in May 1989. That paradigm shift led to the book 'A Caribbean Forest Tapestry: The multidimensional nature of disturbance and response' by Brokaw et al. [4]. In that book, ideas of tropical forest resilience emerged and added new insights into forest conservation in the face of extreme disturbance events.

The recognition by geologists of the onset of the Anthropocene epoch [5] again changed the emphasis of the research program at the Institute and this shift in emphasis is reflected in this Special Issue. The Anthropocene presents new challenges to forest conservation that research programs must address [6,7]. Amongst the challenges, the most perplexing is the uncertainty of conditions faced by both ecosystems and those who study and conserve them, and the response of forests through species composition changes and novelty [8]. Thus, the title of this Special Issue, *Tropical Forest Ecology and Management in the Anthropocene* on the one hand reflects the continuity of the Institute's research focus on forest ecology and management, while on the other hand it recognizes its application and innovations in relation the challenges posed by the new epoch of the Anthropocene. We asked our contributing scientists and collaborators to review progress on their lines of research in light of the conditions of the Anthropocene.

The resulting contributions to this Special Issue illustrate some of the principal elements of an adaptive research and development program for the conservation of tropical forests in the Anthropocene, which includes the sustainable management of forests. These elements include:

- A long-term focus, required to develop perspective and insight into time-dependent ecological processes.
- Attention to all lands and all species because all have a role to play (social and/or ecological) in an uncertain and changing world.
- Science at many scales because the functioning of ecosystems involves hierarchical processes operating from molecular to global scales.
- Monitoring of changes in biodiversity as essential for adaptive conservation and for maintaining a pulse on the response of biotic systems to changing environmental conditions.
- Experimentation as a way of seeking causality and improving understanding of social and ecological phenomena.
- Understanding novelty in ecosystems, to verify its adaptive role in light of environmental uncertainty.
- Attention to climate and environmental change, which are drivers of biodiversity changes and novelty.
- Studying urban systems because most of the human population increasingly depends on these environments for their habitation and quality of life.
- A social-ecological focus because the production and application of human knowledge in the Anthropocene transcends disciplines and interdisciplinary action. Addressing the wicked problems of the Anthropocene requires transdisciplinary approaches, which incorporate multiple ways of knowing when addressing problems.
- Fomenting collaboration among many social sectors to optimize the use of available resources in support of human activities and their adaptation to future climate and environmental change.
- Developing novel policies for effective governance because many of the policies of the Holocene are outdated and ineffective under Anthropocene conditions.
- Improving institutions and their knowledge systems to make them learning and adaptive organizations sufficiently nimble to be capable of adjusting and transforming in light of changing social and ecological environments.

In Table 1 we relate these elements to Special Issue contributions. We see these elements as evolving notions of the lines of research that help us deal with the uncertainty of the Anthropocene. We see the Institute as a learning and evolving research and development organization that strives to develop knowledge that helps forests and people adapt and transform in the Anthropocene. We look forward to our 100th anniversary when the program will likely look as different from this one as this one itself is different from what we were doing during the 50th anniversary.

Table 1. List of research and conservation elements or activities that contribute to an effective forest research and development program relevant to addressing the uncertainties of the Anthropocene epoch, and example manuscripts in this Special Issue or in recent Special Issues produced by Institute scientists.

Element of Research or Action	Contributed Manuscripts*
A long-term focus	Brown and Lugo [9], González and Lodge [10], Heartsill-Scalley [11]
Attention to all lands and all species	Gould et al. [12], Jacobs [13]
Science at many scales	Fonseca da Silva et al. [14], Medina et al. [15]
Experimentation as a way of seeking causality	Wood et al. [16], Shiels and González [17,18], Shiels et al. [19], Kimball et al. [20]
Attention to climate and environmental change	Henareh et al. [21], Gould et al. [22], Feng at al. [23], Jennings et al. [24], Van Beusekom et al. [25,26]
Monitoring of changes in biodiversity	Campos-Cerqueira et al. [27], Wunderle and Arendt [28], González et al. [29], Heartsill-Scalley and González [30]
Understanding novelty	Lugo and Erickson [31]
Attention to urban environments and their functioning	Muñoz-Erickson et al. [32]
A social-ecological-technological focus	Lugo and Alayón [33], Lugo [34]
Fomenting collaboration among many sectors of society	González and Heartsill-Scalley [35]
Development of novel policies for effective governance	McGinley [36], Rudel [37]
Institutional improvement	M. Rains [38]

* Manuscripts are part of this Special Issue or are recent products of the Institute's program.

2. Conclusions

Long-term ecological research is required to support tropical forest conservation, including active management. Such research needs to be trans-disciplinary with a focus on the social, ecological, and technological aspects of forest conservation.

Acknowledgments: This study was conducted in collaboration with the University of Puerto Rico. The Luquillo Critical Zone Observatory (EAR-1331841) and Grant DEB 1239764 provided additional support for G. González from the U.S. National Science Foundation to the Institute for Tropical Ecosystem Studies, University of Puerto Rico, and to the International Institute of Tropical Forestry USDA Forest Service, as part of the Luquillo Long-Term Ecological Research Program. The research is part of the Institute's contribution to the San Juan ULTRA program through the International Urban Field Station. We thank Tischa Muñoz-Erickson and Tamara Heartsill-Scalley for their review of the manuscript.

Conflicts of Interest: The authors declare no conflict of interest.

References

1. Robinson, K.; Bauer, J.; Lugo, A.E. *Passing the Baton from the Tainos to Tomorrow: Forest Conservation in Puerto Rico*; FS-862; U.S. Department of Agriculture Forest Service, International Institute of Tropical Forestry: San Juan, Puerto Rico, 2014.
2. Lugo, A.E.; Scatena, F.N.; Waide, R.B.; Greathouse, E.A.; Pringle, C.M.; Willig, M.R.; Vogt, K.A.; Walker, L.R.; Gonzalez, G.; McDowell, W.H.; et al. Management implications and applications of long-term ecological research. In *A Caribbean Forest Tapestry: The Multidimensional Nature of Disturbance and Response*; Brokaw, N., Crowl, T.A., Lugo, A.E., McDowell, W.H., Scatena, F.N., Waide, R.B., Willig, M.R., Eds.; Oxford University Press: New York, NY, USA, 2012.
3. Lugo, A.E.; Lowe, C. *Tropical Forests: Management and Ecology*; Springer-Verlag: New York, NY, USA, 1995.

4. Brokaw, N.; Crowl, T.A.; Lugo, A.E.; McDowell, W.H.; Scatena, F.N.; Waide, R.B.; Willig, M.R. *A Caribbean Forest Tapestry: The Multidimensional Nature of Disturbance and Response*; Oxford University: New York, NY, USA, 2012.

5. Waters, C.N.; Zalasiewicz, J.; Summerhayes, C.; Barnosky, A.D.; Poirier, C.; Gałuszka, A.; Cearreta, A.; Edgeworth, M.; Ellis, E.C.; Ellis, M.; et al. The Anthropocene is functionally and stratigraphically distinct from the Holocene. *Science* **2016**, *351*, aad2622. [CrossRef] [PubMed]

6. Lugo, A.E. Evolving conservation paradigms for the Anthropocene. In *Forest Conservation and Management in the Anthropocene: Adaptations of Science Policy and Practices*; USDA Forest Service: Fort Collins, CO, USA, 2014; Volume RMRS-P-71, pp. 47–59.

7. Lugo, A.E. Forestry in the anthropocene. *Science* **2015**, *349*, 771. [CrossRef] [PubMed]

8. Lugo, A.E. Novel tropical forests: Nature's response to global change. *Trop. Conserv. Sci.* **2013**, *6*, 325–337. [CrossRef]

9. Brown, S.; Lugo, A.E. Trailblazing the carbon cycle of tropical forests from Puerto Rico. *Forests* **2017**, *8*, 101. [CrossRef]

10. González, G.; Lodge, D.J. Soil biology research across latitude, elevation and disturbance gradients: A review of forest studies from Puerto Rico during the past 25 years. *Forests* **2017**, *8*, 178. [CrossRef]

11. Heartsill-Scalley, T. Insights on forest structure and composition from long-term research in the Luquillo Mountains. *Forests* **2017**, *8*, 204. [CrossRef]

12. Gould, W.A.; Wadsworth, F.H.; Quiñones, M.; Fain, S.J.; Álvarez, N.L. Land use, conservation, forestry, and agriculture in Puerto Rico. *Forests* **2017**, *8*, 242. [CrossRef]

13. Jacobs, K.R. Teams at their core: Implementing an "All LANDS approach to conservation" requires focusing on relationships, teamwork process, and communications. *Forests* **2017**, *8*, 246. [CrossRef]

14. Fonseca da Silva, J.; Medina, E.; Lugo, A.E. Traits and resource use of co-occuring introduced and native trees in a tropical novel forest. *Forests* **2017**, 8. [CrossRef]

15. Medina, E.; Cuevas, E.; Lugo, A.E. Substrate chemistry and rainfall regime regulate elemental composition of tree leaves in karst forests. *Forests* **2017**, *8*, 182. [CrossRef]

16. Wood, T.E.; González, G.; Silver, W.L.; Reed, S.C.; Cavaleri, M.A. On the shoulders of giants: Continuing a legacy of large-scale ecosystem manipulation experiments in Puerto Rico. *Forests*. In press, 8.

17. Shiels, A.B.; González, G. Tropical forest responses to large-scale experimental hurricane effects. *For. Ecol. Manag.* **2014**, *332*, 1–136. [CrossRef]

18. Shiels, A.B.; González, G. Tropical forest responses to large-scale experiments. *BioScience* **2015**, *65*, 839–840. [CrossRef]

19. Shiels, A.B.; González, G.; Willig, M.R. Responses to canopy loss and debris deposition in a tropical forest ecosystem: Synthesis from an experimental manipulation simulating effects of hurricane disturbance. *For. Ecol. Manag.* **2014**, *332*, 124–133. [CrossRef]

20. Kimball, B.A.; Alonso-Rodríguez, A.M.; Cavaleri, M.A.; Reed, S.C.; González, G.; Wood, T.E. Infrared heater system for warming tropical forest understory plants and soils. *Ecol. Evol.* **2018**, *8*, 1932–1944. [CrossRef] [PubMed]

21. Henareh Khalyani, A.; Gould, W.A.; Harmsen, E.; Terando, A.; Quiñones, M.; Collazo, J.A. Climate change implications for tropical islands: Interpolating and interpreting statistically downscaled GCM projections for management and planning. *J. Appl. Meteorol. Climatol.* **2016**, *55*, 265–282. [CrossRef]

22. Gould, W.A.; Díaz, E.L.; Álvarez-Berríos, N.L.; Aponte-González, F.; Archibald, W.; Bowden, J.H.; Carrubba, L.; Crespo, W.; Fain, S.J.; González, G. et al. U.S. Caribbean. In *Impacts, Risks, and Adaptation in the United States: Fourth National Climate Assessment*; Reidmiller, D.R., Avery, C.W., Easterling, D.R., Kunkel, K.E., Lewis, K.L.M., Maycock, T.K., Stewart, B.C., Eds.; U.S. Global Change Research Program: Washington, DC, USA, 2018; Volume II, pp. 809–871.

23. Feng, X.; Uriarte, M.; González, G.; Reed, S.; Thompson, J.; Zimmerman, J.K.; Murphy, L. Improving predictions of tropical forest response to climate change through integration of field studies and ecosystem modeling. *Glob. Chang. Biol.* **2018**, *24*, e213–e232. [CrossRef]

24. Jennings, L.N.; Douglas, J.; Treasure, E.; Gonzalez, G. *Climate Change Effects in El Yunque National Forest, Puerto Rico, and the Caribbean Region*; U.S. Department of Agriculture Forest Service: Asheville, NC, USA, 2014.

25. Van Beusekom, A.E.; González, G.; Rivera, M.M. Short-term precipitation and temperature trends along an elevation gradient in northeastern Puerto Rico. *Earth Interact.* **2014**, *19*, 1–33. [CrossRef]

26. Van Beusekom, A.E.; González, G.; Scholl, M.A. Analyzing cloud base at local and regional scales to understand tropical montane cloud forest vulnerability to climate change. *Atmos. Chem. Phys.* **2017**, *17*, 7245–7259. [CrossRef]

27. Campos-Cerqueira, M.; Arendt, W.J.; Wunderle, J.M.; Aide, T.M. Have bird distributions shifted along an elevational gradient on a tropical mountain? *Ecol. Evol.* **2017**, *7*, 9914–9924. [CrossRef]

28. Wunderle, J.M., Jr.; Arendt, W.J. The plight of migrant birds wintering in the Caribbean: rainfall effects in the annual cycle. *Forests* **2017**, *8*, 115. [CrossRef]

29. González, G.; Willig, M.R.; Waide, R.B. *Ecological Gradient Anlyses in a Tropical Landscape*; John Wiley & Sons: Oxford, UK, 2013.

30. Heartsill-Scalley, T.; González, G. Introduction: Caribbean forest dynamics and community and regional forestry initiatives. *Caribb. Nat.* **2016**, *1*, 1–12.

31. Lugo, A.E.; Erickson, H.E. Novelty and its ecological implications to dry forest functioning and conservation. *Forests* **2017**, *8*, 161. [CrossRef]

32. Muñoz-Erickson, T.; Miller, C.A.; Miller, T.R. How cities think: knowledge co-production for urban sustainability and resilience. *Forests* **2017**, *8*, 203. [CrossRef]

33. Lugo, A.E.; Alayón, M. Understanding the vulnerability and sustainability of urban social-ecological systems in the tropics: Perspectives from the city of San Juan. *Ecol. Soc.* **2014**, *19*, 2.

34. Lugo, A.E. *Social-Ecological-Technological Effects of Hurricane María on Puerto Rico: Planning for Resilience under Extreme Events*; Springer International Publishing. Switzerland AG: Basel, Switzerland, 2019; 112 p. [CrossRef]

35. González, G.; Heartsill-Scalley, T. Building a collaborative network to understand regional forest dynamics and for the advancement of forestry initiatives in the caribbean. *Caribb. Nat.* **2016**, *1*, 245–256.

36. McGinley, K.A. Adapting tropical forest policy and practice in the context of the Anthropocene: Opportunities and challenges for the El Yunque National Forest in Puerto Rico. *Forests* **2017**, *8*, 259. [CrossRef]

37. Rudel, T.K. The dynamics of deforestation in the wet and dry tropics: A comparison with policy implications. *Forests* **2017**, *8*, 108. [CrossRef]

38. Rains, M.T. A Forest Service vision during the Anthropocene. *Forests* **2017**, *8*, 94. [CrossRef]

forests

MDPI

Article

Adapting Tropical Forest Policy and Practice in the Context of the Anthropocene: Opportunities and Challenges for the El Yunque National Forest in Puerto Rico

Kathleen A. McGinley

USDA Forest Service, International Institute of Tropical Forestry, Jardín Botánico Sur, 1201 Calle Ceiba, Río Piedras, PR 00926, USA; kmcginley@fs.fed.us

Received: 7 June 2017; Accepted: 15 July 2017; Published: 20 July 2017

Abstract: Tropical forest management increasingly is challenged by multiple, complex, intersecting, and in many cases unprecedented changes in the environment that are triggered by human activity. Many of these changes are associated with the Anthropocene—a new geologic epoch in which humans have become a dominating factor in shaping the biosphere. Ultimately, as human activity increasingly influences systems and processes at multiple scales, we are likely to see more extraordinary and surprising events, making it difficult to predict the future with the level of precision and accuracy needed for broad-scale management prescriptions. In this context of increasing surprise and uncertainty, learning, flexibility, and adaptiveness are essential to securing ecosystem resilience and sustainability, particularly in complex systems such as tropical forests. This article examines the experience to date with and potential for collaborative, adaptive land and resource management in the El Yunque National Forest (EYNF)—the only tropical forest in the U.S. National Forest System. The trajectory of EYNF policy and practice over time and its capacity for learning, flexibility, and adaptiveness to change and surprise are analyzed through an historical institutionalism approach. EYNF policies and practices have shifted from an early custodial approach that focused mostly on protection and prevention to a top-down, technical approach that eventually gave way to an ecosystem approach that has slowly incorporated more flexible, adaptive, and active learning elements. These shifts in EYNF management mostly have been reactive and incremental, with some rarer, rapid changes primarily in response to significant changes in national-level policies, but also to local level conditions and changes in them. Looking to the future, it seems the EYNF may be better positioned than ever before to address increasing uncertainty and surprise at multiple scales. However, it must be able to count on the resources necessary for implementing adaptive, collaborative forest management in a tropical setting and on the institutional and organizational space and flexibility to make swift adjustments or course corrections in response to system changes and surprises.

Keywords: adaptive management; tropical forest; Anthropocene; U.S. Forest Service Planning Rule; El Yunque National Forest; Luquillo Experimental Forest

---

1. Introduction

Humans use and value tropical forests for a range of objectives, from the preservation of biodiversity to the production of wood products, but these objectives increasingly are challenged by multiple, complex, intersecting, and in many cases unprecedented changes in the environment that are triggered by human activity. Because land use change, habitat fragmentation, pollution, and other anthropogenic processes have spurred new environmental conditions and novel habitats, many in the scientific community have come to agree that we now are living in the Anthropocene—a new geologic epoch in which humans are a dominating factor in shaping the biosphere [1–3]. As human activity further affects the environment in continued, new, and unprecedented ways, changes in social-ecological systems are expected to become less predictably cyclical, while system responses become less certain [4]. Change is nothing new in nature, which has always produced surprises. Likewise, societies have and will continue to change in terms of their interests in, needs from, and demands on tropical forests and other natural resources. Nevertheless, unpredictability and surprise can present significant challenges to those who depend on forests for their livelihood and for those who manage them for multiple purposes, particularly if management aims to maintain current conditions or restore them to some ideal from the past.

People have managed tropical forests for millennia, intentionally manipulating them for desired composition, goods, and services, and in some cases successfully adapting to shifts and even surprises in them. For example, the Amazon Basin, once thought to have been dominated by 'virgin' forest prior to European arrival, was inhabited by sizable, 'sedentary' societies that cleared areas of interior forest for agriculture and managed other forested areas for the optimal distribution of useful species [5]. In Central America, the Mayan civilization developed a complex system of land use that cycled from closed canopy forest, to open field cropping (*milpa*), to tree gardens, and, eventually back to diverse, hardwood forests shaped in part by purposeful plantings and species selection [6]. The presence of many economically important plant species (e.g., mahogany (*Swietenia macrophylla*), Spanish cedar (*Cedrela odorata*), chicle (*Manilkara* spp.)) in the Maya forest today reflect their careful selection and management by the Maya of the past [7]. Some of these and other traditional forest practices have demonstrated a holistic understanding of complex forest processes and an ability to adapt to environmental change for exceptions [8,9], as do many contemporary forest management approaches. However, the changes that are likely to occur in tropical forests in the context of the Anthropocene will test further our capacity to cope with uncertainty and surprise, particularly in already complex systems.

Acknowledging that the tropical forests of the future may differ in terms of composition, structure, and even function from those of the past and present, pushes us to consider how policies and practices shape human–environment interactions and to determine what adjustments, if any, are necessary to resist, respond, or adapt to system shifts and surprises. Some control may be exerted over the projected and unexpected changes associated with the Anthropocene through existing land and resource management strategies, but these are bound to need reconfiguring in the least, to better address sustainability and other societal goals, particularly as these also may shift over time. So far, there is no magic bullet or exact science for managing forests in the context of the increasing environmental variability associated with modern times. Management strategies range from passive to reactive to anticipatory and may be combined in a tool-box approach that depends as much on probabilities and predictions as societal interests and demands [10,11]. While these may differ, for example by system, scale, or objectives; flexibility, adaptiveness, and active, ongoing learning emerge as common attributes among many strategies designed to address uncertainty and surprise [12–15].

This paper examines the practice of and prospects for flexibility, adaptiveness, and learning in the El Yunque National Forest (EYNF) in Puerto Rico to understand better its capacity to address the uncertainty and surprise expected to increase in the context of the Anthropocene. Tropical forests, like the EYNF, are on the front lines of global change, forcing scientists and practitioners to grapple with critical questions, such as the degree to which these ecosystems can persist in human-modified landscapes, and which management strategies will be most effective at maintaining their structures and functions at different spatial and temporal scales. Evaluating the management policies and practices of the EYNF and their potential to effectively address the expected and unknown changes that are likely to occur in this new geologic epoch provides important feedback for the EYNF and its stakeholders, and for the U.S. Forest Service, the federal agency to which it pertains. Outcomes of this study also provide important policy and practical inputs for other tropical forest dwellers, managers, and decision-makers confronted with the imminent changes, uncertainty, and surprise associated with the Anthropocene.

2. Study Setting and Approach

The EYNF is located in northeastern Puerto Rico—a region that has seen significant changes in land use over the past 100 years, similar to the rest of the island. In 2010, the region was covered mostly by forest (43 percent), followed by pasture (36 percent), urban area (10 percent), shrubland (6 percent) and wetland (3 percent) [16]. The EYNF encompasses much of the forested area of northeastern Puerto Rico and protects one of the largest remnants of primary forest on the island. It extends across 11,735 ha and ranges in elevation from 120 to 1074 m above sea level [17]. It is the most biologically diverse forest in the USFS National Forest System, harboring more than 800 native species of plants and wildlife [18]. The EYNF also is highly valued for its recreation and water resources, receiving more than 600,000 visitors per year and producing about 20 percent of the island's total municipal water supply [18,19].

Average temperatures in the EYNF range from about 22° Celsius in the winter to about 30° Celsius in the summer [20]. Average annual rainfall in the forest is about 3000 mm, ranging from about 2500 mm at its lower elevations to more than 4500 mm at the peaks [21]. Records show decreasing annual rainfall and increasing temperatures in and around the EYNF over the past 65 years or so [22]. Statistical models of Puerto Rico's future climate vary, but generally predict increasing average annual temperatures (ranging from 4 to 9° Celsius) along with slightly decreasing total rainfall and increasing extreme weather events by the end of the 21st century [22–24].

The documented and predicted changes in climate in the EYNF are expected to affect its structure and function, possibly leading to shifts in species composition and their distribution along the elevational gradient, as well as leading to effects on water supplies and flows [25–28]. Changes in climate and weather patterns also may affect recreational activities in the forest and lead to changes in visitor use and visitation patterns [29]. Moreover, climate change and the associated effects are likely to intersect with projected and unexpected changes in demographics, economies, land use, and other issues, which may lead to compounding effects on the EYNF and surrounding area.

Land and resource management policies and practices of the EYNF were examined through an historical institutionalist approach, which aims to understand and explain a specific real-world policy process or outcome by studying the historical legacy of related institutional structures and feedbacks [30]. This approach can be used to study the creation, persistence, and change in institutions through time, focusing on pathways of institutional development, patterns of institutional path dependence, and critical junctures of institutional evolution [30–33]. This paper focuses on the extent to which learning, flexibility, and adaptiveness to changes in local and larger conditions are incorporated in EYNF policy and practice and the factors that influence stasis and change in policy and practice over time. Common aspects or characteristics of active, ongoing learning (e.g., monitoring, analysis, and feedback on management effects), flexibility (e.g., proclivity to stakeholder collaboration and coordination, responsiveness to new information), and adaptiveness (e.g., to changes and surprises in environmental and social conditions) were drawn from the literature and considered throughout the analysis [12–15]. Examining EYNF policies and practices, with a specific focus on learning, flexibility, and adaptiveness through an historical institutionalist lens sheds light on the ways and means through which this national forest has dealt with change, uncertainty, and surprise in the past. It also provides a frame of reference for assessing the EYNF's prospects for active learning, adaptiveness, and flexibility under increasing uncertainty and surprise anticipated in the context of the Anthropocene.

3. Shifting Approaches to Land and Resource Management in the EYNF

Over time, the land and resource management policies and practices of the EYNF have shifted from an initial custodial approach that focused mostly on protection and prevention to top-down, technical/scientific management that eventually gave way to an ecosystem management approach that has slowly incorporated more flexible, adaptive, and active learning elements. These shifts in EYNF management mostly have been reactive and incremental, with some rarer, rapid changes primarily in response to significant changes in national-level policy, but also to local level conditions and trends. The trajectory of EYNF land and resource policy and practice is presented in the following sections, which focus on the ways and means through which this forest has confronted system changes, uncertainty, and surprises.

3.1. Custodial Management

Forest reserves in northeastern Puerto Rico and other parts of the island predate political association with the U.S., having been established by the Spanish government in 1876, mostly to protect remaining timber supplies and water sources following decades of forest conversion to agriculture and intensive harvest of timber species that occurred primarily in the lowlands [18]. These forest reserves were ceded to the U.S. government after the Spanish–American War in 1898. Shortly thereafter, in 1903, President Theodore Roosevelt proclaimed the reserved lands in northeastern Puerto Rico as the Luquillo Forest Reserve (eventually renamed the EYNF) and placed them under the direction of the USFS [18].

As part of the National Forest System, early work in the EYNF was guided by the Forest Reserve Act of 1891 (P.L. 51–561), which allowed for lands in the public domain to be set aside as forest reserves (later, renamed as national forests). Also, the Forest Service Organic Administration Act of 1897 (16 U.S.C. § 473 et seq.) provided direction for national forest management, including provisions to secure water flows and permit timber harvests. In its early years, the USFS worked primarily in a custodial approach to resource management, focusing on the acquisition of national forests lands and the protection and later development of resources within their boundaries [34].

Land managers in Puerto Rico followed suit, embarking on early endeavors to survey, map, and mark the EYNF along its boundaries, divert water for downstream communities and towns, and install forest roads, trails, and other construction projects, but did not allow for timber extraction in its earliest years [35]. After Hurricane San Felipe passed over Puerto Rico in 1928—the first of five major hurricanes that have significantly affected the EYNF under USFS administration, land managers shifted part of their focus, permitting the extraction of downed trees and initiating periodic surveys of the forest to determine timber stocking and regeneration in a turn towards timber management [18]. Extensive reforestation in areas surrounding the EYNF that had been deforested prior to U.S. association began in the early 1930s. Shortly thereafter, timber stand improvements (TSI) were initiated, which were later complemented by the first study plots established in 1938 to monitor TSI results. Long-term research in permanent plots to study species composition, stand characteristics, and timber production potential began in earnest in the early 1940s [18].

In these early years, the EYNF implemented policies and practices for a range of activities including livestock grazing, mining, recreation, water supply, and timber production, largely through a custodial approach to land and resource management. Important aspects of active learning were established fairly early in the life of the EYNF through research and monitoring and a close association with the USFS Tropical Forest Experiment Station (now known as the International Institute of Tropical Forestry), which was established in Puerto Rico under the McSweeny–McNary Act in 1939. Throughout much of the early 1900s, decisions on land and resource use in the EYNF followed agency policy and guidance, but were made locally, at the discretion of the forest supervisor. During this period, though there was limited outside input or coordination in forest-related decision making, the EYNF demonstrated some measurable flexibility in management practices and approaches and in its capacity to adapt to new and changing conditions, particularly those associated with the uniqueness of being the only tropical forest in the National Forest System.

3.2. Scientific Management

Through the mid-20th century, the management of the EYNF increasingly focused on multiple resources and services, including the provision of recreation, conservation of watersheds, protection of parrot habitat, research, and wood production, which was mostly for fuelwood and charcoal production to meet the energy demands of a rapidly growing post-war population in Puerto Rico [18]. By 1955, logging within the EYNF had ceased, due mainly to decreasing demands for timber and fuelwood given rising imports of kerosene and mahogany [36]. Then, in 1956 the EYNF was officially dually designated as the Luquillo Experimental Forest (LEF) throughout its geographic extension. Experimental forests and ranges were established throughout the NFS to address large scale problems of forest, range, and watershed management through a broad range of basic and applied studies with short- to long-terms planning horizons. Designating the EYNF as an experimental forest across the entirety of its range was unique in the system and reflected the recognition of the complexity inherent in tropical forests and their management.

The passage of the Multiple Use Sustained Yield Act (MUSYA) in 1960 officially expanded the USFS mandate, giving the agency permissive and discretionary authority to administer national forests for outdoor recreation, range, timber, watersheds, wildlife, and fishing (P.L. 86–517). Following MUSYA and other agency guidance, the EYNF continued to plan and practice forest restoration, recreation, research, and other activities. Though timber was no longer extracted from the EYNF, line plantings of mahogany and additional timber stand improvements were conducted throughout the 1960s and 1970s anticipating future wood harvests from forest areas designated for timber production.

In 1976, the National Forest Management Act (NFMA) (P.L. 94–588) was passed, requiring a systematic approach to land and resource management for all national forests and grasslands, setting standards for timber sales, and providing criteria for timber harvests. The NFMA also required the development of forest planning regulations, which were first issued in 1979 and later superseded by revised regulations in 1982. The 1982 USFS Planning Rule focused on the maximization of multiple public benefits and for the first time required public input on forest planning, but also prescribed a complicated and elaborate planning process that was not easy for the public to access [37]. Under these directives, the EYNF began work on a comprehensive forest management plan in the early 1980s, which was developed internally, largely based on prescriptive guidelines that included the use of scientific and technical information. The final plan adopted a multiple use approach to land and resource management, encompassing several management alternatives, including timber production in suitable areas [38]. This plan was approved by the regional forester in 1986 and submitted to the public for comment along with an environmental impact statement and record of decision.

The 1986 EYNF land and resource management plan was strongly contested by local communities and conservation organizations, who had been largely excluded from the decision process. Major concerns were associated with the proposed timber harvests and new roads and trails, as well as with the fact that the plan and related documents had not been made available in Spanish [18]. On 19 November 1986, thousands of people marched on the grounds of the EYNF in protest of the plan. Soon thereafter, the plan was appealed in court by 12 environmental and recreational organizations, culminating in a court order to provide a Spanish translation of official EYNF documents submitted to the public for comment and to reconsider logging throughout the plan area [39]. Subsequently, the EYNF began a process of plan amendments and revisions that included increasing public input and involvement over the course of the next decade.

In the latter half of the 20th century, decisions about the EYNF increasingly were shaped by agency direction, which increasingly dictated a scientific approach to management that focused on the objective application of scientific methods and technical information to control natural systems and changes in them [40]. However, the scientific management approach also often discounted or excluded other types of nonscientific information, knowledge, and input that made prescriptions messy or more difficult, but which were inextricable from the natural system and its management [40,41]. In its first forest plan submitted for public input, the EYNF had incorporated considerable scientific and technical information, some of which was produced from ongoing research and monitoring throughout the forest, demonstrating important aspects of active learning. However, the relative exclusion of local communities and other stakeholders from internal decision-making resulted in a major setback for the EYNF and its land and resource practices and projects. During this time, there was limited flexibility in decision-making, particularly in terms of incorporating outside perspectives, which only were addressed after mandated by a court order.

3.3. Ecosystem Management

As the EYNF worked to revise its land and resource management plan during the late 1980s and early 1990s, ecosystem management had emerged as a new management approach centering on the conservation of multiple resources and the preservation of structures and processes across multiple scales through the integration of ecological, economic, and social information; collaboration and coordination with stakeholders; and adaptation of management through continuous learning or experimentation [42,43]. The ecosystem approach represented a paradigm shift in natural resource management in the U.S., triggered in part by ongoing conflicts over clearcutting and other environmental issues on federal lands and growing demands from scientists, practitioners, and other stakeholders to move away from a traditional focus on a single species or deliverable, towards a more integrated focus on the ecosystem as a whole [44]. In June 1992, the USFS became the first federal agency in the U.S. to adopt (on paper at least) an ecosystem management approach [45]. Related agency guidance laid out specific means and ends for advancing forest sustainability through integrated scientific information (i.e., ecological, economic, social), collaborative stewardship, interagency cooperation, and adaptive management [44–46]. However, there was no related statutory mandate requiring the use of ecosystem management or its components parts. Consequently, there was very limited funding to fully implement ecosystem management, for example through intensified monitoring or adaptive management, and slow organizational uptake throughout much of the USFS [47,48].

By the late 1990s, the EYNF had adopted an ecosystem approach to land and resource management, which influenced decisions about projects and practices, as well as the forest plan revision process, particularly in terms of plan components and public input and involvement. During this time, the EYNF continued to confront environmental and anthropological processes, pressures, and surprises, further shaping local policies and practices. For example, Hurricane Hugo passed very near to the EYNF in 1989, causing significant loss of standing biomass and wildlife, short-term changes in the water regime, and road and infrastructure damage among other effects throughout many parts of the forest, affecting its functions and services [49]. Additionally, the island's population expanded throughout the latter half of the 20th century (e.g., Pop. in 1950: 2.22 million, Pop. in 2000: 3.81 million), producing increasing pressures on the EYNF, particularly for water and recreation, leading to increasing water withdrawals and visitation rates. Land uses around the EYNF also were shifting during this time, with significant reversion of agriculture and pasture to forest, but also conversion of pasture, agriculture, and even forest to urban and suburban development, in some cases in violation of local zoning rules aimed at protecting a buffer zone around the EYNF [50–52]. Land use changes and their effects on forest connectivity within the region prompted the EYNF to become ever more engaged in local level planning efforts and community outreach.

Ultimately, agency direction on ecosystem management, conditions and trends within and outside the EYNF, and more than a decade of appeals and subsequent public consultation on the previous forest plan all contributed to shaping the 1997 Revised EYNF Land and Resource Management Plan [53]. Reflecting stakeholder priorities, the 1997 forest plan focused largely on forest protection (e.g., wilderness, wild and scenic river segments, research natural area) and human activities (e.g., recreation, environmental education, research) and eliminated opportunities for commercial timber harvesting except within a relatively restricted area designated for the demonstration of sustainable timber production. Several key aspects of the broader ecosystem management approach, such as collaborative stewardship, interagency cooperation, and the integration of scientific information are reflected in this plan and its implementation. Conversely, there was comparatively limited integration of economic and social information relevant to the EYNF. Moreover, though forest monitoring was a requisite part of the plan; in practice, monitoring of forest conditions and responses to management and other effects was limited, as were mechanisms for continuous learning and adaptation through established feedback loops. These disparities in management approach and practice limited EYNF capacity for adaptiveness to environmental and socioeconomic shifts and surprises. Furthermore, though the EYNF gradually opened to stakeholder collaboration and horizontal and vertical inter-agency and inter-sectoral coordination, its flexibility to respond to new conditions or trends was limited, in part, by an expanding legal framework affecting all federal lands, as well as by an inherent aversion to risk and organizational change, which was common throughout much of the agency [47,48].

3.4. Towards Adaptive, Collaborative Management?

With the approval of its 1997 Land and Resource Management Plan, the EYNF set on a course of action that deviated very little for the next 15 years or so. Policies and practices focused mostly on conservation, for example of at-risk species, wilderness, and wild and scenic rivers, often in collaboration with stakeholders and local partners, and on human demands and needs, such as recreation, environmental education, and water supplies. Timber harvesting remained taboo for most stakeholders and was not reactivated in the EYNF, even in the designated demonstration sites, but long-term research continued to expand through the International Institute of Tropical Forestry and the NSF-funded Luquillo Long-term Ecological Research program, as did community engagement and environmental education [19].

By 2012, after multiple attempts to revise agency regulations, the USFS had issued its National Forest System Land Management Planning Rule (36 CFR Part 219) (hereafter, the 2012 Planning Rule) and selected the EYNF as one of eight "early adopter" forests to revise its existing forest plan under the new directives and guidance [54]. The 2012 Planning Rule was developed in accordance with the NFMA, and for the first time, codified requirements for collaboration, integrated scientific information, sustainability, climate change considerations, and adaptiveness in federal land management planning. It prescribes a landscape-scale approach to land and resource management that takes into account conditions and trends beyond management unit boundaries and transfers some decision-making authority back to the local level, designating the forest supervisor with signatory authority for final plan approval and the record of decision (as opposed to the Regional Forester under the previous Planning Rule). Collaboration is required with local communities and other key stakeholders throughout the planning process versus the *ex post facto* consultation of previous Planning Rules. Additionally, the new Planning Rule requires monitoring of forest conditions and responses to management, as well as ongoing evaluation to determine whether new information from monitoring or other sources warrant changes in management direction. Overall, the prescribed process of 'Plan, Monitor, Assess, Repeat' is intended to make national forest plans dynamic and their management adaptive to existing and unforeseen conditions, outcomes, risks and stressors (Figure 1) [54].

In 2012, the EYNF initiated a collaborative, interactive process to revise its land and resource management plan under the new USFS Planning Rule. The first major step in the process was a comprehensive assessment of ecological, economic, and social conditions and trends in and around the EYNF. This assessment was led by an interdisciplinary team of resource managers, specialists, and scientists selected by the EYNF Forest Supervisor and based on the 'best available scientific information' and stakeholder input, as required under the new Rule. The assessment integrates data and findings from nearly a century of research in and around the forest, as well as other research and information sources, including local knowledge and stakeholder input. The EYNF assessment was published in 2014 as a living document (i.e., to be updated with new information as it becomes available and assessed) and providing critical information for determining a proposed action and need for change in existing EYNF management direction [19]. The assessment also informed the development of a draft revised land and resource management plan and environmental impact statement for the EYNF, which incorporated extensive and ongoing public participation and interagency coordination. The EYNF draft revised land and resource management plan and environmental impact statement were officially submitted to the public for comment in September 2016 and are expected to published in final form in the fall of 2017 [17].

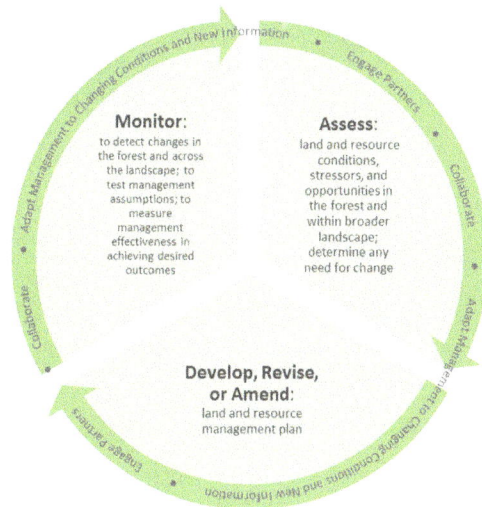

Figure 1. U.S. Forest Service Land Management Planning Process (USFS 2012) [54].

EYNF's revised forest plan focuses on the provision of sustainable ecological processes and socioeconomic benefits for local communities and other key stakeholders, recognizing the important role of people in nature and of their values, beliefs, and customs for sustainable forest management (Table 1) [17]. This plan integrates a systems perspective that focuses on species, communities, and ecosystems through a landscape scale approach that promotes management strategies to work beyond the forest's administrative boundaries. It provides for interagency coordination and stakeholder collaboration through an "all-lands" approach to land and resource management and adopts the concept of "*co-manejo*" (or shared stewardship), which developed through interactions with the public and key stakeholders during the planning process. In this context, *co-manejo* does not involve a delegation of decision-making authority from the USFS to partners or participants, but instead involves the strategic and site-specific engagement of the agency and active partners "who work together in general forest operations, conservation, and restoration activities with a practical sense of shared responsibilities to achieve the mission" [17].

Components of the 2016 draft revised EYNF land and resource management plan focus on the protection of at-risk species; the sustainability of water production and quality; opportunities for sustainable recreation and other ecosystem services; protection of wetlands and riparian areas; and mitigation of and adaptation to climate change. This plan differs from the previous EYNF land and resource management plans in the integrated view of ecological, economic, and social sustainability, explicit connections to local communities, and adaptive approach to land and resource management. The adaptive aspects of the plan include a monitoring program that focuses on management effects and outcomes, as well as management strategies that promote partnerships with scientists, practitioners, decision-makers, and other stakeholders "who learn and work together to support a management system resilient to changes in social, economic, and ecological conditions" [17].

Table 1. El Yunque National Forest Core Management Themes (Adapted from USFS 2016) [17].

• Healthy ecosystems
– Conserve and restore Forest ecosystems.
– Protect and conserve functional wetlands and primary forest.
– Maintain and improve watershed conditions throughout the Forest while monitoring, adapting to, and mitigating the effects of climate change.
• Sustainable recreation, access, and tourism
– Provide for sustainable recreation throughout the Forest in harmony with the natural setting and with historical and cultural resources.
– Develop public support and partnerships to improve recreation facilities and services on the Forest.
• "All-lands" management approach
– Consider the ecological, social, and economic needs of the broader landscape in Forest plans and projects.
– Provide for forest-community interface management area at the lower elevations of the Forest that is sustainably managed in accessible locations suitable for multiple-use management, including forest products.
• Collaborative, adaptive management
– Sustain and develop partnerships and regional collaboration efforts engaged in conservation, management, land use, and research.
– Shift priorities from primarily top-down to more collaborative land and resource management.
– Integrate agencies, local landowners, and other key stakeholders in conservation efforts through the facilitation and coordination of a collaboration network.
– Provide opportunities for research and monitoring and develop related initiatives with agencies and stakeholders.
• Environmental literacy and education
– Connect the surrounding communities to the Forest's natural landscapes, identifying and overcoming barriers that inhibit their participation.
– Provide opportunities to develop community capacity for participation in management activities, including interpretation, recreation, monitoring, etc.

Overall, this most recent EYNF management planning process demonstrates increased organizational and strategic flexibility in decision-making through its openness to extensive and ongoing public participation and interagency coordination, including listening sessions with key stakeholders; focus groups with local, state, and federal governments, regional protected area managers, youth groups, tourism operators, and others; and multiple public meetings, fora, and information exchanges. Following agency direction, the EYNF also has incorporated key aspects of adaptive, ecosystem management and the potential for active, ongoing learning important for dealing with complexity and uncertainty. In particular, the monitoring component of the draft revised plan outlines monitoring questions and associated indicators designed to inform management decisions by testing assumptions, tracking relevant changes, and measuring management effectiveness and progress toward desired conditions and objectives. These are significant developments that differentiate this plan from the ways in which the EYNF has been managed in the past and if put into practice will enhance learning and adaptiveness. Still, the monitoring program stops short of defining thresholds or triggers that signal when to adapt or change course, which are important for sustaining forest structures and processes.

4. Discussion and Implications

Since its inception, the EYNF has had to contend with complexity and uncertainty in the management of land, resources, and people. Scientific understanding of tropical forests as complex social-ecological systems certainly has increased since its establishment. Yet, so too has the associated variability in and across systems and related processes—dynamics which are expected to increase in the context of the Anthropocene. So, how will the EYNF cope with the uncertainty and surprise associated with modern times? Are useful tools and arrangements already available? Or, are changes in policy or practice required for dealing with the Anthropocene?

Historically, the custodial and scientific approaches to EYNF land and resource management integrated important aspects of learning through the early establishment of long-term forest research and monitoring. However, these early policies and practices were not entirely conducive to the learning, flexibility, and adaptiveness needed for managing a complex system in increasingly unpredictable times [12–15]. A significant shift in overall agency direction was set into motion in the 1990s under the ecosystem management approach, but was only recently codified statutorily and expanded through the 2012 USFS Planning Rule. Under this new rule and related guidance, the USFS prescribes and promotes a more collaborative, landscape-level, learning approach to land and resource management that ultimately may permit more flexibility and adaptiveness to system changes and surprises, if adequately implemented on the ground. These changes are likely to be particularly important for managing tropical forests, like the EYNF, which as noted by Lugo (1995) [55] echoing Holling (1973) [56] "does not require a precise capacity to predict the future, but only a qualitative capacity to devise systems that can absorb and accommodate future events in whatever unexpected form they may take".

Nevertheless, this major shift in approach to the management of the ENYF that embraces adaptive, collaborative, ecosystem management will require significant organizational change, which is no easy feat, particularly since institutions tend to be path dependent and "sticky", changing mostly through slow, incremental shifts [57]. Hence, it is not surprising, for example, that it took 20 years or more for the USFS to move from the adoption of ecosystem management as a policy to its codification and implementation through the new Planning Rule. Though the USFS has experimented some with adaptive management over the past few decades, managers and line officers alike have revealed persisting challenges to implementation, including the significant institutional changes that are required, the high costs and limited funding for monitoring, and the lack of public and private support, particularly for the risk involved in experimentation [58]. These obstacles are likely to be a challenge for the EYNF as well and will require innovation, experimentation, and collective learning locally and throughout the agency, especially since the effects of the Anthropocene occur and interact in different ways at different scales, requiring flexibility in policy and practice at all levels [15,59,60]. Ultimately, the 2012 USFS Planning Rule has codified for the first time many of the processes necessary for managing complex systems, like the EYNF, in the context of the Anthropocene—specifically, ongoing learning, flexibility, and adaptiveness. What remains to be seen is if critical elements like monitoring and feedback loops will be adequately developed, funded, and implemented on the grounds, and if aversion to the risks associated with experimentation and flexibility in management decisions will be overcome at local and larger scales.

5. Conclusions

Historically, land and resource managers of the EYNF in Puerto Rico focused mostly on the protection of resources and prevention of harm through a custodial approach to land and resource management. Research and learning on forest characteristics, functions, and processes were established early on in the EYNF. The custodial approach to land and resource management gave way to an increasingly scientific, top-down management approach focused on technically-founded decision making, often at the exclusion of outside input. This eventually opened up to a more holistic management approach focused on the forest as an ecosystem and incorporating collaboration and adaptive elements in forest management and decision-making. These changes in the EYNF management approach mostly have been reactive, slow, and incremental, with some rarer, rapid shifts primarily in response to major changes in USFS policy and guidance that have promoted increasing adaptiveness and flexibility, but also to significant processes or impacts on the ground, such as hurricanes and changes in land use.

Most recently, the EYNF has expanded its potential for learning, flexibility, and adaptiveness in land and resource decisions and practices, in large part through its early adoption of the 2012 National Forest System Land and Resource Management Planning Rule. With sufficient, committed, and consistent resources, this policy shift could provide a robust framework for managing the EYNF into the future—particularly, given the certain uncertainty and surprise that are expected to increase for this tropical forest in the context of the Anthropocene. Building on decades of incremental change, the processes prescribed in the new Planning Rule provide the EYNF with a significant opportunity to make significant changes and blaze the trail for collaborative and adaptive forest management in the tropics. The EYNF is particularly well positioned to do so given its long history of forest research and close relations with scientists and research programs at the ready to engage in long term monitoring, experimentation, and feedback loops. It also has developed an inclusive and stable structure for meaningful collaboration with local communities and other active partners to engage in resource assessments, management decisions and applications, monitoring questions and collections, and analysis and interpretation of new information, within an adaptive management framework.

Collaborative relationships with local level stakeholders, productive ties to research and monitoring, and political support for adaptiveness and flexibility are critical elements in managing for the future resilience and sustainability of tropical forests, like the EYNF. Though the EYNF increasingly counts these and other important elements within its toolbox, this forest is likely to confront challenges as it shifts towards an organizational learning culture ready for and responsive to system changes and surprises [44,57,59]. Likewise, the development of a robust monitoring system with reliable and effective feedback loops will be critical, but not easily or inexpensively established or maintained [60,61]. The EYNF, like other forests in the National Forest System and throughout the tropics, must be afforded the necessary financial, human, and technical resources and the flexibility to put this new management approach into practice, to learn while doing, and adapt when necessary. Considering progress to date in their land and resource management planning process, it seems quite probable that the EYNF will be well positioned to collaboratively and adaptively manage its land and resources for resilience and sustainability even in the context of the changes and surprises certain to come as we continue to move forward through the Anthropocene.

Acknowledgments: Many thanks to the El Yunque National Forest, which has supported various projects that have been instrumental in the development of the ideas in this article. Thanks also to Ariel Lugo, Grizelle González, and two anonymous reviewers for their input on earlier versions of this manuscript, which contributed to its improvement. The findings, conclusions, and views expressed in this manuscript are those of the author and do not necessarily represent the views of the USDA Forest Service.

Conflicts of Interest: The author declares no conflict of interest.

References

1. Crutzen, P.J.; Stoermer, E.F. Global change newsletter. *Anthropocene* **2000**, *41*, 17–18.
2. Waters, C.N.; Zalasiewicz, J.; Summerhayes, C.; Barnosky, A.D.; Poirier, C.; Gałuszka, A.; Cearreta, A.; Edgeworth, M.; Ellis, E.C. The Anthropocene is functionally and stratigraphically distinct from the Holocene. *Science* **2016**, *351*, aad2622. [CrossRef] [PubMed]
3. Davies, J. *The Birth of the Anthropocene*; University of California Press: Oakland, CA, USA, 2016.
4. Millar, C.I.; Stephenson, N.L.; Stephens, S.L. Climate change and forests of the future: Managing in the face of uncertainty. *Ecol. Appl.* **2007**, *17*, 2145–2151. [CrossRef] [PubMed]
5. Heckenberger, M.J.; Kuikuro, A.; Kuikuro, U.T.; Russell, J.C.; Schmidt, M.; Fausto, C.; Franchetto, B. Amazonia 1492: Pristine forest or cultural parkland? *Science* **2003**, *301*, 1710–1714. [CrossRef] [PubMed]
6. Ford, A.; Nigh, R. The milpa cycle and the making of the Maya forest garden. *Res. Rep. Belizean Archeol.* **2010**, *7*, 183–190.
7. Nigh, R. Trees, fire, and farmers: Making woods and soil in the Maya forest. *J. Ethnobiol.* **2008**, *28*, 231–243. [CrossRef]
8. Berkes, F.; Colding, J.; Folke, C. Rediscovery of traditional ecological knowledge as adaptive management. *Ecol. Appl.* **2000**, *10*, 1251–1262. [CrossRef]

9. Diamond, J. *Collapse: How Societies Choose to Fail or Succeed*; Penguin: New York, NY, USA, 2005.

10. Reyer, C.P.O.; Brouwers, N.; Rammig, A.; Brook, B.W.; Epila, J.; Grant, R.F.; Homgren, M.; Langerwisch, F.; Leuzinger, S.; Lucht, W.; et al. Forest resilience and tipping points at different spatio-temporal scales: Approaches and challenges. *J. Ecol.* **2015**, *103*, 5–15. [CrossRef]

11. Keenan, R.J. Climate change impacts and adaptation in forest management: A review. *Ann. For. Sci.* **2015**, *72*, 145–167. [CrossRef]

12. Holling, C.S. (Ed.) *Adaptive Environmental Assessment and Management*; John Wiley and Sons: London, UK, 1978.

13. Millar, C.I.; Swanson, C.W.; Peterson, D.L. Adapting to Climate Change. In *Climate Change and the United States Forests*; Peterson, D.L., Vose, J.M., Patel-Weynad, T., Eds.; Springer: Dordecht, The Netherlands, 2014; pp. 183–222.

14. Vose, J.; Peterson, D.; Patel-Weynand, T. (Eds.) *Effects of Climatic Variability and Change on Forest Ecosystems: A Comprehensive Science Synthesis for the U.S. Forest Sector*; General Technical Report PNW-GTR-870; U.S. Department of Agriculture, Forest Service Pacific Northwest Research Station: Corvallis, OR, USA; Washington, DC, USA, 2012; p. 265.

15. Bhagwat, S.A.; Humphreys, D.; Jones, N. Forest governance in the Anthropocene: Challenges for theory and practice. *For. Policy Econom.* **2017**, *79*, 1–7. [CrossRef]

16. López-Marrero, T.; Hermansen-Báez, L.A. *Land Cover within and around El Yunque National Forest. [Fact Sheet]*; U.S. Department of Agriculture, Forest Service Southern Research Station: Gainesville, FL; Knoxville, TN, USA, 2011; p. 4.

17. USDA Forest Service. *Draft Revised Land and Resource Management Plan: El Yunque National Forest*; USDA Forest Service: Rio Grande, Puerto Rico, 2016.

18. Weaver, P.L. *The Luquillo Mountains: Forest Resources and Their History*; General Technical Report IITF-44; USDA Forest Service, International Institute of Tropical Forestry: Rio Piedras, Puerto Rico, 2012.

19. USDA Forest Service. *Forest Plan Assessment El Yunque National Forest. Draft*; USDA Forest Service: Rio Grande, Puerto Rico, 2014.

20. Scatena, F.N. An assessment of climate change in the Luquillo Mountains of Puerto Rico. In Proceedings of the Tropical Hydrology and Caribbean Water Resources, Third International Symposium on Tropical Hydrology and Fifth Caribbean Islands Water Resources Congress, San Juan, Puerto Rico, 12–16 July 1998; Segarra-Garca, R.I., Ed.; American Water Resources Association: Herndon, VA, USA, 1998; pp. 193–199.

21. Briscoe, C.B. *Weather in Luquillo Mountains of Puerto Rico*; Res. Paper; ITF-3; Institute of Tropical Forestry: Rio Piedras, Puerto Rico, 1996; p. 250.

22. Waide, R.B.; Comarazamy, D.E.; González, J.E.; Hall, C.A.S.; Lugo, A.E.; Luvall, J.C.; Murphy, D.J.; Ortiz-Zayas, J.R.; Ramírez-Beltran, N.D.; Scatena, F.N.; et al. Climate variability at multiple spatial and temporal scales in the Luquillo Mountains, Puerto Rico. *Ecol. Bull.* **2013**, *54*, 21–41.

23. Henareh Khalyani, A.; Gould, W.A.; Harmsen, E.; Terando, A.; Quiñones, M.; Collazo, J.A. Climate change implications for tropical islands: Interpolating and interpreting statistically downscaled GCM projections for management and planning. *J. Appl. Meteorol. Climatol.* **2016**, *55*, 265–282. [CrossRef]

24. Hayhoe, K. *Quantifying Key Drivers of Climate Variability and Change for Puerto Rico and the Caribbean*; Final Report to the Southeast Climate Science Center; Agreement No.: G10AC00582; Southeast Climate Science Center: Raleigh, NC, USA, 2013; p. 241.

25. Lasso, E.; Ackerman, J.D. Flowering phenology of (*Werauhia sintenisii*), a bromeliad from the dwarf montane forest in Puerto Rico: An indicator of climate change? *Selbyana* **2003**, *24*, 95–104.

26. Wunderle, J.M.; Arendt, W.J. Avian studies and research opportunities in the Luquillo Experimental Forest: A tropical rain forest in Puerto Rico. *For. Ecol. Manag.* **2011**, *262*, 33–48. [CrossRef]

27. Schellekens, J.; Scatena, F.N.; Bruijnzeel, L.A.; Van Dijk, A.I.J.M.; Groen, M.M.A.; Van Hogezand, R.J.P. Stormflow generation in a small rainforest catchment in the Luquillo Experimental Forest, Puerto Rico. *Hydrol. Process.* **2004**, *18*, 505–530. [CrossRef]

28. Jennings, L.N.; Douglas, J.; Treasure, E.; González, G. *Climate Change Effects in El Yunque National Forest, Puerto Rico, and the Caribbean Region*; Gen. Tech. Rep. SRS-GTR-193; USDA-Forest Service, Southern Research Station: Asheville, NC, USA, 2014; p. 47.

29. Prideaux, B.; Coghlan, A.; McNamara, K. Assessing tourists' perceptions of climate change on mountain landscapes. *Tour. Recreat. Res.* **2010**, *35*, 187–199. [CrossRef]

30. Hall, P.A. Historical Institutionalism in Rationalist and Sociological Perspective. In *Explaining Institutional Change: Ambiguity, Agency and Power*; Mahoney, J., Thelen, K., Eds.; Cambridge University Press: New York, NY, USA, 2009.

31. Hall, P.; Taylor, R.C.R. Political science and the three new institutionalisms. *Polit. Stud.* **1996**, *44*, 936–957. [CrossRef]

32. Thelen, K. *How Institutions Evolve*; Cambridge University Press: Cambridge, UK, 2004; pp. 208–240.

33. Mahoney, J.; Thelen, K. (Eds.) *Explaining Institutional Change: Ambiguity, Agency and Power*; Cambridge University Press: New York, NY, USA, 2009.

34. Steen, H.K. *The U.S. Forest Service: A History*; University of Washington Press: Seattle, WA, USA, 1976.

35. Wadsworth, F.H. Forest management in the Luquillo Mountains, II. Planning and multiple use. *Caribb. For.* **1952**, *13*, 49–61.

36. Wadsworth, F.H. Review of past research in the Luquillo Mountains of Puerto Rico. In *A Tropical Rain Forest: A Study of Irradiation and Ecology at El Verde, Puerto Rico*; Odum, H.T., Pigeon, R.F., Eds.; U.S. Department of Commerce: Springfield, VA, USA, 1970; Chapter B-2; pp. 33–46.

37. Cubbage, F.W.; O'Laughlin, J.; Peterson, M.N. *Natural Resource Policy*; Waveland Press, Inc.: Long Grove, IL, USA, 2017.

38. USDA Forest Service. *Final Land Use and Resource Management Plan*; Caribbean National Forest: Palmer, Puerto Rico, 1986.

39. USDA Forest Service. *Luquillo Forest Reserve Centennial 1903–2003*; USDA Forest Service: Rio Grande/Rio Piedras, Puerto Rico, 2005. Available online: https://www.fs.usda.gov/detail/elyunque/about-forest/?cid=fsbdev3_042988 (accessed on 19 July 2017).

40. Nelson, R.H. The religion of forestry: Scientific management. *J. For.* **1999**, *97*, 4–8.

41. Brunner, R.D.; Steelman, T.A.; Coe-Juell, L.; Crowley, C.M.; Edwards, C.M.; Tucker, D.W. *Adaptive Governance: Integrating Science, Policy, and Decision Making*; Columbia University Press: New York, NY, USA, 2005.

42. Grumbine, R.E. What is ecosystem management? *Conserv. Biol.* **1994**, *8*, 27–38. [CrossRef]

43. Christensen, N.L.; Bartuska, A.; Brown, J.H.; Carpenter, S.; D'Antonio, C.; Francis, R.; Franklin, J.F.; MacMahon, J.A.; Noss, R.F.; Parsons, D.J.; et al. The report of the Ecological Society of America Committee on the scientific basis for ecosystem management. *Ecol. Appl.* **1996**, *6*, 665–691. [CrossRef]

44. Butler, K.F.; Koontz, T.M. Theory into practice: Implementing ecosystem management objectives in the USDA Forest Service. *Environ. Manag.* **2005**, *35*, 138–150. [CrossRef] [PubMed]

45. Robertson, D. *Memo from Dale Robertson to Regional Foresters and Station Directors Entitled "Ecosystem Management of the National Forests and Grasslands"*; Washington Office, USDA/Forest Service: Washington, DC, USA, 1992.

46. Thomas, J.W. *Statement Concerning Implementation of Ecosystem Management Strategies before the Subcommittee on Agricultural Research, Conservation, and Forest and General Legislation*; Committee on Agriculture, U.S. Senate: Washington, DC, USA, 1994.

47. Killen, J.R. *Federal Ecosystem Management Its Rise, Fall, and Afterlife*; University Press of Kansas: Lawrence, KS, USA, 2015.

48. Koontz, T.; Bodine, J. Implementing ecosystem management in public agencies: Lessons from the U.S., Bureau of Land Management and the U.S. Forest Service. *Conserv. Biol.* **2008**, *22*, 60–69. [CrossRef] [PubMed]

49. Besnet, K.; Likens, G.E.; Scatena, F.N.; Lugo, A.E. Hurricane Hugo: Damage to a tropical rain forest in Puerto Rico. *J. Trop. Ecol.* **1992**, *8*, 47–55. [CrossRef]

50. Lugo, A.E.; Lopez, delM, T.; Ramos, O.M. *Zonificación de Terrenos en la Periferia de El Yunque*; GTR-IITF-16; USDA Forest Service, IITF: Rio Piedras, Puerto Rico, 2000; p. 20.

51. Ramos Gonzalez, O.M. Assessing vegetation and land cover changes in Northeastern Puerto Rico: 1978–1995. *Caribb. J. Sci.* **2001**, *37*, 95–106.

52. Lugo, A.E.; Lopez delM, T.; Ramos Gonzalez, O.M.; Velez, I.I. *Urbanización de los Terrenos en la Periferia de El Yunque*; GTR-WO-66; USDA Forest Service: Washington, DC, USA, 2004; p. 29.

53. USDA Forest Service. *Revised Land Use and Resource Management Plan: Caribbean National Forest—Luquillo Experimental Forest, Puerto Rico*; Management Bulletin R8-MB 80G; Caribbean National Forest: Palmer, Puerto Rico, 1997.

54. USDA Forest Service. Web. Planning Rule. Available online: https://www.fs.usda.gov/detail/planningrule/ (accessed on 19 July 2017).

55. Lugo, A. Management of tropical biodiversity. *Ecol. Appl.* **1995**, *5*, 956–961. [CrossRef]
56. Holling, C.S. Resilience and stability of ecological systems. *Ann. Rev. Ecol. Syst.* **1973**, *4*, 1–23. [CrossRef]
57. North, D. *Institutions, Institutional Change and Economic Performance*; Cambridge University Press: Cambridge, UK, 1990; p. 152.
58. Innes, J.L.; Joyce, L.A.; Kellomaki, S.; Louman, B.; Ogden, A.; Parrotta, J.; Thompson, I. Management for adaptation. In *Adaptation of Forests and People to Climate Change*; Seppälä, R., Buck, A., Katila, P., Eds.; IUFRO World Series; International Union of Forest Research Organizations (IUFRO): Vienna, Austria, 2009; Volume 22, pp. 135–169.
59. Lawrence, A. Adapting through practice: Silviculture, innovation and forest governance for the age of extreme uncertainty. *For. Policy Econnom.* **2017**, *79*, 50–60. [CrossRef]
60. Larson, A.J.; Belote, R.T.; Williamson, M.A.; Aplet, G.H. Making monitoring count: Project design for active adaptive management. *J. For.* **2013**, *111*, 348–356. [CrossRef]
61. Verburg, P.H.; Dearing, J.A.; Dyke, J.G.; van der Leeuw, S.; Seitzinger, S.; Steffen, W.; Syvitski, J. Methods and approaches to modelling the Anthropocene. *Glob. Environ. Chang.* **2016**, *39*, 328–340. [CrossRef]

forests

MDPI

Article

The Dynamics of Deforestation in the Wet and Dry Tropics: A Comparison with Policy Implications

Thomas K. Rudel

Departments of Human Ecology and Sociology, Rutgers University, New Brunswick, NJ 08901, USA;
rudel@aesop.rutgers.edu; Tel.: +1-84-8932-9238

Academic Editors: Grizelle Gonzalez and Ariel Lugo
Received: 4 March 2017; Accepted: 1 April 2017; Published: 5 April 2017

Abstract: Forests in the dry tropics differ significantly from forests in the humid tropics in their biomass and in their socio-ecological contexts, so it might be reasonable to assume that the dynamics that drive deforestation in these two settings would also differ. Until recently, difficulties in measuring the extent of dry tropical forests have made it difficult to investigate this claim empirically. The release of high resolution LANDSAT satellite imagery in 2013 has removed this impediment, making it possible to identify variations in the extent of wet and dry forests within countries by measuring variations in the canopy cover of their forests. These metrics have in turn made it possible to investigate human differences in the dynamics of deforestation between dry forested and wet forested nations in the tropics. Cross-national analyses suggest that international trade in agricultural commodities plays a more important role in driving deforestation in the wet tropics than it does in the dry tropics. The variable salience of international trade as a driver has important implications, described here, for the success of policies designed to slow deforestation in the dry tropics and the wet tropics. Curbing dry forest losses, in particular, would appear to require locally focused and administered policies.

Keywords: dry tropical forests; humid tropical forests; tropical deforestation

1. Introduction

Over the past ten years a consensus has gradually emerged about the chief drivers of tropical deforestation worldwide. A series of studies [1–6] have identified large scale, commercial agriculture, frequently engaged in the export of agricultural products, as a primary driver of deforestation in both Latin America and Southeast Asia. A different pattern seems to have characterized deforestation in sub-Saharan Africa. In this region, the small-scale production of agricultural commodities like charcoal, millet, and cassava for local consumption has played a larger role in deforestation [7,8].

Recognition of these regional differences in the dynamics of deforestation has no doubt made investigators more aware of the diverse, conjunctural nature of the forces that have destroyed tropical forests in different places. In many ways though, the emphasis on regional differences just raises a new set of questions. To what do we attribute the regional differences in deforestation? One conjecture might attribute these differences to variations in climate and associated forest types that, in interaction with rural societies, have produced distinctive regional deforestation dynamics. I explore this possibility here with particular attention to possible differences between wet and dry forests in the dynamics of tropical deforestation. This analysis, if convincing, would help us 'unpack' the continental differences in deforestation dynamics observed by many analysts.

Differences in rainfall, in particular, may have had a cascading set of effects on forests and deforestation processes, so, following this logic, the dynamics of deforestation might differ from dry to wet forests. It would be useful to identify these different deforestation dynamics because, once identified, they might suggest efficacious, location specific policies for reducing rates of deforestation. An investigation of these wet and dry forest dynamics takes on added importance right now because global models of climate change project an expansion in the extent of subtropical dry zones and, conceivably, dry forests in the coming decades [9].

To this end, this paper uses newly available, high resolution remote sensing data [10] of recent forest cover changes in the tropics to analyze and compare the dynamics of deforestation in dry and wet forested countries in the tropics. The prevalence of different types of canopy cover in a country, calculated from LANDSAT data, serves as a proxy for the prevalence of dry and wet forests in a country. This canopy cover measure makes it possible to identify predominantly dry and predominantly wet forested countries, analyze the dynamics of deforestation in each set of countries, and then compare these dynamics. I pursue this intellectual agenda through the following steps. I begin by assessing the forest canopy measure for the prevalence of dry forests. I then offer a theory about differences in the dynamics of deforestation between dry and wet forests, describe the data and methods for examining the theory, present the results of the quantitative analyses, and finally, discuss the results and their policy implications.

1.1. Dry Forests and Canopy Cover

Dry forests and woodlands constituted around 42% of the world's tropical forests during the 1980s [11]. Dry forests include a range of different land covers. Shrublands, thickets, open woodlands, and wooded grasslands would all be classified as dry forests. Large expanses of dry forests exist in the Brazilian northeast, in the Paraguay-Parana river basin of South America, in southern and eastern Sub-Saharan Africa, and in South Asia. The inhabitants of dry forest dominated countries in the tropics are among the world's poorest peoples [11,12]. Women in these communities rely to an extraordinary extent for their sustenance on the non-timber forest products (NTFPs) that they can collect in forests [13]. The NTFPs supplement small-scale agricultural and livestock production for both household consumption and for sale in nearby urban centers [14]. The poverty of the dry forest's inhabitants stems from a range of historical conditions, some of which, like the status of these peoples as colonial subjects of European countries during the 19th and 20th centuries, have little or nothing to do with the dry forests that cover the land where they live.

Despite the large extent of the dry forests and the clarion calls to protect them from destruction at the hands of humans [15], their status has been largely neglected by land change scientists [12,16,17]. Analysts have identified threats to forests like population growth in rural areas, but no one has investigated empirically at a global scale whether or not a particular dry forest dynamic of deforestation exists. This paper uses cross-national data on deforestation in the wet and dry tropics to identify this dynamic.

The paucity of studies of the dry tropics most likely reflects the difficulties of defining and therefore delimiting dry forests. Most deforestation data sets, like those of the Food and Agriculture Organization of the United Nations' (FAO) Forest Resource Assessment (FRA) report, do not distinguish between dry and wet forests [12]. The measures for dry forests and woodlands changed with the introduction of new measurement technologies during the 1990s. Prior to the advent of remote sensing, definitions of dry forests emphasized the seasonality of the climate and the stature of the trees in a place. Dry forests occurred in places that experienced pronounced dry seasons and had trees that were shorter in stature than the trees in places with more humid climates [11]. Canopy cover became a more salient measure of dry forests with the increased use of remote sensing technologies after 1990 [16].

Canopy cover has long been a defining feature of forests. For most of the post-WWII era, the FAO defined a landscape as a forest if the canopy cover provided by trees exceeded 15%. More recently, analysts working at the global scale began to use the extent of canopy coverage to distinguish between wet and dry forests. In 2006, Miles and her colleagues [16] used the higher resolution (250 m) of the then newly available MODIS (Moderate Resolution Imaging Spectroradiometer) data to estimate the extent of dry forests globally. They categorized landscapes with canopy coverage between 40% and 80% as dry forests and forests with canopies between 80% and 100% as wet forests. Landscapes with less than 40% canopy coverage were not considered to be forests. In effect, open canopy forests became dry forests, and closed canopy forests became wet forests. I take the same approach here, using a new, global scale data set of LANDSAT images, with a still higher resolution than the MODIS data used by Miles and her associates [16]. The 30 m resolution of the LANDSAT data makes it possible to reliably discriminate between landscapes with differing amounts of canopy cover [10]. Countries in which most canopy cover fell between 25% and 75% would be categorized as dry forest countries while countries in which most canopy cover fell between 75% and 100% would be categorized as wet forest countries. Countries in which most canopy cover fell between 0% and 25% would be classified as non-forested countries.

Data on precipitation and canopy cover by nation supports this approach. Nations in which dry forests exceeded wet forests in extent in the LANDSAT data averaged 881 mm in annual precipitation compared to 1513 mm per year for those nations in which wet forests exceeded dry forests in extent (See Table 1 for data sources). Creating a binary in the tropics of dry forest nations and wet forest nations sacrifices ecological detail about these places, but the use of nations as the unit of analysis makes it possible to bring together in a single, cross-national data set ecological data on forests and socio-economic data on people. It then becomes possible to use this data set to investigate the human drivers of deforestation in wet and dry forested countries. To implement this analytic strategy, I created separate subsamples of (1) tropical nations in which humid forests exceeded dry forests in extent; and (2) tropical nations in which dry forests exceeded humid forests in extent. Analyses of each subsample should reveal whether or not different social and economic forces appear to drive deforestation in wet forest predominating and dry forest predominating countries.

1.2. The Dynamics of Deforestation in Dry and Wet Tropical Forests

While there are important differences, noted below, between the dynamics that drive deforestation in dry and wet tropical forests, there are also some important similarities in the two deforestation processes. In both wet and dry forests, roads provide links to urban areas, so in both places growth in the numbers and affluence of urban consumers would indirectly increase economic pressures to exploit forests at unsustainable rates [16,18]. In both settings, corridors of deforestation emerge along newly constructed or improved roads. Shifting cultivation and charcoal production clusters in corridors along highways in the dry forests of Zambia [19]. Similarly, smallholders, ranchers, and farmers have created corridors of cleared land along roads in the humid forests of the Brazilian Amazon [20]. Pepper farmers in the outer islands of Indonesia established their smallholdings on roads built by the loggers who first exploited these regions [21].

Stagnant economic conditions in both wet and dry forest countries would contribute to deforestation in both settings by making people reluctant to abandon long practiced agricultural livelihoods that entail slashing and burning forests on a regular basis [22]. The same stagnation in economic activity would slow the climb up the energy ladder [23] from wood to charcoal to natural gas, which in turn could contribute to the persistence of high deforestation rates in dry forest regions. In both wet and dry regions, migration to cities would encourage shifts in fuels because more compact fuels, like charcoal as opposed to firewood, would reduce the transportation costs incurred in getting the fuel to the end user.

Differences in the dynamics of deforestation between the two types of regions would begin with differences in building farm to market roads in wet and dry zones. In the wet zones, it takes considerable capital to build roads: to chop down the trees, dig drainage ditches, surface the road with gravel or tar, and construct bridges across the numerous streams and rivers. Private, not public enterprises have built most of the recently constructed or improved roads in the Amazon [24], so the deforestation associated with the recent construction of the new roads in wet zones like the Amazon has usually been carried out by highly capitalized, private enterprises. Roads in dry forests extend out in myriad directions from cities, creating cutover zones around cities that are readily visible in satellite imagery [25]. These roads usually take the form of 'tracks' caused by the repeated passage of vehicles. The tracks open up areas for exploitation, but their construction does not require the capital expenditures necessary to build roads in the humid tropics. Roads cross fewer streams, so they require fewer bridges. In addition, roads do not have to be constructed with gravel or paved surfaces to resist the mires that often occur on roads in regions with frequent rains. For this reason, among others, expenditures on roads to gain access to forests are less in dry zones, and enterprises that exploit dry forests might be expected to have less capital, on average, than those that exploit wet forests.

The clearing of dry forests should also differ from the clearing of wet forests because the agro-ecological productivities of these lands differ so much. Insufficient rainfall limits the productivity of lands in dry forest tropical biomes [26]. Dry tropical forests contain much lower levels of biomass than do humid tropical forests because the low levels of moisture in the soils of dry forests inhibit plant growth [11]. The insufficient rainfall reduces crop yields and diminishes the amounts of standing wood that loggers can cut [27]. The lower levels of agricultural productivity in dry biomes affect the economic geography of agricultural activities that people pursue in these places. The lower yields imply that cultivators will only be able to profit if they can minimize their input and transportation costs. These cost constraints contribute to small-scale and locally oriented agricultural economies in dry zones. The owners of small enterprises do not increase the scale of their operations because they cannot afford to pay the higher wages to workers outside of the family. The same logic applies to costs of transportation. To earn livelihoods from small harvests, cultivators in the dry forest zones must minimize their transportation costs, and this constraint makes it advantageous to sell products in local markets. Limited networks of penetration roads in dry forest regions, particularly in Sub-Saharan Africa [28], place an additional constraint on the long-distance transport of agricultural commodities. For these reasons, international trade would drive trade in the dry tropics to a lesser extent than it does in the wet tropics. In the dry tropics, the size of nearby consuming populations, as well as the continuing economic importance of agricultural livelihoods among rural peoples, would promote the rapid depletion of forests. Given these localized dynamics and the urban influences mentioned above, the loss of dry forests might occur primarily in peri-urban zones.

A different dynamic would appear to apply in humid tropical forests. The higher yields in the more humid zones enable cultivators to pay the relatively high costs of transporting their product to distant, sometimes overseas markets and still make a profit on the lumber that they extract, the soybeans that they grow, or the beef cattle that they raise. In this setting, a large volume of international trade, supported by a worldwide expansion in the size of markets, would play an important role in providing the impetus for the rapid destruction of wet forests [29]. International trade would also encourage the growth of large agricultural enterprises because their large scale would make it possible for the owners of these enterprises to take advantage of the economies of scale provided by the expansion in trade.

These dynamics suggest that different types of enterprises drive deforestation in the dry and wet tropics. Most deforestation in the dry tropics takes an artisanal form, in which small groups, sometimes extended families, work in labor intensive ways with few tools, cutting down trees in dry forests and woodlands in order to produce foods, wood-based fuels, and construction poles for local and regional markets. In the humid tropics, a contrasting dynamic, industrial in form, would prevail. It would consist of large agricultural enterprises that use machinery to clear land and produce for distant, urban markets. These hypotheses can be put to a partial test through statistical analyses of variations in the dynamics of deforestation in wet and dry nations in the tropics.

2. Materials and Methods

Table 1 provides information on the measures and sources for the data used in the univariate and multivariate analyses reported in Tables 2 and 3. The countries in the wet tropical forest and dry tropical forest subsets are listed in Appendix A (Table A1). The data set used for the analyses reported in Tables 2 and 3 is attached in Appendix B (Figure A1).

Table 1. Data Sources for Variables in Tables 2 and 3.

Agricultural Exports, Value in 2000:
World Bank. Agricultural raw materials exports (% of merchandise exports) [30].
Cereal Production, 2000:
World Bank. World Development Indicators [31].
% of People Economically Active in Agriculture, 2000:
Food and Agricultural Organization of the United Nations (FAOSTAT). Food and agriculture data [32].
Forest Area, 2000:
The total area in km² containing trees with canopy coverage that exceeded 25% in 2000. Supplementary materials—Table S3 in Hansen et al. [10].
Forest Losses, 2000–2012:
Removal of a tree canopy at a pixel (30 m) scale. Supplementary materials—Table S3 in Hansen et al. [10].
Forest Gains, 2000–2012:
Appearance of tree cover in more than 50% of a pixel between time 1 and time 2. Supplementary materials—Table S3 in Hansen et al. [10].
GDP per Capita, 2000:
World Bank. Gross Domestic Product per Capita [33].
Precipitation:
World Bank. Average precipitation is the average in depth (over space and time) precipitation in a country [34].
Urban Population, %: 2000:
United Nations, Department of Economic and Social Affairs. World Urbanization Prospects, the 2014 Revision [35].

Table 2. Differences between Nations with Closed and Open Canopy Forests [*],[†].

Tropics (90 Nations)	Wet Forest Countries Closed Canopy Forests Predominate (More than 50% of Forested Land)	Dry Forest Countries Open Canopy Forests Predominate (More than 50% of Forested Land)
% Pop. Active in Ag., 2000	37.1	60.4
% Econ. Activity in Ag., 2000	17.8	30.6
Value ($) PC of Ag. Exports, 2000	34.75	9.96
Cereal Yields (kg/hectare)	2502	1552
% Urban, 2000	51.0	34.8
% Change in Pop., 2000–2005	8.9	11.7
GDP, 2000 per capita	3243	725
% of Forests Lost, 2000–2012	2.82	1.46
% Reforested, 2000–2012	0.79	0.25

[*] All of the mean differences reported in this table are statistically significant at $p < 0.05$ or lower. [†] Open canopy forests (25%–75% canopy closure) predominate if they exceed in extent closed canopy forests (75%–100% canopy closure). Similarly, closed forests predominate when they exceed open forests in extent. With these data it was difficult to draw a forest–non-forest dividing line which affected the calculation of the total area in forests, which in turn led to some very skewed results for deforestation rates in arid countries with very little forest. To reduce the influence of these extreme values, we calculated the areal extent of forest losses over the 12-year period (2000–2012) as a proportion of a country's land area. Pop., Population; Ag., Agriculture; Econ., Economic; PC, Per Capita.

Table 3. Forest Losses (logged), 2000–2012, across Wet and Dry Forested Countries.

Variables	Robust Regression Coefficients (with Std. Errors)	
	(#1) Nations with More Closed Canopy (Humid) Forests	(#2) Nations with More Open Canopy (Dry) Forests
Area of Forests with Canopy >25%, Logged	0.980 *** (0.129)	0.661 *** (0.089)
Value of Ag. Exports, 2000 Logged	0.091 * (0.044)	0.052 (0.038)
% Urban Population, 2000	−0.024 (0.089)	0.033 * (0.018)
GDP Per Capita, 2000	−0.090 (0.065)	−0.057 * (0.029)
N of Cases	41	37
F	35.34	24.48
Prob. of F	0.000	0.000

$p < 0.001 = $ ***, $p < 0.10 = $ *.

High resolution LANDSAT data enabled this analysis by making it possible to reliably measure the extent of dry forests in a country, classify those countries into sets of predominantly dry forested and wet forested countries, and then carry out multivariate analyses of the social and economic dynamics of deforestation in each set of countries. Where to draw the cut points between the wet, dry, and no forest categories remains a source of some uncertainty in the analysis. I used the 0%–25%, 26%–75%, 76%–100% categories of canopy coverage for no forests, dry forests, and wet forests, respectively, because Hansen et al. [10] used them. Previous researchers [16] have used a somewhat different set of canopy cover categories for the same forest designations, 0%–39% for no forest, 40%–80% for dry forests, and 81%–100% for wet forests.

The dependent variables in these analyses are the forest losses for a twelve-year period, from 2000 to 2012. The remote sensing analysts used a conservative decision rule to determine whether or not forest cover change occurred in a pixel. For a forest loss to have occurred, forest cover would have to have declined below 50% of the area in a pixel between times A and B. Under this rule a pixel with a closed canopy could experience selective logging and, because the overall canopy coverage did not decline below 50%, the pixel would not register as having experienced deforestation. Similarly, a dry forest with an open canopy could experience a decline in forest cover from 45% to 30%, and it would not register as deforested. This distorting effect applies as much to open as to closed forests, so it would not seem to bias an analysis like this one that compares deforestation processes across the dry and wet tropics.

This study only looks at forests in the tropics, defined as countries with land areas located between the Tropics of Cancer and Capricorn. Because large forest losses can only occur in places with large forests, the equations in Table 3 contain a forest area variable that serves in effect as a control variable for the size of countries. The demographic and economic variables come from compendia of data on nation states published by the United Nations and the World Bank. To correct for skewed distributions, I logged three of the five variables in the multivariate analyses, the dependent variable, forest losses, and two independent variables, the size of forests in 2000 and the value of agricultural exports in 2000. The other two independent variables, GDP per capita and % urban were not sufficiently skewed in their distributions to warrant logging.

Cross-national data sets like this one that contain data from very large countries like Brazil and very small countries like Brunei frequently exhibit problems of heteroscedascity (unequal variances) and influential outlier cases that can produce misleading results when analyzed using ordinary least squares (OLS) approaches. A Breusch-Pagan test confirmed the presence of heteroscedascity in OLS regressions on these data, so, to limit the magnitude of these disturbances, I employed robust regression techniques in the multivariate analyses reported in Table 3. The equations do not exhibit problems of multi-collinearity. The variance inflation factors for the independent variables are all 1.41 or lower.

3. Results

In Hansen's post-2000 data, dry forests constitute 43% of the forests in the tropics, close to the 42% estimate in Murphy and Lugo's [11] work in the 1980s. The descriptive statistics assembled in Table 2 indicate some surprisingly large differences between nations with predominantly dry forests and nations with predominately wet forests. The rates of deforestation in Table 2 were higher in wet forest countries than they were in dry forest countries. The value of per capita agricultural exports was much higher in the wet forested countries than it was in the dry forested counties. In 2000, the value of agricultural exports from the wet forested countries was eight times greater than the value of agricultural exports from the dry forested countries. As might be expected from the differences in rainfall, regrowth occurred less frequently in the dry forest nations. As noted above, countries with more dry forests than wet forests have very poor, predominately rural populations with very large numbers of people earning their livelihoods from agriculture. Populations in the dry forest countries are increasing in size more rapidly than the populations in countries with humid tropical forests. Consistent with the line of reasoning offered above about the agro-ecology of dry forests, the agricultural productivity of cereal crops is lower in the dry forest countries than it is in the wet forest countries.

The results from the multivariate analyses in Table 3 largely support the line of reasoning presented above about the differences in the dynamics of land clearing in wet and dry forested countries. The dry forest equation (#2 in Table 3) suggests that consumer demands from large and growing populations of urban residents spur forest losses. This pattern suggests that deforestation may concentrate in the more accessible rural areas with unprotected forests [36]. The association between lower levels of economic activity and higher deforestation in the dry forest countries in Table 3 underscores how poverty and economic stagnation reinforce a people's dependence on the agricultural sector for their livelihoods and increase the pressure that they put on forests and other natural resources.

A primary difference between the two types of deforestation reported in the multivariate analyses in Table 3 has to do with the geographic scope of the associated agricultural economies. Exports of large volumes of agricultural products to distant, overseas markets spurs wet forest losses, as indicated by the findings in column 1 of Table 3. Malaysia, Indonesia, and Brazil exemplify this humid tropical pattern of voluminous agricultural exports and relatively high rates of deforestation. Agricultural exports have no discernible effect on forest losses in countries where dry forests predominate, as reported in column 2 of Table 3. In these places, significant numbers of people exploit the forests to produce goods for nearby consumers.

4. Discussion

The finding about the variable influence of agricultural exports on deforestation, a driver in countries with humid tropical forests but not in countries with dry tropical forests, is consistent with the depiction of the organization of wet tropical deforestation as 'industrial' and dry tropical deforestation as 'artisanal'. Deforestation in the humid tropics involves large-scale producers of agricultural commodities for distant markets, while deforestation in the dry tropics mostly involves small-scale producers who produce in labor intensive ways for local markets. The finding about deforestation in the wet tropics identifies one of the same drivers, agricultural exports, as did earlier, worldwide, remote sensing based analyses of tropical deforestation [1,6].

The binary association of dry forest losses with an artisanal organization of work and wet forest losses with an industrial organization of work can be carried too far. The most rapid rates of deforestation between 2000 and 2012, in analyses of the LANDSAT data, occurred in the dry forest regions of Paraguay and Argentina, where a wave of large-scale, industrialized agricultural expansion occurred [10,37]. Similarly, very detailed remote sensing analyses of deforestation in the wet forests of the Congo River basin between 1990 and 2010 demonstrated that small-scale cultivators drove much of the deforestation by opening up new fields adjacent to old fields close to the villages where they resided [38]. These admittedly large exceptions aside, the binary of artisanal production in dry tropical forests and industrial production in wet tropical forests finds some empirical support in this analysis, and could serve some useful heuristic purposes in future policy making and research on tropical deforestation.

In thinking about the characteristics of efforts to reduce deforestation in wet and dry tropical biomes, it is important to note the disjuncture between research on tropical deforestation in Latin America and Southeast Asia [29,39] and research on forest governance in Sub-Saharan Africa and South Asia [40–42]. Researchers interested in tropical deforestation almost never reference research on forest governance. This disjuncture reflects, more than anything else, the different dynamics that shape processes of land cover change in wet and dry forests. One often involves worldwide trade of agricultural commodities, while the other usually stems from trade in local markets for food and wood products.

The different loci of these market-driven deforestation processes have led policymakers to focus on different institutions. Wet forest analysts have focused on centralized controls from federal governments, as in Brazil [43], or on the governance of global flows of commodities to places where a significant fraction of consumers want to purchase green certified products [44]. These policy instruments make sense given the industrial scale of many producers in the wet forests and the long-distance flows of the commodities that they produce. Dry forest analysts have attended to the decentralization of political controls over forests and the clarification of smallholder tree tenure [40–42]. This focus follows from the localized circuits of production and trade in and around dry forests. In these smaller scale, more localized trading networks, local authorities would have distinct advantages in crafting and enforcing rules for the artisanal exploitation of the forests. They would be more likely to have the detailed knowledge necessary to discover the small, but cumulatively significant, transgressions in managing dry forests that frequently occur [45].

REDD+ (Reducing Emissions from Deforestation and Degradation) systems could be used to reduce either kind of deforestation, but the organizations through which REDD+ programs would be implemented would look quite different in dry forests compared to wet forests. Where small groups of local people deplete dry forests, community based organizations would play an important role in implementing REDD+. Where the agents of deforestation are larger enterprises, more centralized administering structures, including international certification groups and national governments with remote sensing tools, would probably play a more crucial role in administering REDD+.

5. Conclusions: Research Agendas for Wet and Dry Deforestation

The preceding analyses of deforestation in wet and dry tropical forests suggest several common and several different foci for further research. Some questions about deforestation dynamics seem important to investigate wherever they occur. First, landowners and land managers, wherever they are, must manage their forests for multiple uses, but how does one do that sustainably [46,47]? Two, the role of fire in forests remains poorly documented. How does it interact with climate change and the challenges of limiting greenhouse gas emissions? Finally, climate change has almost certainly begun to spawn coupled natural and human feedback effects in tropical biomes. Climate change induced droughts would accelerate forest losses, not only directly through a lack of rain, but also indirectly through shifts in rural livelihoods. In one recently documented instance in Madagascar [48], farmers, after suffering through crop failures caused by droughts, decided to become charcoal producers. In so doing, the farmers accelerated the rates of deforestation in Madagascar's dry forests. How large in magnitude are these feedback effects, and what do they portend for continued deforestation in tropical biomes?

Several research questions seem particularly important to investigate in one type of forest. Forest losses in the humid tropics now seem tied to long, transnational commodity (value) chains. Further research might focus on the organization of these commodity chains and the way that subcontracting has allowed some companies to endorse compacts for sustainability at the same time that they purchase products harvested from humid tropical forests in unsustainable ways by subcontractors. Deforestation in the dry tropics, with its artisanal organization of work, would seem likely to produce a more fragmented forest than the industrial-scale deforestation in the humid tropics. This circumstance, coupled with the presumed deleterious effect of forest fragmentation on biodiversity, would make it particularly important to investigate the deforestation–biodiversity crisis in dry forest as well as wet forest settings in the tropics.

Acknowledgments: The author of this paper benefited from participation the PARTNERS Research Coordination Network grant from the U.S. NSF Coupled Natural and Human Systems Program. #1313788.

Conflicts of Interest: The author declares no conflicts of interest.

Appendix A.

Table A1. Tropical Nations with Predominantly Closed Canopy (Wet) Forests or Open Canopy (Dry) Forests: Ordered beginning on left in column, then top to bottom, by Largest to Smallest Forest Losses, 2000–2012 in km^2 (Top to Bottom): Source: Hansen et al. [10].

Closed Canopy Forests > 50%	Open Canopy Forests > 50%
Brazil, East Timor	Paraguay, Reunion
Indonesia, Brunei	Mozambique, Mauritius
DRCongo, Trinidad Tobago	Tanzania, Cape Verde
Malaysia, Puerto Rico	Angola
Bolivia, Bhutan	Cote D'Ivoire
Colombia, Vanuato	Madasgascar
Mexico, Martinique	Zambia
Peru, Gaudeloupe	Nigeria
Myanmar	India
Venezuela	Ghana
Cambodia	CAR
Vietnam	Liberia
Laos	Guinea
Thailand	Zimbabwe
Guatemala	Uganda
Nicaragua	Benin
PNG	Chad
Philippines	Kenya
Ecuador	Ethiopia
Honduras	Burkina Faso
Cameroon	Sierra Leone
Republic of Congo	Mali
Panama	Malawi
Dominican Republic	Senegal
Gabon	Togo
Cuba	Guinea Bissau
Costa Rica	El Salvador
Belize	Bangladesh
Sri Lanka	Haiti
Guyana	Burundi
Suriname	Rwanda
Solomon Islands	Namibia
French Guinea	Gambia
Equatorial Guinea	Pakistan
Nepal	Somalia
Jamaica	Sudan
Fiji	Botswana

Appendix B.

Package

Figure A1. Data and Variable Definitions—SPSS v. 24.

References

1. DeFries, R.; Rudel, T.; Uriarte, M.; Hansen, M. Deforestation driven by urban population growth and agricultural trade in the twenty-first century. *Nat. Geosci.* **2010**, *3*, 178–181. [CrossRef]
2. Rudel, T. Changing agents of deforestation: From state initiated to enterprise driven processes, 1970–2000. *Land Use Policy* **2007**, *24*, 35–41. [CrossRef]
3. Jepson, P.; Jarvie, J.; Mackinnon, K.; Monk, K. The end for Indonesia's lowland forests. *Science* **2001**, *292*, 859–861. [CrossRef] [PubMed]
4. Margono, B.; Potapov, P.; Turubanova, S.; Stolle, F.; Hansen, M. Primary forest cover loss in Indonesia over 2000–2012. *Nat. Clim. Chang.* **2014**, *4*, 730–735. [CrossRef]
5. Aide, T.M.; Clark, M.; Grau, H.R.; Lopez-Carr, D.; Levy, M.; Redo, D.; Bonilla-Moheno, M.; Riner, G.; Andrade-Nunez, M.; Muniz, M. Deforestation and reforestation of Latin America and the Caribbean (2001–2010). *Biotropica* **2013**, *45*, 262–271. [CrossRef]
6. LeBlois, A.; DaMette, O.; Wolfersberger, J. What has driven deforestation in the developing countries since the 2000s: Evidence from new remote sensing data. *World Dev.* **2017**, *92*, 82–102. [CrossRef]
7. Fisher, B. African exception to drivers of deforestation. *Nat. Geosci.* **2010**, *3*, 375–376. [CrossRef]
8. Chidumayo, E. Environmental Impacts of Charcoal Production in Tropical Ecosystems of the World. In Proceedings of the Meetings of the Association of Tropical Biology and Conservation, Arusha, Tanzania, 12–16 June 2011.
9. Norris, J.; Allen, R.; Amato, E.; Zelinka, M.; O'Dell, C.; Klein, S. Evidence for climate change in the satellite cloud record. *Nature* **2016**, *536*, 72–75. [CrossRef] [PubMed]
10. Hansen, M.; Potapov, P.; Moore, R.; Hancher, M.; Turubanova, S.; Tyukavina, A.; Thau, D.; Stehman, S.; Goetz, S.; Loveland, T.; et al. High resolution global maps of 21st century forest cover change. *Science* **2013**, *342*, 850. [CrossRef] [PubMed]
11. Murphy, P.G.; Lugo, A. Ecology of tropical dry forest. *Annu. Rev. Ecol. Syst.* **1986**, *17*, 67–88. [CrossRef]
12. Blackie, R.; Baldauf, C.; Gautier, D.; Gumbo, D.; Kassa, H.; Parthasarathy, N.; Paumgarten, F.; Sola, P.; Pulla, S.; Waeber, P.; et al. *Tropical Dry Forests: The State of Global Knowledge and Recommendations for Future Research*; Discussion Paper; Center for International Forestry Research (CIFOR): Bogor, Indonesia, 2014.
13. Shackleton, S.; Gumbo, D. Contributions of non-wood forest products to livelihoods and poverty alleviation. In *The Dry Forests and Woodlands of Africa: Managing for Products and Services*; Chidumayo, E., Gumbo, D., Eds.; Earthscan: London, UK, 2010; pp. 63–91.
14. Gambiza, J.; Chidumayo, E.; Prins, H.; Fritz, H.; Nyathi, P. Livestock and wildlife. In *The Dry Forests and Woodlands of Africa: Managing for Products and Services*; Chidumayo, E., Gumbo, D., Eds.; Earthscan: London, UK, 2010; pp. 179–203.
15. Janzen, D. Tropical dry forests. In *Biodiversity*; Wilson, E.O., Ed.; National Academies Press: Washington, DC, USA, 1988; pp. 130–137.
16. Miles, L.; Newton, A.; DeFries, R.; Ravilious, C.; May, I.; Blyth, S.; Kapos, V.; Gordon, J. A global overview of the conservation status of dry tropical forests. *J. Biogeogr.* **2006**, *33*, 491–505. [CrossRef]
17. Chidumayo, E.; Marunda, C. Dry forests and woodlands in Sub-Saharan Africa: Contexts and Challenges. In *The Dry Forests and Woodlands of Africa: Managing for Products and Services*; Chidumayo, E., Gumbo, D., Eds.; Earthscan: London, UK, 2010; pp. 1–9.
18. Gumbo, D.; Chidumayo, E. Managing Dry Woodlands for Products and Services: A Prognostic Synthesis. In *The Dry Forests and Woodlands of Africa: Managing for Products and Services*; Chidumayo, E., Gumbo, D., Eds.; Earthscan: London, UK, 2010; pp. 261–279.
19. Moore, H.; Vaughan, M. *Cutting down Trees: Gender, Nutrition, and Agricultural Change in the Northern Province of Zambia, 1890–1990*; Heinemann: Portsmouth, NH, USA, 1994.
20. Smith, N. *Rainforest Corridors: The Transamazon Colonization Scheme*; University of California Press: Berkeley, CA, USA, 1982.
21. Vayda, A.P.; Sahur, A. Forest clearing and pepper farming by Bugis migrants in East Kalimantan: Antecedents and impacts. *Indonesia* **1985**, *39*, 83–110. [CrossRef]
22. DeBroux, L.; Hart, T.; Kaimowitz, D.; Karsenty, A.; Topa, G. *La Foret en la Republique Democratique du Congo Post-Conflit: Analyse d'un Agenda Prioritaire*; Rapport Collectif par des Equipes de la Banque Mondiale,

du Center for International Forestry Research (CIFOR), du Centre Internationale de Recherche Agronomique pour le Developpement (CIRAD): Paris, France, 2007.

23. DeFries, R.; Pandey, D. Urbanization, the energy ladder, and forest transitions in India's emerging economy. *Land Use Policy* **2010**, *27*, 130–138. [CrossRef]

24. Perz, S.; Overdevest, C.; Arima, E.; Caldas, M.; Walker, R. Unofficial road building in the Brazilian Amazon: Dilemmas and models of road governance. *Environ. Conserv.* **2007**, *34*, 112–121. [CrossRef]

25. Malaisse, F.; Binzangi, K. Wood as a source of fuel in upper Shaba (Zaire). *Commonw. For. Rev.* **1985**, *64*, 227–239.

26. Chamshama, S.; Savadogo, P.; Marunda, C. Plantations and woodlots in Africa's Dry Forests and Woodlots. In *The Dry Forests and Woodlands of Africa: Managing for Products and Services*; Chidumayo, E., Gumbo, D., Eds.; Earthscan: London, UK, 2010; pp. 205–230.

27. Sitoe, A.; Chidumayo, E.; Alberto, M. Timber and wood products. In *The Dry Forests and Woodlands of Africa: Managing for Products and Services*; Chidumayo, E., Gumbo, D., Eds.; Earthscan: London, UK, 2010; pp. 131–153.

28. Foster, V. *Africa Infrastructure Diagnostic—Overhauling the Engine of Growth: Infrastructure in Africa*; World Bank: Washington, DC, USA, 2008.

29. Nepstad, D.; Stickler, C.; Almeida, O. Globalization of the Amazon soy and beef industries: Opportunities for conservation. *Conserv. Biol.* **2006**, *20*, 1595–1603. [CrossRef] [PubMed]

30. World Bank. Agricultural Raw Materials Exports (% of Merchandise Exports). Available online: http://data.worldbank.org/indicator/TX.VAL.AGRI.ZS.UN (accessed on 8 September 2016).

31. World Bank. World Development Indicators. Available online: http://databank.worldbank.org/data/reports.aspx?source=world-development-indicators (accessed on 8 September 2016).

32. Food and Agricultural Organization of the United Nations (FAOSTAT). Food and Agriculture Data. Available online: http://faostat.fao.org/site/375/default.aspx (accessed on 9 September 2016).

33. World Bank. Gross Domestic Product Per Capita. Available online: http://data.worldbank.org/indicator/NY.GDP.PCAP.CD (accessed on 9 September 2016).

34. World Bank. Average Precipitation Is the Average in Depth (over Space and Time) Precipitation in a Country. Available online: http://data.worldbank.org/indicator/AG.LND.PRCP.MM (accessed on 9 September 2016).

35. United Nations, Department of Economic and Social Affairs. World Urbanization Prospects, the 2014 Revision. Available online: https://esa.un.org/unpd/wup/ (accessed on 14 July 2016).

36. Ickowitz, A.; Slayback, D.; Asanzi, P.; Nasi, R. *Agriculture and Deforestation in the Democratic Republic of the Congo: A Synthesis of the Current State of Knowledge*; Occasional Paper 119; CIFOR: Bogor, Indonesia, 2015.

37. Gasparri, N.I.; Grau, R. Deforestation and Fragmentation of Chaco Dry Forests in NW Argentina (1972–2007). *For. Ecol. Manag.* **2009**, *258*, 913–921. [CrossRef]

38. Mayaux, P.; Pekel, J.; Desclée, B.; Donnay, F.; Lupi, A.; Achard, F.; Clerici, M.; Bodart, C.; Brink, A.; Nasi, R. State and evolution of the African rainforests between 1990 and 2010. *Philos. Trans. R. Soc. B* **2013**, *368*. [CrossRef] [PubMed]

39. Curran, L.; Trigg, S.; MacDonald, A.; Astiani, D.; Hardiono, Y.; Siregar, P.; Caniago, I.; Kasischke, E. Lowland forest lost in protected areas of Indonesian Borneo. *Science* **2004**, *303*, 1000–1003. [CrossRef] [PubMed]

40. Agrawal, A.; Chhatre, A. Explaining success on the commons: Community forest governance in the Indian Himalaya. *World Dev.* **2006**, *34*, 149–166. [CrossRef]

41. Ribot, J.; Treue, T.; Lund, J. Democratic decentralization in Sub-Saharan Africa: Its contribution to forest management, livelihoods, and enfranchisement. *Environ. Conserv.* **2010**, *37*, 35–44. [CrossRef]

42. Reij, C.; Tappan, G.; Smale, M. *Agri-Environmental Transformation in the Sahel: Another Kind of Green Revolution*; Discussion Paper 00914; International Food Policy Research Institute (IFPRI): Washington, DC, USA, 2009.

43. Boucher, D.; Roquemore, S.; Fitzhugh, E. Brazil's success in reducing deforestation. *Trop. Conserv. Sci.* **2013**, *6*, 426–444. [CrossRef]

44. Sikor, T.; Auld, G.; Bebbington, A.; Benjaminsen, T.; Gentry, B.; Hunsberger, C.; Izac, A.; Margulis, M.; Plieninger, T.; Schroeder, H.; et al. Global land governance: From territory to flow. *Curr. Opin. Environ. Sustain.* **2013**, *5*, 522–527. [CrossRef]

45. Lund, J.; Treue, T. Are we getting there?: Evidence of decentralized forest management in the Tanzanian Miombo woodlands. *World Dev.* **2008**, *36*, 2780–2800. [CrossRef]

46. Pretty, J. Sustainable intensification in Africa. *Int. J. Agric. Sustain.* **2011**, *9*, 3–4. [CrossRef]
47. Montpellier Panel. *Sustainable Intensification: A New Paradigm for African Agriculture*; Agriculture for Impact: London, UK, 2013.
48. Onishi, N. Africa's Charcoal Economy Is Cooking. The Trees Are Paying. Available online: https://www.nytimes.com/2016/06/26/world/africa/africas-charcoal-economy-is-cooking-the-trees-are-paying.html?_r=0 (accessed on 25 June 2016).

![forests logo] *forests*

MDPI

Communication

How Cities Think: Knowledge Co-Production for Urban Sustainability and Resilience

Tischa A. Muñoz-Erickson [1,*], Clark A. Miller [2] and Thaddeus R. Miller [2]

[1] USDA Forest Service, International Institute of Tropical Forestry, Río Piedras, PR 00926, USA
[2] School for the Future of Innovation in Society, Arizona State University, Tempe, AZ 85287, USA;
    clark.miller@asu.edu (C.A.M.); thad.miller@asu.edu (T.R.M.)
* Correspondence: tamunozerickson@fs.fed.us; Tel.: +1-928-600-1613

Academic Editors: Grizelle González and Ariel E. Lugo
Received: 1 April 2017; Accepted: 6 June 2017; Published: 10 June 2017

Abstract: Understanding and transforming *how cities think* is a crucial part of developing effective knowledge infrastructures for the Anthropocene. In this article, we review knowledge co-production as a popular approach in environmental and sustainability science communities to the generation of useable knowledge for sustainability and resilience. We present knowledge systems analysis as a conceptual and empirical framework for understanding *existing* co-production processes as preconditions to the design of *new* knowledge infrastructures in cities. Knowledge systems are the organizational practices and routines that make, validate, communicate, and apply knowledge. The knowledge systems analysis framework examines both the workings of these practices and routines and their interplay with the visions, values, social relations, and power dynamics embedded in the governance of building sustainable cities. The framework can be useful in uncovering hidden relations and highlighting the societal foundations that shape what is (and what is not) known by cities and how cities can co-produce new knowledge with meaningful sustainability and resilience actions and transformations. We highlight key innovations and design philosophies that we think can advance research and practice on knowledge co-production for urban sustainability and resilience.

Keywords: knowledge co-production; idiom of co-production; knowledge infrastructures; knowledge systems; knowledge systems analysis; cities; land use governance; Anthropocene

1. Introduction

Cities are increasingly leaders in the creation and transition to more sustainable and resilient pathways. From more efficient transportation and building technologies to green infrastructure solutions that protect people from flood hazards, cities are on the front line of implementing sustainable strategies and building new infrastructures to enhance resilience to climate change [1,2]. Yet, cities also face great challenges to sustainability transformations. Cities exhibit obduracy because of existing social, economic, political, and physical structures that are difficult to change, even when the vision and actions needed are known [3]. Why is it that, even when agreeing on what needs to be done, city institutions and infrastructure are resistant to change towards more sustainable pathways?

We believe that part of the answer lies in the way that urban knowledge systems—the social practices through which knowledge, ideas, and beliefs are produced, circulated, and put into action—keep certain patterns of thinking in place. Events like Hurricane Katrina and Superstorm Sandy, for instance, have exposed failures in the knowledge systems that engineers, designers, and decision-makers used to design hurricane protection infrastructures and limited the abilities of cities like New Orleans and New York to reduce the vulnerability of their populations to various stresses and shocks, including extreme climate and weather variability [4,5]. Addressing the changing conditions of the Anthropocene will thus require innovations in not only how we design cities' built infrastructures but also in how we upgrade and design their knowledge infrastructures as well [6]. In other words, sustainability demands transformations in ways of thinking—or *how cities think*.

This article examines *knowledge co-production*, an idea which is increasingly popular within the environmental and sustainability research communities, as a promising approach to generate and apply usable knowledge for complex sustainability challenges [7–12]. In its most robust form, knowledge co-production refers to linked practices of knowledge production and application where diverse science, practice, and policy actors collectively identify problems, produce knowledge, and put that knowledge into action through collaboration, integration, and learning processes [13–15]. Knowledge co-production re-thinks the relationship between knowledge and decision-making beyond conventional notions of the 'science–policy interface' that assume that knowledge production and decision-making happen independently from one another [16,17]. This approach is deemed promising for building knowledge systems for cities because it acknowledges the diversity of actors, knowledge systems, social relations and networks involved in creating and applying knowledge relevant to sustainability [18,19].

Too often, however, experiments in knowledge co-production suppose that the construction and use of new knowledge can simply happen de novo, independent of what has come before. Yet, as illuminated in detail by Sheila Jasanoff [20], in regulatory settings, the construction and use of knowledge is deeply intertwined with arrangements and practices of governance—and cities are no different [21]. How cities know and how they design social and policy arrangements go hand-in-hand; they get made and produced together. Knowledge both is an outcome of governance and creates the conditions for it. It contributes to, comes to be embedded in, and helps to construct shared beliefs, discourse, practices, policies, and visions. Thus, the city transformations envisioned by advocates of knowledge co-production cannot be understood as mere exercises in creating and applying knowledge, however broadly sourced across diverse participants; rather, they are exercises in reconfiguring the relationships between and institutional configurations of both how cities think and how they act. They are thus social and political exercises at least as much as they are epistemic ones.

We propose in this article that an analysis of the co-production of existing knowledge-governance dynamics and conditions, as defined by Jasanoff, can help cities to understand and improve their ability to create and deploy new knowledge effectively in service of sustainability and resilience. Large investments are currently being directed towards knowledge co-production experiments in support for sustainability and resilience in cities. The project we are currently involved in, for instance, the Urban Resilience to Extreme Weather-related Events Sustainability Research Network (UREx SRN), is a $12 million dollar investment by the National Science Foundation to co-produce new knowledge and new strategies to improve the resilience of urban infrastructures among researchers, cities, and urban stakeholders. This effort engages urban governance institutions that already know in well-defined ways—and through well-defined practices and routines—that shape how they design and implement infrastructure projects and plans. Understanding how city knowledge systems and dynamics construct and shape what decision-makers already know and wish to know, vis-à-vis infrastructure in their cities, is thus a crucial prior step to investing in new organizations and policy arrangements for knowledge co-production in cities. To put it differently, analyzing how cities think is a necessary precondition to building capacities and designing institutions for knowledge co-production for sustainability.

We present *knowledge systems analysis* as a conceptual and empirical framework to understand how cities think. Following a review of the definitions of and approaches to co-production found in the literature, we describe knowledge systems analysis and how it can be used by both researchers and practitioners to analyze the contexts in which new efforts to co-produce sustainability and resilience knowledge and action are situated. In particular, knowledge systems analysis emphasizes the structured social and institutional processes within which knowledge and information are produced, evaluated, circulated, and applied in governance and decision-making [22,23]. We then highlight key innovations and design philosophies that we think can advance efforts to co-produce knowledge and action for urban sustainability and resilience. We conclude with suggestions for future research directions for analyzing urban knowledge systems and applying these to improving future knowledge co-production efforts.

2. Knowledge Co-Production for Sustainability and Resilience

There are two main interpretations and uses for the term "co-production" [9,21,24]. Within the sustainability science community, knowledge co-production, as van Kerkhoff and Lebel [9] define it, is a prescriptive and instrumental form as it invokes an agenda where relationships can and should be deliberatively designed and managed for improving the scientific basis of decision-making at the project and program scale. This instrumental use of the concept involves shared or collaborative knowledge production to link knowledge to action. Specifically, this literature focuses on how to make knowledge systems—or the institutions to harness science and technology for sustainability—more effective [15]. A key finding of this line of research has shown that knowledge systems are most likely to be effective in influencing action if they are perceived to be salient, credible and legitimate by the larger stakeholder community [15]. This idea of knowledge co-production has taken hold most notably in the contemporary literature exploring science–policy interactions in part as a response to failed conventional science–policy models that assumed that if you get the science right and put it in the hands of the right people, it will be used automatically to inform decision-making. Examples of these conventional models include the loading dock model, where science is transferred to the policy 'dock' through a one-way loading truck, or the bridge model, wherein academia and policy engage in a two-way interaction by building bridges between the two [17]. By giving a new look into how science–policy interfaces are organized, the literature is moving away from looking at the relationship between science and society as a one- or two-way interaction to more of a complex relationship in terms of multiple actors and knowledges, multiple interactions, and multiple mechanisms (see for instance [10–19,24–26]).

The recent popularity of organized arrangements, such as "boundary organizations" in sustainability science [26], reflects the growing importance and social investment given to these institutional approaches as a way to effectively link knowledge systems with user demands [27]. Other examples of knowledge co-production ideas put into practice include joint knowledge production [13], collaborative adaptive management [14], transdisciplinary research [28], and communities of practice [12]. Throughout each of these flavors of knowledge co-production there are several common themes. Building trust between and amongst both researchers and stakeholders and developing a common sense of project goals is fundamental to the process. Collaborating with a broad and relevant range of stakeholder groups [29] with different skills and assets (e.g., knowledge brokers, assessment teams, implementers, and bridging agents) across project elements, including the articulation and identification of knowledge needs and questions, is also crucial to maximize knowledge co-production [30]. These practices and an open, deliberative, transparent setting that promotes trust help to promote mutual, social learning—a goal as important as more specific project specific deliverables [10].

The other form of the concept of co-production has a long lineage as an analytical lens in the fields of history of science and science and technology studies (STS), particularly through the work by Sheila Jasanoff on the dynamic interaction between the production of knowledge and social order [20]. According to Jasanoff, the idiom of co-production highlights the mutually constitutive, interactive, and influential arrangements of knowledge-making and decision-making in various aspects of political life—knowledge both shapes and is shaped by social processes. In Jasanoff's words, "the ways in which we know and represent the world (both nature and society) are inseparable from the ways in which we choose to live in it" [20] (p. 2). Therefore, the production and use of knowledge is deeply embedded in all kinds of social, cultural, and political dynamics, such that what we know cannot be separated from how we act and organize the world.

This version of co-production brings into focus underlying knowledge–power dynamics and social practices that can help to explain both how worldviews and ways of thinking remain in place and difficult to change and how they change over time. For Jasanoff [20], co-production emerges from the constant interplay of different cultural domains, including the cognitive, the material, the social, and the normative. Jasanoff and Wynne [31] further argue that these cultural domains can vary across different policy cultures—bureaucratic, civic, economic, scientific (Figure 1). These policy cultures have different knowledge-governance formulations such that they share practices for producing knowledge that also align with how they view and understand how the world works, and more importantly, how it should work. These policy cultures are constantly interacting with each other, but they are also competing forms of rationality that shape social order, within their own domains and across the collective whole.

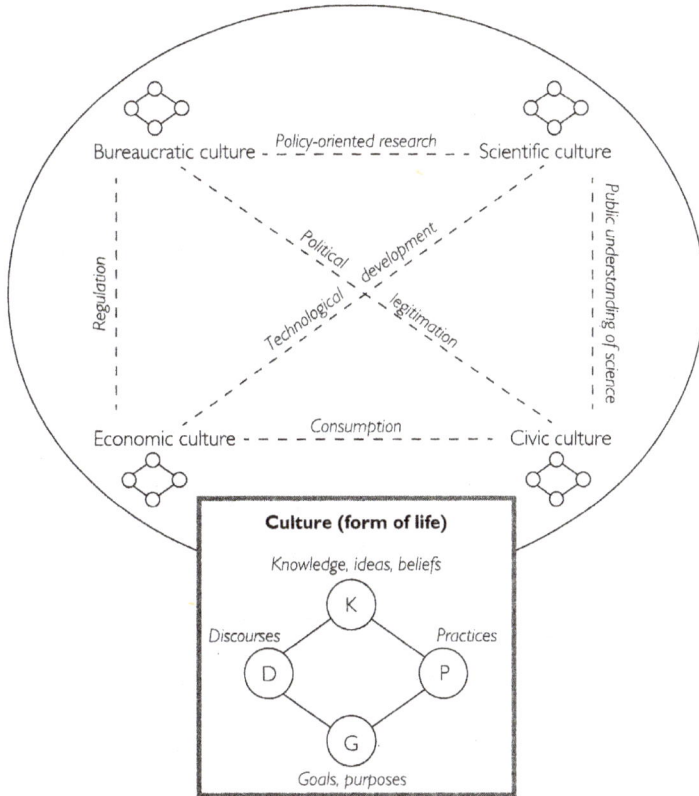

Figure 1. Illustration of the model of co-production of knowledge and society. Derived from Jasanoff and Wynne 1998 (Battelle Press: Columbus, OH, USA).

Research by Carina Wyborn on what she terms "connectivity conservation" offers an empirical example of the application of co-production as analytical lens. Wyborn [24] operationalized Jasanoff's categories of co-production—context (material), knowledge (cognitive), process (social), vision (normative)—as a lens to empirically examine co-production processes in two cases of connectivity conservation in the US (Yellowstone to Yukon Region) and Australia (Habitat 141°). In both cases, researchers and practitioners were attempting to establish knowledge co-production efforts to facilitate the link between conservation science and governance. Wyborn found that, while both cases had similar propositions of the relationship between science and governance, the ways in which the work played out in each case to co-produce context, knowledge, process, and vision of governance determined the different framings and outcomes. In other words, on-the-ground knowledge–power dynamics played out differently in each case, and in each case the dynamics did not correlate with the design principles for linking conservation science and action. Wyborn suggests that highlighting how co-production shapes the relationship between science and governance can be a fruitful contribution to the design of efforts to advance knowledge for adaptive governance.

The analytical form of co-production resonates with the concept of "knowledge governance" that is developing in the sustainability literature. While this concept has a distinct interpretation in organizational economics as an approach to maximizing knowledge transactions to improve organizational efficiency [32], the analytical form of co-production we are discussing here is more closely aligned with the critical lens of socio-political approaches described in van Kerkhoff [33]. Specifically, like knowledge governance approaches, co-production analyzes direct attention to the formal and informal rules, conventions, and networks of actors that shape the ways we approach knowledge processes, such as creating, sharing, accessing, and using knowledge [33–36]. Similarly, knowledge governance focuses on a broader level than the project-based use of knowledge co-production through joint knowledge efforts or boundary management, to what van Kerkhoff describes as the middle layer where the institutional 'rules of the game' shape the possibilities and choices available to decision-makers and organizations. A key distinction, however, is that, by examining how these knowledge governance dynamics are embedded in broader social, political, and cultural dynamics, Jasanoff's co-production goes further to describe the macro-social processes that link how we govern knowledge with how we govern society [20]. At the same time, knowledge governance, like knowledge co-production, tends to focus more on how knowledge gets made and less on the organization of decision-making as an instantiation of particular ways of knowing. Still, there are important similarities, and the co-production and the knowledge systems analysis framework we present in the next section lend themselves to examining existing knowledge governance dynamics and conditions that may enhance or constrain cities' knowledge processes in cities.

3. Knowledge Systems Analysis: A Framework to Design Knowledge Co-Production in Cities

Both variants of the concept of co-production we have discussed are important for urban sustainability. Together, they present a more sophisticated and nuanced view of the relationship between knowledge and action. No longer is the relationship between knowledge and policy seen as a one-way or two-way interaction where knowledge is generated on one side, (the 'knowledge' side of scientists and/or experts that is then transferred to the other side), and 'policy' on the other side (where decision-making bodies use the knowledge). Rather, the interactions of knowledge and decision-making in governance processes are much more complex, especially as we seek to transform both how institutions think and act in pursuit of greater sustainability. Knowledge is rarely singular, for example, in sustainability problems, nor is governance; instead, multiple knowledge institutions intersect across a multiplicity of governing sites that transcend traditional institutional and jurisdictional boundaries [37].

Ideas of co-production particularly highlight the challenges to changing how cities think. They show that the social organization of cities is closely coupled with how cities organize knowledge, such that to re-organize and transform cities requires simultaneously changing how they organize knowledge production and how they put that knowledge to use in formulating policy. At the same time, to re-organize knowledge requires understanding how urban governance and life function socially, politically, and economically, including the factors that enable and constrain the possibility of change in urban knowledge systems. Therefore, in efforts to create and apply new knowledge for urban sustainability and resilience (knowledge co-production), a crucial first step is to understand the complex ways in which epistemic and governance practices are already interlaced across diverse city processes and institutions (the co-production of knowledge and governance).

Cities present a great challenge to the design of knowledge co-production approaches for sustainability. We are concerned that efforts to engage in knowledge co-production in support of urban sustainability and resilience generally lack a thorough examination of how cities think—what local people know about the city, how they know and experience the city, how they envision the city. Cities are more than the physical and institutional infrastructures that service an urban population. Cities are also spaces where a high diversity of organizations and their knowledge systems can come together in networks to catalyze or oppose new ideas and innovations. While urban governance scholars recognize the importance of multiple knowledges or expertise in researching and developing strategies toward the sustainable city [38,39], they may lack a critical analysis of the politics and power dynamics surrounding expertise, of the institutional practices that shape what knowledge is produced and how cities are envisioned, and whether capacities are present to rethink and reconfigure the linkages between knowledge and action.

We present *knowledge systems analysis* as a framework to describe and analyze existing knowledge and governance interactions as pre-conditions to designing knowledge co-production efforts for urban sustainability and resilience (Figure 2). We define knowledge systems as the organizational practices and routines that generate, validate, communicate, and apply knowledge [22,23]. We consider knowledge systems as more than sites where research, data, and information are produced and used in decision-making. They are also where imaginations, ideals, and beliefs of social order are being forged by different social groups [37]. Knowledge systems frame which questions get asked, and which don't, and determine the methods used to answer those questions. They define assumptions, establish burdens of proof, and decide who does the review and how. They lay out how to decide when knowledge is uncertain and what to do if it is and they set limits on the boundaries of relevant expertise. They also set priorities for investments in new knowledge.

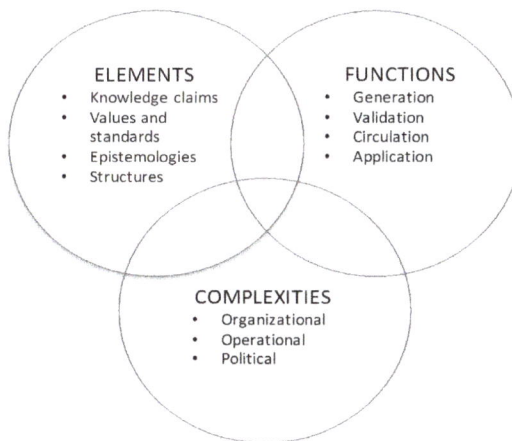

Figure 2. Main components of the knowledge systems analysis framework.

Another important distinction in the way that we view and use knowledge systems is how we define knowledge. We define knowledge as a claim or an idea or belief that someone, whether an individual or a community, takes to be true, or at least relatively more true than other kinds of statements, and therefore of sufficient merit to guide his, her, or their reasoning or, especially for our purposes here, action. This definition of knowledge stems from a sociological perspective that acknowledges the complex judgments, ideas, framings, tacit skills and values that shape what knowledge is, rather than viewing it as just simple statements of truth or fact [31]. As Jasanoff [40] argues, to understand knowledge requires understanding knowledge-in-the-making. Dynamic social processes are involved in the making of knowledge such that its production is a result of the articulation, deliberation, negotiation, and valorization of particular knowledge claims. The structure and dynamics of these social processes determine, in turn, whose knowledge claims matter and how claims are constructed, evaluated, contested, and sanctioned as knowledge [41]. This view of knowledge is a fundamental basis for the "idiom of co-production" that informs our knowledge systems analysis framework.

Our approach to analyzing knowledge systems begins with an examination of the elements of knowledge systems, including the content and types of knowledge being produced, the values, standards, and epistemologies (or ways of knowing) that guide its work, and the social practices and structures involved in creating and applying knowledge. Knowledge systems analysis describes what the knowledge system knows as well as what it doesn't know (e.g., the tacit and explicit uncertainties that surround knowledge claims and the kinds and varieties of knowledge claims that the knowledge system might produce but doesn't for one reason or another), as well as the values, methods, and epistemologies that organize how the system knows what it does (Table 1). In addition, knowledge systems analysis focuses on the people and social practices that make knowledge. The practices used to make knowledge are often hidden from plain view, even to those who are producing or using this knowledge. Even less obvious are the cognitive and cultural dispositions that shape how groups and institutions think. More often than not, organizations take for granted how they know what they know. Much like journalists use a variety of sources to put together a story, or an archeologist uses material and textual tools to 'dig up' evidence from the past, knowledge systems analysis uses different conceptual lenses and approaches to map and describe where within a knowledge system a particular knowledge is located: who knows what; where data is generated and stored; and how, where, and to whom it flows as it is processed, handled, shared, and used. At the same time, knowledge system mapping requires understanding how the people involved in knowledge systems are organized, trained, evaluated, and rewarded for their work [22,23].

Table 1. Elements of knowledge systems.

Framework Concepts	Definition or Use in Knowledge Systems Analysis	Example
Knowledge Claims	Statements or propositions about the world whose relationship to truth cannot be easily or directly ascertained. Whether they are correct or not is always uncertain, at least to some degree.	The statement that "the 2010 Census enumerated 308,745,538 people in the US" is a claim, since the Census cannot obtain an exact count of every single person in the country.
Values and Standards	Define the foundation of knowledge production in the system through a process of simplification, or creating simplified representations of complex social and/or natural processes. Which aspects of reality get simplified, and to what extent, is a value choice.	Standardized methodologies to measure greenhouse emissions defined by normative principles outlined by the Conference of Parties of the UN Framework Convention on Climate Change.
Epistemologies	Ways of knowing and reasoning about the world, including a diversity of elements, such as problem framing, forms of evidence and argumentation, deeper imaginaries that inform them, and the technologies used to produce knowledge.	As knowledge production methods, statistical and experimental epistemologies employ different analytical and conceptual approaches, techniques, standards of evidence, and underlying assumptions of causality.
Structures	The social and organizational arrangements, networks, and institutions of the people that construct knowledge. Involves understanding how the people involved in knowledge systems are organized, trained, evaluated, and rewarded for their work.	The USDA Forest Service Research and Development program is a highly-structured knowledge system, organized into many levels of research science (e.g., GS level), with specified standards and norms that define the expectations of the scientists' work and level of productivity.

Knowledge systems analysis also examines the tasks or functions of knowledge systems, with an emphasis on four key functional areas: knowledge generation, validation, circulation, and application ([22,23]; Table 2). Knowledge generation refers to the act of generating knowledge through research, for instance, whether it is scientific research, market research, or journalism. This process involves problem formulation, data collection and analysis, and reporting of information. The second function of knowledge systems is knowledge validation and involves the practices by which knowledge is subject to review, critique, assessment, and check. A common example is the review process used to publish scientific papers or expectations that journalists check the facts of their story before publishing it. Rules and expectations for reviewing and judging the validity of the information can vary from scientific journals and media outlets, so part of analyzing knowledge systems involves figuring out what the expectations are, who determines them, and how are they put into practice. Knowledge circulation refers to the practices of communicating, exchanging, transmitting, or translating knowledge from one person or organization to another. Other ways people often refer to this activity are knowledge exchange or information flows. Describing this activity involves sorting out who has access to new knowledge claims, through what channels, and what forms of communication are used and whether these are properly communicating knowledge.

Table 2. Functions of knowledge systems.

Framework Concepts	Definition or Use in Knowledge Systems Analysis	Example
Generation	The process and activities of problem formulation, data collection, data analysis, and reporting of information. A common example is research, whether scientific, market, or journalistic research. Activities include the ways these activities are carried out, by whom, with what attention to detail and with what methodologies and resources.	The Census data collection process involves significant fieldwork (e.g., surveyors that travel around communities knocking on doors for people to fill out their forms), but also legal and political work that govern knowledge generation (e.g., Congress writes laws specifying how the Census will be conducted). Agencies must also develop regulatory processes to determine exactly what data to collect and which methods to use.
Validation	The practices, processes, and routines by which knowledge claims are subject to review, critique, assessment, or check. Includes who in a knowledge system is assessing, reviewing, testing, or otherwise checking the knowledge that is being generated.	The National Science Foundation peer review process is known for the rigor of its procedures and the caliber of the scientists that it brings together to evaluate the quality of the research generated by the agency's funding.
Circulation or Communication	The practices by which knowledge claims are exchanged, transmitted, or translated from one location to another. Involves sorting out who has access to new knowledge claims, through what channels, whether those are the right people, whether the forms of communication are properly communicating enough additional information to judge a knowledge claim and its value.	Nutritional labeling in the US is an explicit effort to ensure that knowledge claims are circulated to a wide array of citizens. The standardization of food packaging labels enhances consumer decision-making by making knowledge available and easy to read at the time of purchase.
Application	The social and institutional practices by which knowledge is factored into decisions. This phase is often also referred to as the use, uptake, or consumption of knowledge.	Regulatory agencies, like the EPA, have internal and external processes, such as administrative hearings, to present and review relevant scientific research when constructing a new regulatory rule. The agency must decide how to put the knowledge collected and reviewed to use, typically through formal and informal conversations and deliberations, an official judgment and then formal statement by the Administrator.

A final function of knowledge systems is knowledge application. This phase is usually where most of the literature on linking knowledge to action and knowledge co-production focuses, as it refers to the social and institutional practices by which knowledge is factored into decisions, or put into action by decision-makers and stakeholders. In other words, this is the phase related to users and consumers of knowledge, or knowledge users. For instance, we know from previous research that knowledge systems tend to be more effective when the knowledge is viewed as credible, legitimate, and salient by multiple stakeholders (e.g., [15]). We know less, however, about the nuances of how exactly knowledge is acted upon and how this use of knowledge feeds back into the other functions of knowledge systems (generation, validation, and circulation). Who acts on particular kinds of knowledge? What other knowledge systems do stakeholders already rely on to make their decisions? What expectations do they have about the knowledge system? How is uncertainty about the knowledge being communicated? What do users know about how the knowledge was generated, validated, and circulated in order to evaluate whether the knowledge expressed is credible, legitimate, and salient? These questions raise the point that the functions of knowledge systems are not independent of one another, but rather are tightly coupled, with each facet of the system reinforcing the others.

More often than not, the co-production of knowledge, decisions, and actions around sustainability and resilience involve many diverse institutions. The functions and tasks of knowledge systems are thus often distributed across multiple organizations with varying structures, goals, and degrees of accountability. In boundary organizations, for instance, Guston describes multiple lines of accountability to both scientific and political organizations [26]. In addition, because knowledge co-production efforts attempt to bring together different types of knowledge and expertise, the intertwining of multiple epistemic cultures will likely bring to the fore both epistemic conflict over different assumptions about how the world works and political conflict over whose expertise should count in decision-making [42,43].

In our framework we describe these dynamics in terms of three layers of complexities in knowledge systems: organizational, operational, and political ([22,23]; Table 3). Organizational complexity arises when multiple organizations or networks are involved in knowledge production. Operational complexity refers to instances when the goals and values underlying the collection of knowledge, and the processes needed to generate that knowledge, are not obviously aligned; thus considerable effort needs to be placed to coordinating activities and routines, such as standardizing research protocols, to ensure the credibility of the system. Political complexity arises when the work and products of knowledge systems become entangled with politics or conflicts within or between organizations. The case studies examined by Wyborn, which we reviewed briefly above, offer a good illustration of these complexities [24]. Both cases showed significant organizational and political complexity as they engaged multiple science, management and policy organizations to outline strategies for connectivity conservation based on conservation science, yet neither was effective at actively connecting science with governance. In the US case, the Y2Y conservation proposals in Yellowstone experienced backlash from the local community because the proposals used science to justify a narrow vision of appropriate land-use that did not line up with local normative visions of how the landscape should be managed. While the Habitat 141°'s science vision wasn't in conflict with local governance goals, project leaders couldn't re-organize themselves appropriately because of disagreements over where decision-making power for conservation actions should be located. The leadership was not able to coordinate the organizational and operational complexity involved in a large-scale conservation project involving multiple institutional levels. These examples again highlight the importance of paying close attention to how key actors and stakeholders formulate and re-organize themselves to reconcile tensions between science and governance.

Another illustrative example specifically related to knowledge systems dynamics in cities is the case of land use planning in San Juan, Puerto Rico, described by Muñoz-Erickson [21,44,45]. In 2009, increasing development of the city's green areas, especially in the upper headwaters of the main watershed of the city, exposed many residents to river and urban flood risks. While the Municipality's land use regulatory framework included protection of these green areas as part of the sustainable development of the city, projects were still permitted. In her analysis of the land use governance landscape in San Juan, Muñoz-Erickson [21] found that, in addition to economic and political interests, knowledge systems also played an important role in shaping outcomes. Relevant factors included a lack of organizational capacity to generate and validate site-specific knowledge about proposed projects and power dynamics within the state's planning agency. The latter was key because the state continued to make decisions on land use in San Juan (based on their own knowledge systems and not the Municipality's) even though the Municipality had gained autonomy in 2003. Muñoz-Erickson applied the knowledge-action systems analysis (KASA) framework, a type of knowledge systems analysis that uses social network analysis to map and analyze co-production processes that link knowledge to action [39,45]. She mapped and analyzed the network of organizations producing knowledge on land use, what frames and epistemologies where circulating across the network and how, and which organizations had greater influence over how that knowledge was applied.

Table 3. Complexities of knowledge systems.

Framework Concepts	Definition or Use in Knowledge Systems Analysis	Example
Organizational Complexity	When knowledge systems are in a complex decision-making landscape that involves a multiplicity of interacting actors and viewpoints, and complicated rules of procedure. Oftentimes knowledge and decision-making become tightly coupled to one another, such that integrating new knowledge into this form of closed system can be a very difficult undertaking.	Decisions involving ecosystem services typically involve trade-offs among ecosystem services and multiple stakeholders and organizations. Knowledge of the trade-offs among ecosystem services is often absent from or neglected within disconnected decision-making processes, leading to decisions that have unexpected or problematic outcomes.
Operational Complexity	Conditions under which highly dynamic social work is necessary to carry out the core functions of knowledge systems, involving diverse participants and organizations, and requiring careful coordination across the system's many organizational components.	The UN Framework Convention on Climate Change coordinates across multiple experts and organizations the various tasks of emissions inventories, including defining which emissions to count and allocate to responsible parties, the standardization of those methods, and the review processes by independent experts from other countries to ensure transparency. Boundary work and orchestration are also crucial functions to ensure legitimacy and credibility across multiple institutions and forms of expertise.
Political Complexity	Conditions of high interconnection between knowledge production and the exercise of political power, especially in the presence of conflicts within or between organizations. In the adversarial political context of the US, in particular, the connection of science and expert advice within many facets of decision-making in the US federal government is an illustration of the political complexity of knowledge systems.	The knowledge claims underpinning EPA regulatory decisions have been widely contested by both industry groups and environmental organizations, depending on which group perceived an interest in undermining EPA credibility on any given policy issue. Further layers of organizational complexity, e.g., the presence of the EPA Science Advisory Board, often exacerbate knowledge conflicts rather than mitigate them by presenting another opportunity for divergent views of the proper use of scientific evidence to arise and become subject to critical commentary by policy actors.

The application of knowledge systems analysis revealed various complex dynamics in San Juan's co-production processes that could serve as barriers to the design of knowledge co-production efforts to build urban sustainability and resilience pathways. For instance, while a diverse network of organizations existed to generate, exchange, and use knowledge informing Municipal land use practices, including non-governmental organizations, a significant breakdown in knowledge flow between the Municipality's office of territorial ordinance and the state's planning agency created barriers to communicating knowledge of local conditions to the state agency [23]. In addition, political complexities created distinct power asymmetries that impacted the ways in which diverse knowledge systems and visions were able to inform planning processes. The Municipality's ideas and epistemic cultures, which included social dimensions of urban planning such as quality of life and equity considerations, conflicted with (and often lost out to) the state's hegemonic ideas of the city as a node for regional economic power [44].

The case of San Juan highlights the forms of organizational, operational, and political complexity that knowledge systems can experience. The knowledge systems' tasks and functions around land use planning and decisions in San Juan were carried out by multiple organizations in competition with each other. Still today, although the Municipality has sketched out a pathway towards more sustainable futures through a vision of a Livable City [44], knowledge–power dynamics may keep these ideas from moving into action. In this respect, knowledge systems analysis is useful as a diagnostic tool to examine and make explicit the interplay of values, knowledge, and power that enable and constrain research and decision-making processes underpinning elements of societal stability and transformation. With this context, one can see multiple definitions and applications of the concept of knowledge systems. We interpret these as variations across a knowledge systems spectrum that ranges from specific and tightly closed knowledge systems, such as the US Census, to more complex knowledge–action systems where multiple knowledge systems and organizations interact fluidly with one another across complex social and physical landscapes.

Knowledge systems analysis is a powerful framework to uncover hidden relations and highlight the societal foundations that shape what is (and is not) possible (or, arguably more appropriately, what can be easily accomplished and what will require extensive work) in the creation and application of new knowledge to advance sustainability. The framework can help identify and make explicit the tensions and assumptions informing efforts to design and implement new knowledge-making arrangements such that they can work within, or transform, existing knowledge–power structures, thus increasing the likelihood of knowledge leading to action. In the next section we present general guidelines or design criteria that we view useful for designing knowledge co-production processes in cities from a knowledge systems lens.

4. Design Philosophies for Knowledge Co-Production for Urban Sustainability and Resilience

Understanding the way cities think is necessary to building knowledge infrastructures that transform cultural and institutional barriers to building sustainable and resilient pathways to sustainability. Because every context will present particular barriers and opportunities to linking knowledge and action, analyzing and evaluating existing knowledge–power dynamics can help in designing appropriate architectures for knowledge systems. Simply put, one size does not fit all in the design of knowledge systems. Simplistic assumptions about how knowledge systems work in the real world have led to a plethora of lists of ingredients for 'science–policy interfaces' with outcomes that remain unexamined. Thus, the following are not meant to serve as a 'blueprint', but rather normative and organizational elements of the design of co-production—what we might call design philosophies—that need close attention to ensure the success of knowledge co-production initiatives. Following each, we provide a set of questions and strategies to aid the design and practice of knowledge co-production.

4.1. Context and Inclusiveness

Building knowledge systems that align with the local context entails the use of more inclusive definitions and approaches for defining knowledge and the actors that produce and use it. Breaking down knowledge stereotypes is necessary, removing a priori assumptions about who produces and uses knowledge. For instance, analyzing and evaluating the local epistemic context in San Juan revealed and helped explain not only the knowledge produced and the needs of knowledge users (and gaps between them) but also distributions of power and expertise and perceptions of credibility and legitimacy across actors in the local political context. The investigation showed, for instance, a heterogeneous network of land use and green area knowledge with a variety of sources of knowledge, including organizations not traditionally perceived as experts (i.e., civic groups) [21]. This may be indicative that credibility and legitimacy in San Juan is more widely distributed among a more diverse set of actors than commonly considered in US policymaking (where academic, scientific, or technical government institutions commonly predominate). Researchers and practitioners engaged in knowledge co-production processes should be exposed to and experience the complex social and institutional dynamics shaping knowledge and governance in a place.

In conducting these analyses, focus should be put on the "interactional" elements of the co-production of knowledge and governance. As described above, for Jasanoff, co-production occurs through interactions among diverse elements of and participants within a given political culture. These interactions both maintain stability but also create the potential for structural change. Except in rare circumstances, research on co-production suggests that transformational change occurs more frequently through reconfigurations of existing knowledge and political arrangements than through their replacement with entirely novel alternatives. Thus, understanding the dynamics and structures of existing knowledge systems and the ways that they contribute to larger processes in the co-production of knowledge and society—and situating new knowledge-making initiatives within this context—can help open up the potential for the kinds of major changes in cities necessary to achieve sustainability and resilience.

Key questions and strategies for building context and inclusiveness in knowledge co-production.

- Analyze existing knowledge systems; do not make assumptions about how they work in the city: What do people know or need to know about the city? Who are the key actors producing and using knowledge for urban planning and sustainability? How are their knowledge systems structured and functioning? What epistemic practices inform their visions and expectations of the city? How is their network constituted? How do the credibility and legitimacy of science and other knowledge play out in this context? What actors are perceived as credible and legitimate, and why or why not?

- Expose researchers to these conditions and the complex social–ecological realities of the place. Ethnographic research approaches, such as field work, observations and unstructured interviews can be useful tools to build epistemic context and initiate rapport, and hence trust, with local stakeholders.

- Identify all knowledge-relevant stakeholders (including marginal actors) and engage early to assess their needs, priorities, and existing knowledge systems. Develop trust by engaging in multiple ways, formally and informally, and continuously follow-up and communicate with stakeholders.

4.2. Adaptability and Reflexivity

Building institutional reflexivity is crucial to avoid failures in the future and build more adaptive knowledge systems. Reflexivity is the idea that those who produce and use knowledge are aware of and reflective about how they do so [23]. It implies that the assumptions, framings, values, and practices underpinning knowledge production and use for sustainability be open to scrutiny [46]. In other words, reflexivity calls for knowledge-producing institutions to be self-critical and routinely reflect on how they build knowledge about cities, the assumptions they make about how cities work, and their normative premises for how urban development pathways should be steered in the future. Reflexivity is related to adaptability in that the approach demands awareness of system uncertainty and unintended consequences. It goes further, however, to consider the effects that such reflection has on how we produce or change the production of knowledge, as producers and users come to terms with the impossibility of having full and complete knowledge of system dynamics [47].

From the standpoint of practice, reflexivity involves 'opening up' knowledge production processes for review and critique. In other words, it involves developing institutional mechanisms that allow outside actors, including non-scientists, to be part of the design and review of the research process [48]. Much like the peer review process in science, knowledge systems need an external review body, such as extended peer communities [49] or advisory committees, to provide context and critical assessment of the assumptions, methods, and direction of research in relation to city needs, changes, and expectations. These bodies should not only bring accountability to the knowledge system by integrating various stakeholder or actor groups involved in governance but must also be inclusive of the various ideas, knowledge, and values needed to address and be congruent with the system. A reflexive approach to improving a knowledge production process, however, brings up an 'efficiency paradox' as it implies a balance between opening up and closing it down [50]. Closing down is necessary to do the work and have the ability to act, but the timing of closing may cause rigidity. Voss and Kemp argue that the issue is not a matter of either/or but of doing both throughout the knowledge co-production process [50]. The key to this balancing act is the timing and structure of mechanisms to open up using an iterative process. For instance, broad inclusiveness is crucial in the beginning and final phases of a project, therefore using methods that allow greater representation and deliberation of ideas, viewpoints, and epistemologies. Other points in the stage are more technical and may require a narrower and more specific set of expertise to review and provide critique (but be wary of too glibly assuming this; even minimal checking in with stakeholders can help spot problems early). Finally, mechanisms for monitoring and evaluating the knowledge production process are crucial to assess whether learning is occurring and if both ecological and social outcomes are being met.

Key questions and strategies for building adaptability and reflexivity in knowledge co-production

- Institute an advisory review body, in which both political interests and epistemologies (ways of knowing) are represented, and build accountability in the knowledge production process.
- Be flexible with engagement methods—use a variety of methods with varying frequencies, including consultative (e.g., surveys, rapid appraisals), informal meetings (e.g., office visits, fields trips), and active participation (e.g., engagement in decisions on research) to develop an appropriate framework that fits local context and the diversity of ways that researchers and practitioners are able to engage given different reasoning styles, time, and other capacities.
- Iteratively frame research agenda and process; approach knowledge systems as experiments; evaluate and adapt.
- Monitor knowledge systems through learning indicators and knowledge system analysis and evaluation.
- Account for the 'intangibles', or non-quantifiable elements, of quality of life in a city.

4.3. Knowledge–Action Networks

While the previous two design philosophies related more to the dynamics and functions of knowledge systems, attention to knowledge–action networks focuses on the structure, or architecture, of efforts to design new strategies for creating and applying knowledge for advancing sustainability. We use the term knowledge–action networks to refer to the multiplicity of spaces (i.e., nodes), both physical and organizationally, where knowledge and action interact frequently. In San Juan, for instance, this happens not only in expert organizations that produce knowledge and link it to action through various means for circulating and applying it but also to places where a diversity of ideas about urban sustainability are being constructed and deliberated, such as community meetings, coffee shops, and even churches. As we suggested above, in the first design philosophy, the architecture of new knowledge systems and knowledge co-production processes needs to engage with existing knowledge systems and their relationships to the ecological *and* political landscape of the city, to be most effective. In this way, the new interactions stimulated by knowledge co-production initiatives can help catalyze the transformations necessary for sustainability and resilience. Network theory reinforces this perspective, observing that creativity and innovation are best fostered by diverse and polycentric networks, as opposed to isolated networks composed of siloed entities with similar views and perspectives. A polycentric design entails strengthening existing capacities and connections where there are weak links and building new ones where they are absent. Interventions, such as establishing new knowledge co-production efforts, should take these local network properties into consideration and build on them, enhancing polycentricity and opening up possibilities for change [18].

Following the adaptive and reflexive approach proposed here, this structure needs to reflect the knowledge–power relationships in these networked and complex contexts, while at the same time be adaptive and recognize when re-organization or new institutional arrangements are needed for knowledge production. The structure also should be flexible enough to help link existing knowledges together (and to action) and facilitate knowledge flows where needed, thus allowing local stakeholders to feel ownership of the knowledge co-production process. Monitoring and evaluation of knowledge systems functions and performance is part of designing a reflexive structure. Strong leadership is needed to manage knowledge systems complexities (e.g., organizational, operational, political) and to work with existing capacities/projects so as to not compete or be redundant. Developing and maintaining a network imaginary, as Goldstein and Butler [51] have proposed for the US Fire Learning Network (FLN), is an approach that can provide the cultural and organizational 'glue' that helps balance the social cohesion, yet flexibility, of a distributed knowledge–action network. The authors describe that the FLN is able to maintain an extensive network of research nodes across the US without the need for a hierarchical authority structure, by articulating a network imaginary through technologies, planning guidelines and media. Put differently, a shared-mental schema of a community of diverse interests and knowledge but with a common goal (i.e., manage fire) was created and perpetuated through the communication and research practices of the network such that people working at different locations feel part of this imagined community.

Key questions and strategies for designing knowledge co-production.

- Evaluate and invest existing institutional structure and capacities for co-production: do not assume capacity is already there. Where capacities do exist, work or help transform them, instead of automatically building new structures (e.g., new organization).
- Recognize that in an increasingly networked society, power and knowledge are distributed, thus the knowledge–action networks need to be cognizant of the distribution of expertise in the governance space and the inevitable political, organizational, and operational complexity that this creates.
- Develop epistemic or transdisciplinary consortiums: instead of looking for uniformity or consensus, foster diversity and pluralism of ideas, knowledge and ways of reasoning. Individuals trusted and deemed credible by researchers and stakeholders alike can serve as the 'mediators' between knowledge and action.
- Create a variety of spaces and/or activities or support others in leading them (i.e., field trips, seminars, workshops, retreats, office visits, etc.) to deliberate research questions and outputs such that stakeholders feel ownership of the process.
- Develop a network imaginary as the cultural glue to keep the network together and thus allow actors to have ownership of the process and outcomes of the networked structure.

As we mentioned earlier in this section, these design philosophies are meant to highlight key normative and organizational dimensions of co-production that require close attention. More empirical research is needed to explore how these philosophies can guide innovations in practice and to evaluate the results in advancing urban sustainability and resilience. The set of questions and strategies we present here offer a starting point.

5. Conclusions

Co-production requires a fundamental transformation of both knowledge and governance toward more critical, inclusive and reflexive practices. The social, institutional, and ecological complexities of cities defy simple arrangements that link knowledge producers on one side and knowledge users on the other. Instead, institutional arrangements that are able to meaningfully engage the institutional and ecological complexity and dynamism of cities are more likely to be effective in generating useful and innovative strategies for sustainability and putting them to work to create long-term transformation. A lack of awareness of how these existing knowledge systems work can have unforeseen consequences on the resilience of cities.

In this study we discussed knowledge systems analysis as a conceptual and empirical framework to understand how cities think. This framework is useful to both scientists and practitioners interested in designing knowledge co-production efforts to produce better knowledge and facilitate successful implementation of sustainable outcomes. It provides a way to understand existing institutional conditions, as well as to build reflexivity and change through its long-term application to evaluate how existing and new knowledge co-production processes perform over time. Future research should apply this framework to understand co-production in multiple cities and for multiple resource domains (e.g., water, energy, etc.) to develop more robust assessments of how these systems work in multiple sustainability contexts. Experimenting with different institutional configurations could also provide a way to test the design propositions recommended here. Doing so will create new insights into the arrangements and stakeholder engagement processes most useful to tackle urban sustainability issues.

We also hope that future research in this area can broaden the scope of how knowledge systems are addressed in sustainability science and science and technology studies (STS) by acknowledging the complexity of these systems, especially in cities, and presenting ways to tackle this complexity analytically. The use of multiple, interdisciplinary concepts and methods can highlight important institutional and epistemological aspects of knowledge systems that are more difficult to assess through a single analytical approach. From a practical perspective, understanding the complex workings of knowledge systems has important implications for how we design and build them in practice. Thus, linking knowledge to action is not as simple as building 'interfaces' or other institutional arrangements drawn from theoretical designs. Rather, it requires that we first assess how knowledge gets made, vetted, circulated, and applied within complex political and institutional terrains, such that whatever intervention we design not only makes sense within that place but also has the interactional capabilities to create necessary change. The knowledge systems analysis framework challenges researchers and practitioners in cities to ask themselves: are the social and institutional conditions of the system they are working in conducive to knowledge co-production efforts? If not, why not? What needs to change to build an urban knowledge infrastructure for sustainability and resilience? If yes, what kinds of capabilities are necessary to transform knowledge co-production from a new way of thinking about knowledge to a force for effective change? Ultimately, the goal of understanding how cities think is not only to help produce better outcomes for knowledge co-production efforts but also to provide a window into the adaptive capacity and transformation potential of cities.

Acknowledgments: We are thankful for the support of the National Science Foundation, which has supported numerous projects that have been instrumental in the development of the ideas in this article. Specific projects include the Urban Resilience to Extreme Weather-Related Events Sustainability Research Network (1444755), the San Juan Urban Long-Term Research Area (0948507), the Urban Sustainability Research Coordination Network (1140070), and several interdisciplinary research and educational projects focused on urban climate resilience (1462086, 1441352, 1360509, 1043289).

Author Contributions: Clark A. Miller and Tischa A. Muñoz-Erickson conceived and designed the knowledge systems framework presented in this paper. Thaddeus Miller contributed to the review on knowledge co-production processes. Tischa A. Muñoz-Erickson wrote the paper.

Conflicts of Interest: The authors declare no conflict of interest.

References

1. Rosenzweig, C.; Solecki, W.; Hammer, S.A.; Mehrotra, S. Cities lead the way in climate-change action. *Nature* **2010**, *467*, 909–911. [CrossRef] [PubMed]
2. Wheeler, S.M.; Beatly, T. *The Urban Sustainable Development Reader*; Routledge: New York, NY, USA, 2010.
3. Hommel, A. *Unbuilding Cities: Oduracy in Urban Sociotechnical Change*; MIT Press: Cambridge, MA, USA, 2008.
4. American Society for Civil Engineering. The New Orleans Hurricane Protection System: What Went Wrong and Why: A Report, 2007. Available online: http://www.pubs.asce.org (accessed on 7 December 2013).
5. New York City. *Hurricane Sandy after Action Report and Recommendations to Mayor Michael R. Bloomberg*; New York City: New York, NY, USA, 2013.
6. Edwards, P.N. Knowledge infrastructures for the Anthropocene. *Anthr. Rev.* **2017**, *4*. [CrossRef]
7. Nel, J.L.; Roux, D.J.; Driver, A.; Hill, L.; Maherry, A.C.; Snaddon, K.; Petersen, C.R.; Sminth-Adao, L.B.; van Deventer, H.; Reyers, B. Knowledge co-production and boundary work to promote implementation of conservation plans. *Conserv. Biol.* **2016**, *30*, 176–188. [CrossRef] [PubMed]
8. Cvitanovic, C.; Hobday, A.J.; van Kerkhoff, L.; Wilson, S.; Dobbs, K.; Marshall, N.A. Improving knowledge exchange among scientists and decision-makers to facilitate the adaptive governance of marine resources: A review of knowledge and research needs. *Ocean Coast. Manag.* **2015**, *112*, 25–35. [CrossRef]
9. Van Kerkhoff, L.E.; Lebel, L. Coproductive capacities: Rethinking science-governance relations in a diverse world. *Ecol. Soc.* **2015**, *20*, 14. [CrossRef]
10. Fazey, I.; Evely, A.C.; Reed, M.S.; Stringer, L.C.; Kruijsen, J.; White, P.V.L.; Newsham, A.; Jin, L.; Cortazzi, M.; Phillipson, J.; et al. Knowledge exchange: A review and research agenda for environmental management. *Environ. Conserv.* **2013**, *40*, 19–36. [CrossRef]
11. Edelenbos, J.; van Buuren, A.; van Schie, N. Co-producing knowledge: Joint knowledge production between experts, bureaucrats and stakeholders in Dutch water management projects. *Environ. Sci. Pol.* **2011**, *14*, 675–684. [CrossRef]

12. Roux, D.J.; Rogers, K.H.; Biggs, H.C.; Ashton, P.J.; Sergeant, A. Bridging the science-management divide: Moving from unidirectional knowledge transfer to knowledge interfacing and sharing. *Ecol. Soc.* **2006**, *11*, 4. [CrossRef]

13. Hegger, D.; Lamers, M.; Van Zeijl-Rozema, A.; Cieperink, C. Conceptualising joint knowledge production in regional climate change adaptation projects: Success conditions and levers for action. *Environ. Sci. Pol.* **2012**, *18*, 52–65. [CrossRef]

14. Berkes, F. Evolution of co-management: Role of knowledge generation, bridging organizations and social learning. *J. Environ. Manag.* **2009**, *90*, 1692–1702. [CrossRef] [PubMed]

15. Cash, D.W.; Clark, W.C.; Alcock, F.; Dickson, N.M.; Eckley, N.; Guston, D.H.; Jäger, J.; Mitchell, R.B. Knowledge systems for sustainable development. *Proc. Natl. Acad. Sci. USA* **2003**, *100*, 8086–8091. [CrossRef] [PubMed]

16. Vogel, C.; Moser, S.C.; Kasperson, R.E.; Dabelko, G.D. Linking vulnerability, adaptation, and resilience science to practice: Pathways, players, and partnerships. *Glob. Environ. Chang.* **2007**, *17*, 349–364. [CrossRef]

17. Van Kerkhoff, L.; Lebel, L. Linking Knowledge and action for sustainable development. *Annu. Rev. Environ. Resour.* **2006**, *31*, 445–477. [CrossRef]

18. Grove, M.; Childers, D.L.; Chowdhury, R.R.; Galvin, M.; Hines, S.; Muñoz-Erickson, T.A.; Svendsen, E. Linking Science and Decision Making to Promote an Ecology for the City: Practices and Future Opportunities. *Ecol. Health Sustain.* **2016**, *2*, 1–10.

19. Campbell, L.K.; Svendsen, E.S.; Roman, L.A. Knowledge co-production at the research-practice interface: Embedded case studies from urban forestry. *Environ. Manag.* **2016**, *57*, 1262–1280. [CrossRef] [PubMed]

20. Jasanoff, S. The idiom of Co-Production. In *States of Knowledge: The Co-Production of Science and Social Order*; Jasanoff, S., Ed.; Routledge: London, UK, 2004; pp. 1–13.

21. Muñoz-Erickson, T.A. Co-production of knowledge-action systems in urban sustainable governance: The KASA approach. *Environ. Sci. Pol.* **2014**, *37*, 182–191. [CrossRef]

22. Miller, C.; Muñoz-Erickson, T.A. Designing Knowledge. In *The Rightful Place of Science, Book Series*; Arizona State University: Tempe, AZ, USA, 2017; in progress.

23. Miller, C.; Muñoz-Erickson, T.A.; Monfreda, C. Knowledge Systems Analysis: A Report to the Advancing Conservation in a Social Context. In *CSPO Report 10-05*; Arizona State University: Tempe, AZ, USA, 2010. Available online: http://www.cspo.org/content/knowledge-systems-project-publications (accessed on 5 December 2010).

24. Wyborn, C. Co-productive governance: A relational framework for adaptive governance. *Glob. Environ Chang.* **2015**, *30*, 56–67. [CrossRef]

25. Matson, P. *Linking Knowledge with Action for Sustainable Development*; The National Academies Press: Washington, DC, USA, 2008.

26. Guston, D.H. Boundary organizations in environmental policy and science: An introduction. *Sci. Technol. Hum. Val.* **2001**, *264*, 399–408. [CrossRef]

27. McNie, E. Reconciling the supply of scientific information with user demands: An analysis of the problem and review of the literature. *Environ. Sci.* **2007**, *10*, 17–38. [CrossRef]

28. Wiek, A.; Walter, A. A transdisciplinary approach for formalized integrated planning and decision making in complex systems. *Eur. J. Oper. Res.* **2009**, *197*, 360–370. [CrossRef]

29. Clark, W.C.; Tomich, T.P.; van Noordwijk, M.; Guston, D.; Catacutan, D.; Dickson, N.M.; McNie, E. Boundary work for sustainable development: Natural resource management at the Consultative Group on International Agricultural Research (CGIAR). *Proc. Natl. Acad. Sci. USA* **2016**, *113*, 4615–4622. [CrossRef] [PubMed]

30. Reyers, B.; Nel, J.L.; O'Farrell, P.J.; Sitas, N.; Nel, D.C. Navigating complexity through knowledge coproduction: Mainstreaming ecosystem services into disaster risk reduction. *Proc. Natl. Acad. Sci. USA* **2015**, *112*, 7362–7368. [CrossRef] [PubMed]

31. Jasanoff, S.; Wynne, B. Science and Decision-Making. In *Human Choice and the Climate Change-Vol. 1: The Societal Framework*; Battelle Press: Columbus, OH, USA, 1998.

32. Foss, N.J. The emerging knowledge governance approach: Challenges and characteristics. *Organization* **2007**, *14*, 29–52. [CrossRef]

33. Van Kerkhoff, L. Knowledge governance for sustainable development: A Review. *Chall. Sustain.* **2013**, *1*, 82–93. [CrossRef]

34. Wyborn, C. Future oriented conservation: Knowledge governance, uncertainty and learning. *Biodivers. Conserv.* **2016**, *25*, 1401–1408. [CrossRef]

35. Gerritsen, A.L.; Stuiver, M.; Termeer, C.J.A.M. Knowledge governance: An exploration of principles, impact, and barriers. *Sci. Publ. Pol.* **2013**, *40*, 604–615. [CrossRef]

36. Manuel-Navarrete, D.; Gallopín, G.C. Feeding the world sustainably: Knowledge governance and sustainable agriculture in the Argentine Pampas. *Environ. Dev. Sustain.* **2011**, *14*, 321–333. [CrossRef]

37. Miller, C.A. Civic epistemologies: Constituting knowledge and political order in political communities. *Sociol. Compass* **2008**, *2*, 1896–1919. [CrossRef]

38. Petts, J.; Brooks, C. Expert conceptualisations of the role of lay knowledge in environmental decisionmaking: Challenges for deliberative democracy. *Environ. Plan.* **2006**, *38*, 1045–1059. [CrossRef]

39. Evans, R.; Marvin, S. Researching the sustainable city: Three modes of interdisciplinarity. *Environ. Plan. A* **2006**, *38*, 1009–1028. [CrossRef]

40. Jasanoff, S. *Designs on Nature: Science and Democracy in Europe and the United States*; Princeton University: Princeton, NJ, USA, 2005.

41. Shapin, S. *A Social History of Truth: Civility and Science in Seventeenth-Century England*; University of Chicago Press: Chicago, IL, USA, 1994.

42. Leach, M.; Scoones, I.; Stirling, A. *Dynamic Sustainabilities: Technology, Environment, Social Justice*; Earthscan: London, UK, 2010.

43. Smith, A.; Stirling, A. The politics of social-ecological resilience and sustainable socio-technical transitions. *Ecol. Soc.* **2010**, *15*, 1. Available online: http://www.ecologyandsociety.org/vol15/iss1/ (accessed on 18 August 2012). [CrossRef]

44. Muñoz-Erickson, T.A. Multiple pathways to the sustainable city: The case of San Juan, Puerto Rico. *Ecol. Soc.* **2014**, *19*, 2. Available online: http://dx.doi.org/10.5751/ES-0647-190302 (accessed on 18 August 2014). [CrossRef]

45. Muñoz-Erickson, T.A.; Cutts, B. Structural dimensions of knowledge-action networks for sustainability. *Curr. Opin. Environ. Sustain.* **2016**, *18*, 56–64. [CrossRef]

46. Hendriks, C.M.; Grin, J. Contextualizing reflexive governance: The politics of Dutch transitions to sustainability. *J. Environ. Pol. Plan.* **2006**, *9*, 333–350. [CrossRef]

47. Leach, M. Pathways to Sustainability in the forest? Misunderstood dynamics and the negotiation of knowledge, power, and policy. *Environ. Plan. A* **2008**, *40*, 1783–1795. [CrossRef]

48. Stirling, A. Precaution, Foresight, and Sustainability: Reflecting and Reflexivity in the Governance of Science and Technology. In *Sustainability and Reflexive Governance*; Voss, J.P., Kemp, R., Eds.; Edward Elgar Publishing: Northampton, MA, USA, 2004; pp. 225–272.

49. Funtowicz, S.O.; Ravetz, J.R. Science for the post-normal age. *Futures* **1993**, *25*, 739–755. [CrossRef]

50. Voss, J.P.; Bauknecht, D.; Kemp, R. *Reflexive Governance for Sustainable Development*; Edward Elgar: London, UK, 2006.

51. Goldstein, B.E.; Butler, W.H. The network imaginary: Coherence and creativity within a multiscalar collaborative effort to reform US fire management. *J. Environ. Plan. Manag.* **2009**, *52*, 1013–1033. [CrossRef]

forests

Communication

Teams at Their Core: Implementing an "All LANDS Approach to Conservation" Requires Focusing on Relationships, Teamwork Process, and Communications

Kasey R. Jacobs

United States Forest Service International Institute of Tropical Forestry, Caribbean Landscape Conservation Cooperative, Río Piedras, Puerto Rico, USA; kaseyrjacobs@fs.fed.us; Tel.: +1-772-486-7561

Academic Editor: Timothy A. Martin
Received: 1 April 2017; Accepted: 7 July 2017; Published: 11 July 2017

Abstract: The U.S. Forest Service has found itself in an era of intense human activity, a changing climate; development and loss of open space; resource consumption; and problematic introduced species; and diversity in core beliefs and values. These challenges test our task-relevant maturity and the ability and willingness to meet the growing demands for services. The Forest Service is now on a transformative campaign to improve abilities and meet these challenges. The "All-Lands Approach to Conservation" brings agencies, organizations, landowners and stakeholders together across boundaries to decide on common goals for the landscapes they share. This approach is part of a larger transformation occurring in the American Conservation Movement where large-scale conservation partnerships possibly define the fourth or contemporary era. The intent of this communication is to present one perspective of what large-scale conservation partnerships should include, namely an emphasis on rethinking what leadership looks like in a collaborative context, relational governance, cooperative teamwork procedures, and communications.

Keywords: landscape conservation; network governance; strategic teams; communications; leadership

1. Introduction

In a speech at the Western States Land Commissioners Association in the United States in 2012, Chief Tom Tidwell of the U.S. Forest Service described the all-lands approach. He said, " . . . We need a common vision. Restoration is predicated on partnerships . . . None of this can happen on a piecemeal scale. It has to be on a scale that supersedes ownership. An all-lands approach brings landowners and stakeholders together across boundaries to decide on common goals for the landscapes they share. It brings them together to achieve long-term outcomes. Our collective responsibility is to work through landscape-scale conservation to meet public expectations for all the services people get from forests and grasslands" [1].

That same year, the Department of the Interior of the United States signed Secretarial Order No. 3289 launching the Landscape Conservation Cooperatives (LCCs) [2], an initiative to better integrate science and management to address climate change and other landscape scale issues. The Department of the Interior adopted several pre-existing cooperatives and formed a network of 22 LCCs that work collaboratively with federal, state, and local governments, Tribes and First nations, non-governmental organizations, universities, and interested public and private organizations. The goal is to identify best practices, connect efforts, identify science gaps, and avoid duplication through conservation planning and design.

The emergence of landscape conservation and the all lands approach to conservation in federal government is indicative of a larger movement that has been building for decades.

Arguably, there are three eras of the American conservation movement. The first major historical stage occurred in the late 19th and early 20th centuries with conservationists such as Theodore Roosevelt, John Muir, Gifford Pinchot and women such as Rosalie Barrow Edge whose lasting legacies include national parks, forests and monuments, and private land trusts. The second major historical stage occurred in the 1960s and 1970s which is known as the launch of the environmental movement, this was largely brought on by Rachel Carson and her book *Silent Spring*, Paul Ehrlich and his book *Population Bomb*, former Interior Secretary Udall's book *The Quiet Crisis* and his work with many others to pass and enact important legislation including the Clear Air Act, Clean Water Act, the Wilderness Act, and the Endangered Species Preservation Act and Wild and Scenic Rivers Act. The third stage could be considered the Grassroots Advocacy stage from the 1980s to the 2000s and still into today. Many non-governmental organizations (NGOs) either had their starts or really picked up speed during this time, they brought environmentalism into the homes and hearts of American citizens and worked very successfully to mobilize and activate many around key environmental issues; often to enforce the very laws passed during the last second era in the conservation movement.

Without the gift of hindsight, views might differ for how to define the contemporary conservation movement. Many would say that if the 20th century was a conservation battle on land, the ocean is a major focus for conservation in the 21st century. There is another way to frame this new era and it includes marine conservation. Possibly, we are now in the era of large-scale conservation partnerships. The U.S. federal government initiatives are just a handful of dozens around the world and in the Caribbean (Table 1).

Table 1. Examples of large-scale partnerships in the United States Caribbean and Non-U.S. Caribbean.

U.S. Partnerships	Scale	Non-U.S. Caribbean Partnerships	Scale
El Yunque National Forest All Lands Planning Process	9 municipalities (plan covers municipal, regional and island-wide)	Global Landscapes Forum	135 countries (committed to restoring 128 million hectares of degraded and deforested landscapes)
Caribbean Regional Ocean Partnership and Regional Planning Body	Exclusive Economic Zones of PR and USVI	Caribbean Challenge Initiative	20% of the Caribbean's marine and coastal ecosystems (by 2020)
U.S. Coral Reef Task Force	All coral jurisdictions of the U.S.	Atlantic Conservation Partnership	U.S., Bermuda, and Wider Caribbean
Co-managed Nature Reserves	Individual Reserves	Wider Caribbean Sea Turtle Network	40 nations and territories
Caribbean Landscape Conservation Cooperative	U.S. and Wider Caribbean	Caribbean Sustainable Development Solutions Network	All countries and territories bordering the Caribbean Sea
Model Forest (Bosque Modelo)	19 protected areas in 31 municipalities	Eastern Caribbean Marine Managed Areas Network	6 Organization of Eastern Caribbean States' countries of St. Kitts and Nevis, Antigua and Barbuda, Dominica, Saint Lucia, St. Vincent and the Grenadines and Grenada
Our Florida Reefs	4 counties		

These initiatives are all using similar methodologies. With the rise of ecosystem-based management, managing species by species in forested landscapes is no longer the norm. Nor by ecosystem type. Recognizing the importance of connectivity across types, the focus has largely become on the panorama of ecosystems at multiple scales. Reed et al., 2016 argues in a research review that other methodologies since at least 1992 are a form of the integrated landscape approach, such as agrolandscape ecology (1992), sustainable landscape approach (1994), integrated resource management (2000), collaborative decision making (2005), landscape ecology (2006), integrated watershed management (2008), ecosystems approach (2010), integrated water resources management (2011), integrated coastal zone management (2011), and participatory land use planning (2004) [3].

These methodologies and initiatives are the new conservation model that is, or at least is becoming, the new identity of the American Conservation Movement—large-scale partnerships. This new model is partner-driven, management-driven, large-scale in that it is bigger spatially (across large geographies) and temporally (across long periods of time). The new model is user-focused, future-focused, revolves around a shared vision, and is rooted in both the natural and social sciences. Most importantly, it steps outside of conservation science and management and requires the engagement of multiple sectors. In the U.S., this is often defined by engaging industry, hazard mitigation agencies, the military, private landowners such as farmers, and other economic interests. In international partnerships, it is defined by engaging those working on poverty reduction, food insecurity, development, and disaster risk reduction.

The language used in all these approaches are different, but have the same intent.

Alternative language you occasionally hear the U.S. Forest Service use is "All Lands-All Hands". The "All Hands" might better convey the approach of All Lands and Landscape Conservation. They connect people. They study people and actions of people. They coordinate people. Furthermore, they empower people.

In my experience, in four years serving as Partnership and Communications Coordinator for the Caribbean Landscape Conservation Cooperative (CLCC) with the U.S. Forest Service International Institute of Tropical Forestry, I have found that when you boil down these approaches to the nuts and bolts of cooperative conservation, what you find yourself doing daily, is facilitating strategic teams. Large-scale conservation partnerships are teams at their core.

This means that the most relevant land and sea managers are not just at the table but are engaged and feel empowered to be part of a team that is working for a shared vision. The new model requires that those whom affect the landscape have a way for understanding the values of the others at the table that also affect the land. Collectively, they understand the stressors on the system based on past observations and future projections and, perhaps most importantly, they see real opportunities for increased coordination.

For the purposes of CLCC partnership efforts, strategic teams are defined as teams that are created for a clear purpose, that are tied to larger regional goals, or to a broader strategy, and that strive to be high performing with well-defined tactics for how team outputs will be delivered and ultimately used in accordance with the team's purpose. Strategic teams also tend to be multidisciplinary or transdisciplinary, though that is not required. The organizational structure of the CLCC is comprised of nested strategic teams that interconnected though autonomous. The teams are staff (and staff with advisory groups), steering committee (and executive team with staff and steering committee with advisory groups), research teams, and conservation action teams. The LCC network is also comprised of multiple teams, such as network staff, science agenda teams, working groups, and the LCC Council (similar to the CLCC Steering Committee).

The CLCC is the newest Landscape Conservation Cooperative, created in 2012, but like all landscape partnerships the CLCC has already experienced certain barriers. Through a literature review, Reed et al., 2016 identified five main barriers to implementation of a landscape approach and the CLCC has experienced all five, namely: (1) Time lags as theory is still evolving; (2) Terminology confusion as different actors are familiar with different terms in different languages; (3) Operating silos as actors and agencies work internally to overcome institutional norms that prevent integration; (4) Internal/External engagement that goes beyond "box-ticking" exercises and moves to true engagement that empowers stakeholders; and (5) Monitoring as this is the least developed area of landscape approach application and finding metrics that inform stakeholders and guide decision-making processes is still difficult [3].

There are four solutions to these implementation barriers that are paramount and are often neglected as institutional norms within federal and state forest agencies traditionally have not emphasized them: (1) rethinking leadership in a cooperative context; and deliberately focusing on (2) relational governance, (3) cooperative teamwork procedures, and (4) communications.

This communication seeks to describe these solutions using examples from how the CLCC works to overcome landscape conservation implementation barriers. To be successful in answering the call of the current era of the American Conservation Movement, we cannot be landscape focused, nor all lands-focused, and not even just people-focused. Forest researchers and managers working at the landscape scales must be deliberately focused on relationships, procedures, and communications, and that starts with how leadership in a network is viewed. To do so requires collecting data and knowledge strategically and using that information to design our path forward. Moreover, it requires us as individual scientists and practitioners as well as agencies and organizations to adopt new skills and step outside comfort zones.

2. Rethinking Leadership and Effectiveness

From the inception of the CLCC, official communications have sought to be "agency neutral" in that no one agency would be attributed as the lead agency. One of the first questions received from an interested stakeholder in 2012 was, "Are we a 'cooperative' in name only or we do follow the principles of the International Cooperative Alliance?" meaning do we have voluntary and open membership, democratic member control, member economic participation, autonomy and independence, education and training, cooperation among cooperatives, and concern for community [4]? For individual accomplishments within the cooperative, staff are frequently asked "Who led it?" or "Who is responsible for making that happen?" and when the response is "everyone in the team" or "it was truly collaborative" we are often met with disbelief. At times, the Cooperative's Steering Committee, Staff and Conservation Action Teams, the main components of the CLCC's organizational structure, are led by one or a few individuals or organizations, allowing the traditional hierarchal views of leadership embedded within partner agencies and organizations to dominate. The truth is we seek to be cooperative but we are *learning-by-doing*, meaning we are adjusting how cooperative we are as we go based on lessons learned.

It is more accurate to describe the CLCC's governance structure as "network governance", in that decision-making is horizontal and not vertical and is characterized by systems of affect, communication, knowledge exchange, and dialogue [5]. Network governance is sometimes confused with the governance of a network and governance networks; those are different types of governance. Jones et al. (1997) define network governance as involving, "a select, persistent, and structured set of autonomous firms engaged in creating products or services based on implicit and open-ended contracts to adapt to environmental contingencies and to coordinate and safeguard exchanges. These exchanges are socially- not legally-binding" [6] (p. 914). The array of purposes for network governance include policy formation and implication, service delivery (in the case of the CLCC, science and conservation strategy delivery) and innovation development. Structures might include formal, or informal, arrangements or tight, or loose, structures [5]. Several typologies of network governance have been introduced but all are non-static, evolving from one form to another due to changes in priorities, changing contexts, as well as the actions of individual actors. Because of the nature of this type of governance, participating agencies and organizations in the CLCC find themselves reflecting on the role they have or would like to have in the network. In network governance, "role" means the type of "leadership".

Imperial et al. (2016) make a strong case well-founded in the literature that large landscape conservation requires three interconnected types of leadership: *collaborative leadership*, in which network members share leadership functions at different points in time; *distributive leadership* in which network processes provide local opportunities for members to act proactively for the benefit of the network; and *architectural leadership*, in which the structure of the network is intentionally designed to allow network processes to occur [7]. In network governance, all members of the partnership are a source of leadership and their roles continually shift to match the challenges the network is addressing. The ability to govern this way depends on whether network members and staff have a collaborative mindset and are willing to share leadership.

The successes and failures in this are very much determined by how well the Cooperative fosters the collaborative mindset and how well defined for each project tackled by a strategic team the type of leadership approach that will serve that task best.

It is true that "network governance does not emerge spontaneously to advance large landscape conservation; someone has to call the initial meeting and decide whom to invite" [7] (p. 128). For the CLCC, the pioneers to create the cooperative to begin with, the sponsors that established credibility and legitimacy, and the thought leaders to provide knowledge and expertise for where to begin and what to tackle together, were the U.S. Forest Service International Institute of Tropical Forestry and the U.S. Fish and Wildlife Service Southeast Region.

Without the collaborative leadership roles of pioneer, sponsor, and thought leader, the CLCC would never have been created nor would it have had certain research accomplishments. Once the cooperative was formed, new thought leaders arose out of the main decision-making body known as the CLCC Steering Committee, namely the Puerto Rico and Virgin Islands Coastal Zone Management Programs, the National Oceanic and Atmospheric Administration, The Nature Conservancy in the U.S. Virgin Islands, Para La Naturaleza (a unit of the Puerto Rico Conservation Trust), and the U.S. Forest Service El Yunque National Forest. Each of the members of the Steering Committee and Staff then took on the networking leadership role to engage people across jurisdictions, conservation sectors, and interests primarily this was done by the Puerto Rico Department of Natural Resources, the Virgin Islands Department of Planning and Natural Resources, the National Oceanic and Atmospheric Administration, the National Park Service in the U.S. Virgin Islands, and the U.S. Forest Service International Institute of Tropical Forestry. During this time, the Steering Committee and Staff continued to build identity, decide what to do, and generate capacity as thought leaders and stewards, coordinating activities, managing research teams and ensuring results.

As in all teams and partnerships, differences and conflicts do arise and the facilitation leadership role has been employed by different agencies and organizations at different points in order to cope with those problems and build agreement. Representatives from the following organizations have been particularly adept in this leadership role: the U.S. Geological Survey, Para La Naturaleza, the U.S. Forest Service El Yunque National Forest, and the National Oceanic and Atmospheric Administration. Categories of differences experienced include political dynamics and differing opinions on leadership roles, roles of the cooperative in the larger conservation community, and decision-making procedures.

Additionally, the staff envisioned a new component of the Cooperative organizational structure that would work on specific resources or implementation activities, Conservation Action Teams (CATs). In 2015, three CATs were approved by the Steering Committee though all three originated by different partners through different activities: (1) Protected Areas Conservation Action Team (PA-CAT); (2) Offshore Cay Systems Conservation Action Team (Cays-CAT); and (3) Dune Building and Stabilization with Vegetation Conservation Action Team (Dune-CAT). CATs are unique to the CLCC, though other cooperatives in the LCC Network have strategic teams. CATs work together on science-based actions that facilitate conservation of land and seascapes for specific resources or conservation mechanisms. Envisioning and creation of the CATs were not spur-of-the-moment occurrences. Each took time (e.g., one to three years) to identify the conservation action needed, the shared priorities, and build the necessary relationships to bring the appropriate partners to the table. Without strong networkers with the ability to identify and engage individuals and organizations with compatible priorities, the CATs would not have been able to form. Many members of the CATs and staff take on facilitator roles, as well as the other types of leadership roles previously discussed. Throughout each of these phases or activities in the CLCC's formation and operation, but specifically for the CATs, champions are necessary to give legitimacy to the strategic teams, promote network governance in general, and complete certain activities.

All seven of these leadership roles (i.e., Pioneer, Sponsor, Thought Leader, Networker, Steward, Facilitator, and Champion) as defined by Imperial et al., 2016 [7] (Table 2) are necessary for the development and coordination of the CLCC. These roles are fluid as distributive leadership assumes there will be multiple opportunities for individuals within the network to lead, as well as influence and support, the network process. For the CLCC, these forms of decentralized collaborative leadership are not well-established within participating entities nor are they employed in other types of conservation partnerships. Furthermore, it is not formalized clearly in our charter, administrative orders, or agreements, but has rather come out of four years of learning-by-doing and has emerged as an informal institution of the cooperative. Borrowing from Helmke and Levitsky's (2004) definition of informal institutions, they are "socially shared rules, usually unwritten, that are created, communicated, and enforced outside of officially sanctioned channels. By contrast, formal institutions are rules and procedures that are created, communicated, and enforced through channels widely accepted as official. This includes state institutions (courts, legislatures, bureaucracies) and state-enforced rules (constitutions, laws, regulations), but also what Robert C. Ellickson calls *organization rules*, or the official rules that govern organizations" [8].

Table 2. Collaborative Leadership Roles and Definitions.

Collaborative Leadership Role	Definition [1]
Pioneer	Catalyzes action and recruits others
Sponsor	Establishes credibility and legitimacy
Thought Leader	Provides knowledge and expertise
Networker	Engages people across jurisdictions, sectors, and interests
Steward	Coordinates activities and ensure results
Facilitator	Bridges differences and builds agreement
Champion	Promotes network governance process throughout development

[1] As defined by Imperial et al., 2016.

The challenge now is formalizing the decentralized governance structure while a traditional governance approach is more familiar for individual partner entities and thus sometimes the default setting we find ourselves slipping back into. The formalization of these rules and norms (the "architectural leadership") is important for the long-term sustainability of the CLCC as partner entity representatives inevitably change and the institutional history is lost. What the formalization looks like will also determine how we measure the efficacy of the partnership. Additionally, decentralization can make it challenging to have a shared history on who did what as well as for crafting compelling communications that accurately reflect the leadership roles partners have taken on in order to reach certain milestones. Inreach within partner entities and outreach to potential new partners rely on such communications.

3. Role of Relational Governance

Relationships matter. This is not a novel statement as it is the conclusion of many teams in research and natural resource management circles alike. This reality comes more into the forefront for strategic teams that are composed of multiple agencies and organizations. How well a team performs is largely determined by how well they recognize and appreciate relational governance in their operations. In natural resource management, teams comprised of top subject matter experts tend to be emphasized. However, in the CLCC, we have learned in three years of partnering that team dynamics are far more indicative of successful outcomes than having top experts or managers. A good example of this is our Protected Areas Conservation Action Team where a mix of top experts and managers work directly with mid-level and entry-level employees and new researchers and have performed exceptionally. Top managers for a few of the most active agencies delegated and only once or twice a year do they meet with the full team to see progress and provide feedback. This team, like the other high performing strategic teams of the Cooperative, have interpersonal relationships based on trust, reciprocity, and mutual goals and have met their goals more efficiently than those without.

This experience is consistent with findings in the literature. "Networks are based on the relational ... it is relationships that give networks their strength and edge over other governance forms" [5] (p. 443).

3.1. Lessons from Google

One well-known company learned the importance of relationships for strategic teams through three years of research. Google put together a team of social scientists tasked with observing, interviewing, and collecting data on hundreds of Google teams [9] in order to answer the research question "What makes a team effective at Google?"[10]. Code-named Project Aristotle—a tribute to Aristotle's quote, "the whole is greater than the sum of its parts" (in reference to Google's belief that employees can do more working together than alone). After defining what a team is and how to measure effectiveness, the data collection and analyses began. They found that how the team worked together (the team culture and interpersonal relationships) and not who was on the team, mattered most. This was contrary to what Google believed prior to the research. They had thought I.Q. or talent would be the primary factor. More specifically, the following factors determined team effectiveness:

1. **Psychological safety**, "an individual's perception of the consequences of taking an interpersonal risk or a belief that the team is safe for risk taking in the face of being seen as ignorant, incompetent, negative or disruptive."
2. **Dependability**, "members reliably complete quality work on time (versus the opposite—shirking responsibilities)"
3. **Structure and clarity**, "an individual's understanding of job expectations, the process for fulfilling these expectations, and the consequences of one's performance ... Goals can be set at the individual or group level, and must be specific, challenging and attainable"
4. **Meaning**, "finding a sense of purpose in either the work itself or the output ... The meaning of work is personal and can vary: financial security, supporting family, helping the team succeed, or self-expression for each individual, for example."
5. **Impact**, "the results of one's work, the subjective judgement that your work is making a difference, is important for teams. Seeing that one's work is contributing to the organization's goals can help reveal impact."

The Project Aristotle team also found the following variables were *not* significantly connected with team effectiveness (for Google; these variables might be significant for other organizations): (1) colocation of teammates; (2) consensus-driven decision making; (3) extroversion of team members; (4) individual performance of team members; (5) workload size; (6) seniority; (7) team size; and (8) tenure [10].

3.2. Large Partnerships on Small Islands

The local discourse in Caribbean islands often emphasizes that relational governance seems more necessary on small islands. The point is often made that because there are smaller numbers in the conservation community or professional circles, maintaining good relationships is even more important for sustaining existing and future teamwork. Moreover, because of the prevalent culture of socializing and "island time", teams need to integrate Caribbean cultural norms into the work environments. Research would be needed to determine whether that is just perception from island practitioners or if in fact a difference between landscape conservation in islands and continents. Emphasis on relationships is critical to determining team success anywhere and not just on small islands. However, consequences of past team failures may influence new partnerships more frequently in island contexts given the smaller pool of professionals and collaboration opportunities.

3.3. The Wicked Problem of Team Formation

The social-ecological problems that the All-Lands Approach to Conservation seeks to address have been described as wicked [11] and super wicked [12]. Norris et al. (2016) make a compelling case for how the teams that employ a transdisciplinary approach to research and management to address those wicked problems experience challenges that make the formation of these teams a wicked problem in itself [13]. The evidence for this is the challenges experienced in team formation are consistent with challenges described by Rittel and Webber (1973) that defined challenges of wicked problems, namely: (1) the process of formulating the problem and of conceiving a solution are identical; (2) it is difficult to ascertain when the work is done; (3) every solution is a "one-shot operation" meaning no opportunities to learn by trial and error (different from learning by doing); and (4) every wicked problem can be considered to be a symptom of another problem [14].

Norris et al. (2016) [13] go on to discuss ways to address the wicked problem of team formation, similar to ones discussed below, and argue that team formation is a planning process. For landscape conservation, this means in essence that we need to go through planning process to form the teams that will develop long-term and all-lands conservation planning processes. As Luis Villaneuva-Cubero, CLCC Spatial Analyst and graduate of the University of Puerto Rico's Graduate School of Planning, stated at a Steering Committee meeting, "it is important to note that planning processes inherently must end in implementation [of said plans] or it is not a true planning process" [15].

Teams that do not go through well thought-out, relationship-based processes are like icebergs [16]. The ordinary or technical problems are at the surface but the governance and political problems and the fundamental and cultural problems are below the surface and only become apparent after your team's ship has struck the iceberg.

The CLCC has learned (sometimes through failures) that it takes considerable investment of time and effort to build and sustain the interactions that are central to the healthy functioning of network relationships. The investment needed for strategic relationship building and monitoring and embedded social relations can limit the actions of members and increase the need for coordination and communications. It also takes skilled network leadership and management in order to navigate the complex sets of relationships and agendas and identify ways to overcome barriers to action. It is critical to keep the perspective alive in the network that the teams are working to ensure a public value is delivered. In my opinion, this is the principal reason why patience is deemed necessary when employing the landscape approach [3], as it can take a lot of time to achieve mutually agreed on actions and outcomes [7] because of relational governance challenges. The collective impact of the collaborative actions are greater than when attempting to work alone as a single entity, but the path is longer and with more speed bumps.

3.4. Role of Cooperative Teamwork Procedures

The CLCC is still developing well-defined rules, incentives, and norms within the network. These are sketched out through a combination of allowing them to develop organically and by formalizing certain necessary procedures. Imperial et al. (2016) say that "structures and processes themselves are theorized to be sources of leadership, separate from the formal or visible leader" [7] (p. 128). This is a dynamic process but the literature provides insights as to why certain rules should be formalized sooner rather than later.

Bennett and Gadlin (2013) describe collaboration and team science along a continuum extending from collaborations with minimal levels of interaction to scientific teams with significant levels of interaction and integration [17]. CLCC Research Teams have occurred in different points on this continuum with some having higher levels of internal interaction and integration than others. A couple research projects were investigator-initiated where the scopes of the work were related to CLCC priorities and so were supported by the Cooperative but until the findings were available for dissemination there was little to no interaction between the rest of the Cooperative and the investigator's research team. On the other end of the spectrum are research projects where the teams worked on the research problem together, each member bringing specific expertise to the table. The projects also used sub-teams that would take on certain tasks conducting research separately but going back to the larger group at key stages. There were regular meetings and discussions during the team formation, during the work, and afterwards during delivery of the results and multiple times communicated with the wider Collaborative via workshops, webinars, and in written form to elicit input used during the research. All teams, despite where they lie on the continuum, are connected to the broader CLCC strategy so in that regard are strategic teams though not necessarily collaborative or consistent with all network governance definitions.

While Bennett and Gadlin's work tends to be more focused on collaborative teams in the biomedical sciences, their perspectives are suitable for large-scale conservation partnerships. They focus on the interpersonal context (relationships among scientists) and borrow from an extensive study on collaborations in physics (Shrum, Geruth, and Chompalov 2007 [18]). From their work, I have crafted a list of factors that contribute to the successful interactions and communications among teams, as appropriate for landscape conservation. Their list comes from in-depth interviews with five National Institutes of Health teams that either were successful, did not succeed in getting fully off the ground, or came to an end due to conflict. Reviewing these variables revealed many CLCC teams and teams of the wider LCC network have experiences consistent with those five teams. Indeed, the Cooperative has experienced the three types of varying degrees of success as consequences of abilities to operationalize procedures that consider these variables. To conceptualize and executive these elements in everyday practice is not easy or intuitive as they may appear:

1. *Self- and Team-Awareness and Abilities to Self-Regulate* (e.g., communications style, conflict management approach, personality types, approach to giving and receiving feedback)
2. *Understanding Team Development* (Forming, Storming, Norming, Performing)
3. *Trust*
4. *Building a Team* (identification of people interested and capacity to work as a cooperative is important; they have synchronicity between team goals and individual interests as they cannot achieve their individual goals on their own)
5. *Creating a Shared Vision* (recognizing vision statements and team objectives are dynamic and will change over time)
6. *Sharing Recognition and Credit* (e.g., review criteria for evaluations should allow for multiple leaders to share status and power, recognition and reward; craft agreed-upon criteria for authorship on products and how decisions are made about gives talks or responds to media inquiries and how intellectual property will be handled; discuss how the team will promote the careers of individual members who depend on taking appropriate credit and receiving proper recognition for their career trajectories)
7. *Communicating about the Science: Promoting Disagreement while Containing Conflict* (taking advantage of the numerous benefits engaging in disagreements provide: new and stronger relationships within the team; keeps problems or issues from simmering thus prevents accumulation of resentment or of disagreements; continued re-evaluation of the team dynamic and the team procedures; strengthened trust; and emergence of creative solutions)
8. *Communicating with Each Other* (see Table 3—Ways to Strengthen Team Relationships and Dynamics)
9. *Share the Excitement of the Process and Discovery* (in theory, everyone in the team loves what they do—do not be afraid to share that passion).

The storming under "Understanding team development" is the point in CLCC teams that some might describe as the most painful or where the most uncertainty lies in whether the team will move forward or fold. This is the phase where team members develop processes often that come out of identification of differences and people opening up to one another. The storming is when individuals move into teamwork and have to shift mentality from being an individual or sole expert to sharing influence and leadership with others. Because of the potential sensitivity in this stage, it is important to develop processes early for managing conflict while creating a safe space for open and honest discussion articulating expectations and defining roles and responsibilities.

Table 3. Ways to Strengthen Team Relationships and Dynamics (compiled from Bennett and Gadlin (2013), Project Aristotle, and author perspectives of Caribbean Landscape Conservation Cooperative (CLCC) practices).

Bennett and Gadlin (2013)	Project Aristotle (2016)	CLCC Practices
Foster a collegial and non-threatening environment.	Establish a common vocabulary about team behaviors and norms you want to foster.	Use of process agendas that define purpose and desired outcomes of meetings and use of meeting ground rules, such as "silence is agreement" and "try to offer a solution if stating a problem".
Openly recognize strengths of all members of the team and note as a team how different strengths contribute to advancing the project.	Create a forum to discuss team dynamics allowing for teams to talk about subtle issues in safe, constructive ways.	Outside formal meetings discuss views on process and progress and make sure members know procedures for speaking up. Or offer to bring up the comment for the team member if some restriction to them doing so.
Take a few minutes at regularly scheduled group meetings to do a check-in: How is everyone doing?	Commit leaders to reinforcing and improving.	Create time outside formal meetings to develop relationships (e.g., coffee before or after or via scheduled coffee breaks; time for socializng before or after; field trips with built in socializing time).
Encourage open and honest discussion by establishing trust.	Actively solicit input and opinions from the group. Share information about personal and work style preferences.	Rotate who moderates or facilitates work sessions and meetings.
Jointly develop a process for bringing issues and disagreements forward for early resolution.	Foster dependability by clarifying roles and responsibilities of team members and developing concrete project plans to provide transparency into every individual's work.	Use of user-centered performance metrics for individual tasks or teams, usually through meeting process agendas or tactical communications plans.
Assure that when decisions are being made that require everyone's input that each person has an opportunity and understands the process for providing comment.	Regularly communicate team goals and ensure members understand the plan for achieving them.	Definition of science delivery that stresses coproduction of knowledge, engagement with users, and working towards informing management, policy, investment, behavior change, community actions, or further research.
Schedule periodic assessments and feedback, including opportunities for collaborators to discuss what is going well, what is not, and what needs to be improved.	Ensure your team meetings have a clear agenda and designated facilitator.	Learning-by-doing; Recognizing that network governance/team procedures need to be dynamic as we learn how to operationalize this new model of collaborative conservation. Trainings and capacity building are key!
	Give team members positive feedback on something outstanding they are doing and offer to help with something they struggle with.	Strive for transparency, open membership and open processes, and use variety of available digital tools to assist.
	Publicly express gratitude for someone who helped you out.	Encourage developing new perspectives and new skills.
	Co-create a clear vision that reinforces how each team member's work directly contributes to the team's and broader organization's goals.	Recognize the eight values that motivate human actions and strive for teams that produce and allocate values (Respect, Influence/Power, Wealth, Well-being/Security, Knowledge, Skill, Affection, Ethics) [1]
	Reflect on the work you are doing and how it impacts users or clients and the organization. Adopt a user-centered evaluation method and focus on the user.	Strive to embedd in processes standpoint clarification, problem orientation, social process mapping, and decision-process mapping [1], sometimes called Structured-Decision-Making [2]
		Look for partner needs you can provide (reciprocity)

[1] Clark, T.W. 2002. The Policy Process: A Practical Guide for Natural Resources Professionals. Yale University. New Haven, CT, USA; [2] Keeney, R. 2004. Making better decision makers. Decision Analysis 1(4) 193–204.

The five factors that Google's Project Aristotle found to set successful teams apart come into play during the storming phase, especially psychological safety and structure/clarity. It is the norming and performing phases where team members begin to work together effectively and efficiently, developing trust and comfort as they learn to rely on each other ultimately working together seamlessly focusing on the shared goals and resolving issues and problems that emerge. These last two phases are highly dependent on what happens in the storming phase. The CLCC has experienced in a couple teams the negative consequences of not taking the time for the intentional development of these procedures. The consequences being less trust and more difficulty transitioning to the norming phase.

Trust just might be what all the other eight elements boil down to. The CLCC has experienced trust and lack of trust in a few areas of work. Trust that team mates will deliver on their assignments or share the necessary data. Trust in the shared vision that we are doing something meaningful that will have positive results for society. Trust in the resources of member organizations to support the work. Trust that each members' contributions are appreciated and will be credited appropriately for the good of the team and the individual. Trust that team members will not try to dominate or make unilateral decisions. Trust that if there is disagreement on process or scientific outcomes that there is a safe space to voice opinions and the team will act constructively. Trust that everyone will be respected and human dignity protected. Unfortunately, trust is often overlooked in collaborative conservation. As Bennett and Gadlin (2013) state, "For many, [trust seems hopelessly subjective and even softheaded" [17] (p. 5). The experiences so far in the CLCC have taught us that trust can make or break a team. Work relationships play a critical role in the teamwork itself. There are risks in teamwork as each individual member forfeits some of their control or influence and when the outcomes of the collaboration affect individual performance the risks are even greater. This dependence (the team's dependence on the individual and the individual's dependence on the team) creates vulnerability. Without trust-filled relationships that vulnerability leads to individuals being protective or defensive rather than collaborative. Time and great care must be taken to build and nurture trust among team members. Furthermore, trust can come from strong personal relationships or created or reflected in written agreements.

As there is little space in this communication to detail all the different tactics to ensure strong team relationships, I have compiled several ways based on the work of Bennett and Gadlin (2013), Google's Project Aristotle, and the most successful CLCC practices based on informal staff monitoring and evaluation. The entries in the table are not exhaustive and many more tools are available. As you read these tactics, it becomes evident that some require certain skill-sets that not everyone possesses. Imperial et al. (2016) review the literature and find that network participants should possess abilities to have a collaborative mindset, link to external resources, mobilize existing assets, be persuasive, deal with changing contexts and challenges, manage group processes, and lead even when not in charge or empower others to lead. Obviously, not everyone can possess all of these skillsets so drawing from those who have those skills and pooling them is one way, but to be most effective collaborative leadership training is crucial. Some may feel they are lacking in these areas and that demonstrates a strong level of self-awareness needed for developing the capacity to work collaboratively [17]. Many natural resource management programs in universities are investing in developing programs that equip students and researchers with these collaborative skills and federal and state agencies in the United States are beginning to provide relevant trainings. The LCC network through the Department of the Interior has benefited from a variety of resources and trainings by the Partnership and Community Collaboration Academy [19] and the National Conservation Training Center [20] but there are many other institutes and centers that can help develop these skills.

4. Role of Communications

It has already been described how team communications are important for rethinking leadership and focusing on relational governance, but communications play other roles in large-scale conservation partnerships as well. It is no secret that agencies, universities, and scientific partnerships do not focus or invest enough in communications. When communications are integrated into a project, it tends to occur at the end of the project or after the research has been done. Sometimes, the project managers do not leave ample room for communications and it is an afterthought. Furthermore, when communications are included, often the communications are not strategically tied to available or generated information on user needs or project goals. For strategic teams to be successful, strategic internal and external communications need to be supported before or during team formation. Bennett and Gadlin (2013) identified this as a factor for internal team procedures [7] and Google's Project Aristotle identified this as a key dynamic for internal and external activities because team members should reflect on the work they are doing and how it impacts users or clients [10]. They put a strong focus on the user.

Similarly, the CLCC and LCC Networks are user-centered in that team outputs utilized by users are how success or effectiveness is determined in landscape conservation delivery efforts. Because landscape conservation partnerships are inherently applied science initiatives, they strive for the products to be used to inform management, investment, policy, behavior change, community actions, etc. [21]. Co-production or actionable science is a science delivery approach increasingly utilized by federal agencies in conservation and communications play a strong role [22]. The term "science delivery" has become a bit of a misnomer as its meaning has evolved as the benefits of co-production are realized, rather than just delivery and dissemination itself. While clear pathways cannot be delineated easily for knowledge-to-action as it is more of a web of interactions that lead to knowledge being applied [23], the co-production approach is showing to be effective [22].

When the LCCs began, communications were focused on translating science in formats useful to decision-makers. Communications staff were brought in primarily after research projects were completed to package the results in a variety of formats based off of presumed needs of the user groups and "delivered" using suitable tactics for each target audience or user. Stakeholder engagement was a strong element of the approach but it was more viewed as a bridge where information was brought from science providers to science users, and from science users to science providers [24]. In 2014, while setting the LCC Network Science Agenda, a sub-team of science coordinators and communicators discussed the positives and negatives of this approach based off of lessons learned up until that point. The recommendation to the larger network was that communications and stakeholder engagement, "science delivery", need to be considered before a project was even envisioned and employed throughout the research project and of course afterwards. The needs of user groups and the tactics employed for delivering the conservation approaches or the science should be designed with the users. As a result, the 2014 Strategic Plan [25] has a series of objectives and tactics that move the network towards co-production and actionable science, including "Encourage communications guidance, policy, training, and support to principle investigators for science delivery regarding outreach strategies and applications of their research and results to end users (e.g., land managers) and assist them in demonstrating the ecosystem services and socio-economic values of their conservation research" [25] (p. 19). To develop this guidance, the network created the LCC Science Delivery Working Group. The guidance has not been finalized yet, but the draft steps of "science delivery" include four larger best practice categories where communications are integrated: Scoping, Analysis, Outreach/Use of Results, Evaluation. Each of these four are broken down into nine elements of science delivery: (1) engagement with decision makers/implementers (collectively: stakeholder engagement); (2) identification of science needs put into a clear policy or research question; (3) Synthesis of scientific and agency information to ensure the research has not already been conducted; (4) conservation and research design; (5) analysis/conducting research; (6) science translation; (7) delivery and deployment; (8) conservation adoption; and (9) measurement and evaluation [21]. Distinctive in this approach is that teams working to co-produce knowledge include the users and managers that will be utilizing the outputs. Additionally, communications are integrated throughout the process and not just at the end of a project when outputs need to be translated and communicated to broader audiences.

Because each team's needs for internal and external communications are dependent on the shared priorities they are tackling together, communicators are also integrated into the teams. Some team members might play a dual role in that they are skilled communicators who also contribute as scientists or users and others serve exclusively for designing stakeholder engagement processes, facilitating meetings or workshops, coordinating internal communications, and for communicating externally to other target users. It is important that the teams are being strategic about their communications in that tactics used are data-driven and have clear linkages to objectives and metrics of success.

In the CLCC, each team has a different approach for how communications, and thus communicators, are integrated. Many of the Table 3 ideas require strong communications and high levels of organization. The organizational and administrative skills have been a factor in the effectiveness of CLCC Conservation Action Teams and the CLCC Steering Committee. Each team employs different digital collaborative tools and operating procedures that are based on team composition, purpose or tasks, collective skill, and the norms that emerge through the team formation processes mentioned above. The Cooperative has had varying levels of success with internal and external communications. Key to improvement is monitoring and evaluation of performance, without this strategic teams do not have strategic communications.

5. Final Thoughts and Conclusions

This communication is just one person's perspective. Other members of the CLCC or others who work in the implementation of the All-lands Approach in forested landscapes might disagree on a number of points. The landscape conservation community does not have all the answers yet on how to operationalize the teamwork of these large-scale partnerships. That is why the work towards this new model of conservation is challenging and exciting for landscape conservation practitioners. We are learning-by-doing and it takes experimentation, creativity, risk taking, and learning from others and each other. There are a few professional challenges we need to overcome in order to see a clear path forward for the All Lands, All Hands approach to conservation. One is patience. Institutional change can be slow. There is a whole social science field around how change happens in institutions. There exist institutional restrictions, baggage from past team efforts, doubt, limited human capital, and varying levels of capacity.

Additionally, much of the social and natural science research needed for the partnerships of the new era in conservation, especially climate studies, take years to complete. If only institutional change and landscape conservation science could work as quickly as our changing climate, then we might be able to avoid the worst effects. This offbeatness might just be the greatest challenge, though there are many. Practitioners of the All-Lands Approach are attempting to do something at a pace institutions and scientists might not be structurally ready for. Sometimes, it seems a lot like trying to speed through a red mangrove stand; lots of starts and stops as you go over prop roots encumbering your path. At least it can feel that way when patience is at its lowest. The strategic teams we interact with everyday and their accomplishments are what keep the new model of conservation moving forward. If we can learn to place a greater emphasis on rethinking leadership in collaborative settings and on relational governance, cooperative teamwork procedures, and communications there is reason to hope we will see long-term success and have more documented case studies by landscape conservation partnerships.

Acknowledgments: The author would like to thank A. Lugo, P. Rios, T. Muñoz-Erickson, I. Parés-Ramos and L. Villanueva-Cubero for reviewing an earlier version of this manuscript. All research conducted at the International Institute of Tropical Forestry is done in collaboration with the University of Puerto Rico.

Conflicts of Interest: The author declares no conflict of interest.

References

1. Tidwell, T. An All-Lands Approach to Conservation. U.S. Forest Service. 2010. Available online: https://www.fs.fed.us/speeches/all-lands-approach-conservation (accessed on 1 March 2017).
2. Secretary of the Interior. Order No. 3289: Addressing the Impacts of Climate Change on America's Water, Land, and Other Natural and Cultural Resources. U.S. Department of the Interior. 22 February 2010. Available online: https://lccnetwork.org/sites/default/files/Resources/DOI_SecretarialOrder_3289A1.pdf (accessed on 1 March 2017).
3. Reed, J.; Van Vianen, J.; Deakin, E.L.; Barlow, J.; Sunderland, T. Integrated landscape approaches to managing social and environmental issues in the tropics: Learning from the past to guide the future. *Glob. Chang. Biol.* **2016**, *22*, 2540–2550. [CrossRef] [PubMed]

4. International Cooperative Alliance. Available online: https://ica.coop/en/what-co-operative (accessed on 20 January 2017).

5. Keast, R. Chapter 36: Network governance. In *Handbook on Theories of Governance*; Ansell, C., Torfing, J., Eds.; Edward Elgar Publishing Limited: Northampton, MA, USA, 2016.

6. Jones, C.; Hesterley, W.; Borgatti, S. A general theory of network governance: Exchange conditions and social mechanisms. *Acad. Manag. Rev.* **1997**, *22*, 911–945.

7. Imperial, M.T.; Ospina, S.; Johnston, E.; O'Leary, R.; Thomsen, J.; Williams, P.; Johnson, S. Understanding leadership in a world of shared problems: Advancing network governance in large landscape conservation. *Front. Ecol. Environ.* **2016**, *14*, 126–134. [CrossRef]

8. Helmke, G.; Levitsky, S. Informal institutions and comparative politics: A research agenda. *Perspect. Politics* **2004**, *2*, 725–740. [CrossRef]

9. Duhigg, C. What Google learned from its quest to build the perfect team. *New York Times Magazine*, 25 February 2016.

10. Google re:Work Team. Guide: Understand Team Effectiveness. Available online: https://rework.withgoogle.com/blog/five-keys-to-a-successful-google-team/ (accessed on 20 January 2017).

11. Carroll, M.S.; Blatner, K.A.; Cohn, P.J.; Morgan, T. Managing fire danger in the forests of the U.S. inland Northwest: A classic wicked problem in public land policy. *J. For.* **2007**, *105*, 239–244.

12. Levin, K.; Cashore, B.; Bernstein, S.; Auld, G. Overcoming the tragedy of superwicked problems: Constraining our future selves to ameliorate global-climate change. *Policy Sci.* **2012**, *45*, 123–152. [CrossRef]

13. Norris, P.A.; O' Rourke, M.; Mayer, A.S.; Halvorsen, K.E. Managing the wicked problem of transdisciplinary team formation in socio-ecological systems. *Landsc. Urban. Plan.* **2016**, *154*, 115–122. [CrossRef]

14. Rittel, H.W.; Webber, M.M. Dilemmas in a general theory of planning. *Policy Sci.* **1973**, *4*, 155–169. [CrossRef]

15. Villanueva-Cubero, L.; (University of Puerto Rico, Río Piedras, Puerto Rico, U.S.). Personal communication, 2016.

16. Clark, T.W. *The Policy Process: A Practical Guide for Natural Resources Professionals*; Yale University: New Haven, CT, USA, 2002.

17. Bennett, L.M.; Gadlin, H. Collaboration and team science: From theory to practice. *J. Investig. Med.* **2012**, *60*, 768–775. [CrossRef] [PubMed]

18. Shrum, W.; Genuth, J.; Chompalov, I. *Structures of Scientific Collaboration*; The MIT Press: Boston, MA, USA, 2007.

19. Partnership and Community Collaboration Academy. Available online: http://www.partnership-academy.net/ (accessed on 20 January 2017).

20. National Conservation Training Center. Available online: https://training.fws.gov/ (accessed on 31 March 2017).

21. LCC Network Science Delivery Working Group. Guidance: Effective Science Delivery (Working Paper), 2015; In preparation.

22. Beier, P.; Behar, D.; Hansen, L.; Helbrecht, L.; Arnold, J.; Duke, C.; Farooque, M.; Frumhoff, P.; Irwin, L.; Sullivan, J.; et al. *Guiding Principles and Recommended Practices for Co-Producing Actionable Science: A How-To Guide for DOI Climate Science Centers and the National Climate Change and Wildlife Science Center*; Report to the Secretary of the Interior; Advisory Committee on Climate Change and Natural Resource Science: Washington, DC, USA, 2015; Available online: https://nccwsc.usgs.gov/sites/default/files/files/HowToGuideforActionableScience_ACCCNRS_2.pdf (accessed on 20 January 2017).

23. Muñoz-Erickson, T.A. Multiple pathways to sustainability in the city: The case of San Juan, Puerto Rico. *Ecol. Soc.* **2014**, *19*, 2. [CrossRef]

24. Jacobs, K.R.; Nicholson, L.; Murry, B.A.; Maldonado-Roman, M.; Gould, W.A. Boundary organizations as an approach to overcoming science-delivery barriers in landscape conservation: A caribbean case study. *Caribb. Nat.* **2016**, *1*, 87–107.

25. Landscape Conservation Cooperative Network. *2014 Strategic Plan*; The US Department of Agriculture (USDA): Washington, DC, USA, 2014.

forests

MDPI

Review

Soil Biology Research across Latitude, Elevation and Disturbance Gradients: A Review of Forest Studies from Puerto Rico during the Past 25 Years

Grizelle González [1,*] and D. Jean Lodge [2]

[1] United States Department of Agriculture, Forest Service, International Institute of Tropical Forestry, Jardín Botánico Sur, 1201 Ceiba St.-Río Piedras, 00926, Puerto Rico
[2] United States Department of Agriculture, Forest Service, Northern Research Station, Luquillo 00773-1377; Puerto Rico; dlodge@fs.fed.us or dlodgester@gmail.com
* Correspondence: ggonzalez@fs.fed.us; Tel.: +1-787-764-7800; Fax: +1-787-766-6302

Academic Editor: Timothy A. Martin
Received: 31 March 2013; Accepted: 20 May 2017; Published: 24 May 2017

Abstract: Progress in understanding changes in soil biology in response to latitude, elevation and disturbance gradients has generally lagged behind studies of above-ground plants and animals owing to methodological constraints and high diversity and complexity of interactions in below-ground food webs. New methods have opened research opportunities in below-ground systems, leading to a rapid increase in studies of below-ground organisms and processes. Here, we summarize results of forest soil biology research over the past 25 years in Puerto Rico as part of a 75th Anniversary Symposium on research of the USDA Forest Service International Institute of Tropical Forestry. These results are presented in the context of changes in soil and forest floor biota across latitudinal, elevation and disturbance gradients. Invertebrate detritivores in these tropical forests exerted a stronger influence on leaf decomposition than in cold temperate forests using a common substrate. Small changes in arthropods brought about using different litterbag mesh sizes induced larger changes in leaf litter mass loss and nutrient mineralization. Fungi and bacteria in litter and soil of wet forests were surprisingly sensitive to drying, leading to changes in nutrient cycling. Tropical fungi also showed sensitivity to environmental fluctuations and gradients as fungal phylotype composition in soil had a high turnover along an elevation gradient in Puerto Rico. Globally, tropical soil fungi had smaller geographic ranges than temperate fungi. Invertebrate activity accelerates decomposition of woody debris, especially in lowland dry forest, but invertebrates are also important in early stages of log decomposition in middle elevation wet forests. Large deposits of scoltine bark beetle frass from freshly fallen logs coincide with nutrient immobilization by soil microbial biomass and a relatively low density of tree roots in soil under newly fallen logs. Tree roots shifted their foraging locations seasonally in relation to decaying logs. Native earthworms were sensitive to disturbance and were absent from tree plantations, whereas introduced earthworms were found across elevation and disturbance gradients.

Keywords: tropical forests; invertebrates; microbiota; soil biota; litter; wood; latitude; elevation; disturbance; gradients

1. Introduction

Although below-ground research has lagged behind above-ground studies of plants and animals, especially in tropical forests, the Luquillo Experimental Forest (LEF) in Eastern Puerto Rico has some of the earliest research on effects of disturbance on fungi and ecosystem processes. Research carried out in the 1960s by the US Department of Energy under the Atoms for Peace program, and largely published in a book edited by H.T. Odum and R.F. Pigeon [1] showed that disturbances that open the forest canopy, whether from cutting or gamma irradiation, induce major shifts in fungal communities (see summary in Lodge [2]). The methods used at the time were limited, so the studies of disturbed vs. undisturbed environments by Holler [3] and Holler and Cowley [4] focused on comparisons of morphospecies of fungi that could be grown on agar. Nevertheless, they showed strong responses of litter fungi to disturbance. Those early results have been largely validated more recently using modern techniques that detect all fungi and bacteria, such as identifying phylospecies using Terminal Restriction Length Polymorphism (TRFLP) and sequences of the Internal Transcribed Spacer (ITS)—A DNA region used as a molecular barcode in fungi, and Extracted Microbial Fatty Acid Methyl Ester (FAME) analysis for overall changes in dominance among microbial groups e.g., Cantrell et al. [5,6]. Early work on whole ecosystem respiration [1] has been superseded by soil gas flux measurements using automated samplers for below-ground fluxes e.g., [7–9].

In the tropics, a great deal of attention has been given to above-ground organisms, while few studies deal with the diversity of invertebrates in the soil, leaf litter, or dead wood [10]. There have been some quantitative studies of below-ground invertebrates in tropical forests [11]. Early faunal inventories at El Verde in the Luquillo Experimental Forest [12,13] showed that about half of the faunal biomass was concentrated in the thin upper soil horizon and litter layer. Given the persistence of both anthropogenic and non-anthropogenic disturbances in the tropics, it is important to study the diversity of its fauna and assess how they can affect ecosystem functioning. For example, recent research has focused on the effects of disturbance on arthropods and their connectivity to other biotic and abiotic factors in forested ecosystems that can have significant effects on detrital processes [14–17].

A recent focus of studies in the tropics has been on changes in biotic communities along elevation gradients [18,19]. These studies shed light on the abiotic and biotic factors that control species distributions, biotic assemblages and ecosystem processes. Further, elevation gradient studies provide baseline data for biota that may be imperiled by climate change, especially for species that are restricted to high-elevation cloud forests. Changes in ecological space along elevation gradients can be used as a proxy for environmental changes across latitudinal gradients [18,19]. While species distributions may respond similarly to changes in ecological space along both elevation and latitudinal gradients, they are also influenced by barriers to dispersal and colonization (i.e., filters) at the longer distances associated with latitudinal gradients.

The aim of this paper is to review the salient results of research on soil and litter biota in forests over the last 25 years in Puerto Rico. This manuscript is part of a Special Issue comprised of presentations at the 75th anniversary of the establishment of the International Institute of Tropical Forestry in Puerto Rico (United States Department of Agriculture, Forest Service).

2. Summary of Results over the Past 25 Years

2.1. Latitudinal Gradients

The status of information on soil animal diversity in the tropics is limited, particularly when compared to other ecosystems such as temperate forests, grasslands and deserts [11]. In Puerto Rico (as in many other tropical regions), when present, earthworms compose the highest biomass among the soil macrofauna [1]; and thus play important roles in regulating soil processes [20]. The density of macroarthropods (such as myriapodans and crustaceans) is higher at LEF than in tropical sites elsewhere; ants are also an important component of the litter invertebrate community with densities ranging from 500 to 1200 m^{-2} [13,21]. In the LEF, millipedes appear to be the predominant taxa in the tabonuco forest litter at mid-elevations in terms of standing stocks (0.6 g·m^{-2}; Pfeiffer, [13]). Yet, the contribution of soil fauna activities to ecosystem processes varies widely along latitudinal gradients because it depends on the confounding effects of the size, abundance, diversity, and functionality e.g., [22–26].

2.1.1. Arthropods Are More Diverse and Accelerate Leaf Litter Mass Loss More in Wet Tropical Forest than in Cold Temperate Forest

Soil fauna can influence decomposition and mineralization processes either directly by modifying litter and soil environments or indirectly via interactions with the microbial community [27–30]. In a study comparing leaf decomposition across a latitudinal gradient between wet tropical forest in Puerto Rico and subalpine forest in Colorado, soil fauna appeared to have greater direct effects via grazing on the microbial community whereas in the tropical forest, soil fauna appeared to have primarily indirect effects via litter comminution [25–27]. Comminution fragments litter and opens fresh surfaces to microbial decomposers. González and Seastedt [25] compared leaf decomposition rates and arthropod higher-order assemblage diversity (expressed as the number of orders) in wet and dry tropical forests of Puerto Rico and temperate subalpine forest in Colorado and found highest diversity and decomposition rates in wet tropical forest. Similarly, Heneghan et al. [23,31] compared decomposition of a common substrate (*Quercus* leaves) among wet neotropical forests of Puerto Rico and Costa Rica and warm temperate forest in North Carolina, and found highest decomposition rates among neotropical forests despite having similar or even lower species diversity. On the other hand, data of Crossley in Coleman et al. [32] showed high magnitude (45%–71%) soil fauna effects in warm temperate forest in western North Carolina, USA. Similarly, soil fauna can account for up to 66% of the total decomposition of tough *Cecropia schreberiana* leaves with high lignocellulose content in the tropical wet forest in Puerto Rico [25], and macroarthropods involved in comminution of litter, such as millipedes, can have the strongest effect on leaf litter mass loss. For example, González et al. [33] found that millipedes can affect leaf litter decomposition both directly and indirectly, but the extent of their effect depends on their density and the quality of the substrate (leaf lignin content). Also, Ruan et al. [34] found that millipede density explained 40% of the variation in leaf decomposition rates, whereas microbial biomass explained only 19% of the variance. Leaf litter comminution by termites is common in the paleotropics where fungal gardening termites occur, but not in neotropical forests where this termite feeding guild is absent. Termites that consume leaf litter are either uncommon or poorly documented in neotropical forests, though there is a specialized grass-feeding termite in the grasslands of the Great Savanna of Venezuela. There are no termite species that consume leaf litter in Puerto Rico.

Wood is the main constituent of tropical forest detritus [35]. Boreal, temperate, tropical, and island ecosystems vary in climate, species composition, decomposer community structure and rates of biomass production, resulting in variable amounts of carbon stored in persistent downed woody debris [36,37]. Thus, González et al. [34] set up a wood decomposition experiment to quantify the decay of aspen stakes (*Populus tremuloides*) in dry and moist boreal, temperate and tropical (Puerto Rico) forest types. They concluded that moisture content is an important control of wood decomposition over broad climate gradients, and that such relationship can be non-linear. Furthermore, they also found that the presence termites significantly altered the decay rates of wood in ways that cannot be predicted solely with climatic factors. These data suggest that biotic controls, rather than abiotic constraints, can better predict wood decay in tropical regions [36].

2.1.2. Tropical Soil Fungi Are Generally More Diverse and Have Smaller Geographic Ranges than Temperate Fungi, and Fungal Diversity Is Related to Rates of Leaf Decomposition

Comparisons of fungal diversity between tropical and temperate forests have previously been generated using a few well-studied plots across latitudinal gradients and then making projections based on fungi to plant species richness ratios [38,39]. For example, Mueller et al. [38] used inventories and name databases to validate macrofungal (those with fruiting bodies visible without magnification) to plant species ratios and arrived at a ratio of 2:1 for temperate regions and 5:1 for tropical regions. Since tropical regions have much higher plant diversity than temperate regions, this led to much higher estimates of macrofungal diversity in the tropics. Further, Mueller et al. [38] indicated that tropical macrofungi had higher rates of regional endemism than temperate fungi. Lodge et al. [40]

also found that in two families of mushrooms, 41% of Hygrophoraceae and 54% of Entolomataceae species were endemic to the Greater Antilles or the Caribbean islands. Animals, including detrital invertebrates, have long been known to have smaller geographic ranges in the tropics than at higher latitudes—A pattern now referred to as part of Rapoport's Rule [41]. Several publications correctly noted the relationship between larger body sizes among detrital invertebrates and more discontinuous and narrower distributions, but they incorrectly assumed that microorganisms including fungi fit into the small body size end of this scheme and had predominantly wide, ubiquitous distributions [42–44]. A more recent global analysis by Tedersoo et al. [45] of all soil fungal diversity based on DNA barcode sampling of natural communities in soils including El Verde in the Luquillo Experimental Forest of Puerto Rico supports the patterns of latitudinal gradients in fungal diversity observed by Mueller et al. [38], and also high abundance of regional endemism found by Mueller et al. [38] and Lodge et al. [40]. Except for ectomycorrhizal fungi that are obligate symbionts of mostly temperate and boreal trees, Tedersoo et al. [45] found that soil fungal diversity was generally greater at low than at high latitude. Furthermore, Tedersoo et al.'s [45] results are consistent with those of Mueller et al. [38] and Rapoport's Rule in showing that geographic ranges of tropical fungi are smaller than those of high latitude fungi. Tedersoo et al.'s [45] study greatly underestimates forest floor fungal diversity in tropical forests, however, as they removed loose leaf litter and duff from the soil before collecting samples of humus and soil. Polishook et al. [46] studied microfungi (not visible without magnification) in decomposing leaves of two tree species that occurred together on the forest floor at El Verde in Puerto Rico and showed they had strong differential abundances. These host 'preferences' were strongest among the dominant microfungi of each leaf species [46]. The segregation of decomposer microfungi among leaf species helps explain the high species richness of microfungi in decomposing litter of wet tropical forests [46]. Further, Polishook et al. [46] showed that while some species are ubiquitous, a large proportion are regionally or locally endemic. Santana et al. [47] later showed that the dominant microfungi on leaves of five tree species decomposed their source leaves faster than dominant microfungi from the other four leaf species. This result indicates that the high species richness of microfungal leaf decomposers is related to rates of decomposition because different leaves have different fungal dominants that are more efficient at decomposing their preferred hosts. The Santana et al. [47] study published in 2005 was among the first to show that diversity of primary decomposers was related to ecosystem function, and that there is more complementarity and less redundancy in below-ground ecosystems than previously thought. A subsequent review by Eisenhauer [48], supports the theoretical basis for diversity effects on ecosystem function through complementarity in below-ground systems, whereas Bardgett and van der Putten [49] have argued that species richness only has effects in very simple systems because there is much redundancy. Most recently, analyses of European grasslands by Soliveres et al. [50] showed that multitrophic diversity strongly predicted ecosystem functions, and that diversity of microbial decomposers has a particularly strong effect.

2.2. Elevation Gradients

Changes in biotic assemblages along elevation gradients can reveal sensitivities in particular groups of organisms to environmental variation that is correlated with elevation [18,19], or dependencies on other organisms that respond strongly to climate [51]. High turnover in microbial species assemblages along an elevation gradient in eastern Puerto Rico [5] resembles patterns found on mountains in Malaysia, Mexico and Peru where high turnover of plant and fungal assemblages along elevation gradients are also found [52–55]. The study of phylogenetic origins of endemic species of plants, animals and fungi on a young mountain in Malaysia by Merckx et al. [52] showed strong niche conservatism that results in high species turnover along the elevation gradient [55,56]. A subsequent analysis by Geml et al. [57] from the same mountain showed that the peak in species richness of ectomycorrhizal fungi at lower-middle elevation was primarily tied to narrow environmental niches and not the result of broad-range species overlapping in the middle of the gradient (known as the mid-domain effect). Sensitivities of tropical montane organisms to changes in environmental factors is important in the context of climate change [18,19,51,58], but not all changes in biota are direct responses to environment [51]. For example, restriction to neotropical cloud forests of certain Ascomycete species in the genus *Xylaria* was related to their specificity to endemic cloud forest plants rather than to the environment per se [51].

2.2.1. Invertebrate Diversity and Abundance along an Elevation Gradient in Puerto Rico

The Luquillo Mountains represent an ideal setting to study dramatic changes in climatic characteristics over a short distance inland (25 km), as an elevation gradient spanning about 1000 m and differences in temperature and precipitation of about 5 °C and 2600 mm respectively, can be found going from the coast to the top of the mountain. Even though the pattern of tree species abundance along this elevation gradient shows species with narrow as well as wide ranges [59], four forest types have been recognized within LEF [60]. In an effort to relate species richness and abundance of litter-based invertebrate communities to forest productivity along elevational/ecological gradients, Richardson et al. [21] controlled for forest types by comparing mixed forest stands with adjacent areas under palm vegetation at different elevations within the Luquillo Mountains. In forest floor litter communities, using palm litter as a control for forest type, they found that although overall net primary productivity (NPP) declined with increasing elevation and rainfall, animal abundance, biomass, and species richness were remarkably similar along the gradient. In non-palm litter, all community parameters declined with increasing elevation, along with NPP and litter nutrient concentrations [61]. Therefore, they found differences observed in animal abundance and species richness, and the uniformity of communities along the increasing elevational gradient were better explained by the contribution of forest composition to the chemical and physical nature of litter and forest heterogeneity, rather than to direct effects of temperature and rainfall differences [19]. Likewise, Willig et al. [62] have shown that abundances of most species of terrestrial gastropod decrease with increasing elevation, as do metrics of taxonomic biodiversity (i.e., species richness, species rarity, species diversity). González et al. [63] found that the number of earthworm species significantly increases as elevation and annual rainfall increase and temperature decreases. The highest numbers of native earthworm species were found in the elfin and palo colorado forests (10 and five species, respectively). Introduced earthworms were also widespread. *Pontoscolex corethrurus* (pantropical introduced worm) was found in all but the dry, *Pterocarpus* and mangrove forests. The exotic *Ocnerodrilus occidentalis* was found in all but the palo colorado, *Pterocarpus* and mangrove forests. The lowland moist forest had the highest presence of exotic worms. Richardson et al. [64] studied the effects of nutrient availability and other elevational changes on bromeliad populations and their invertebrate comminuters. They found that animal abundance in bromeliads peaked at intermediate elevations.

Woody debris is an important component of the carbon pool and a potential carbon sink in terrestrial ecosystems globally [65–67]. Yet, most surveys of amounts and properties of woody debris have been performed within temperate systems as well as the mainland tropics where these collections are often limited to a few forest types encompassing large land areas [68,69]. In Puerto Rico, González and Luce [35] characterized coarse woody debris (CWD) and fine woody debris at 24 sites (encompassing eight distinct forest types) along an elevation gradient in northeastern Puerto Rico. They found that the contribution of different groups of decomposers to the decay of CWD varies among different forest types located along elevation and environmental gradients [35]. For example, they found in the elfin forest (on peaks in the LEF), the decay class of CWD was most strongly correlated with white rot fungi. Termites were most abundant in dry forests at low elevation. Fungal white-rot was positively correlated with mean annual precipitation and was most abundant at high elevation whereas brown-rot was most abundant at middle elevations [35].

2.2.2. Microbial Diversity, Abundance and Turnover along an Elevation Gradient in Puerto Rico

Microbial biomass and diversity are often correlated [70,71]. Zalamea and González [71] found a decline in total microbial biomass with increasing elevation and moisture from dry coastal forest to mid-elevation wet forest in the Luquillo Mountains of Eastern Puerto Rico using a substrate induced respiration method. Similarly, Cantrell et al. [5] found an overall decline in microbial fatty acid diversity from dry coastal forest to montane rain forest along the same gradient using FAME and TRFLP analyses. The peak in soil fungal abundance and fungal to bacteria ratios occurred in mid-elevation wet forest [5]. Correspondingly, abundance of decomposer basidiomycete (macrofungi that cause white rot) mycelia in leaf litter (% forest floor cover) also declined with elevation from wet mid-elevation forest to montane rain forest at the peak of the Luquillo Mountains [5,72]. White-rot in wood, however, increased with elevation and annual precipitation, whereas brown-rot (caused by different species of basidiomycete macrofungi) was most abundant in the middle of the elevation gradient [35]. Diversity and abundance of Mycetozoa ('slime molds') decreased with increasing elevation in the Luquillo Mountains [5,73–76]. In contrast, bacteria abundance, especially among G- bacteria, was greatest at the two ends of the elevation gradient and lowest in mid-elevation wet forest [5]. The highest diversity of sulfidogenic bacteria and Chrenarchaeota was in the frequently waterlogged soils at high elevation where rainfall is highest [5]. Waterlogging reduces soil oxygen and redox potential, which also favors growth and activity of methaogenic bacteria, so methane production was found to increase with elevation in the Luquillo Mountains [7,8].

Similar to patterns found elsewhere in tropical forest elevation gradients, the turnover of fungal assemblages along the elevation gradient in the Luquillo Mountains of Puerto Rico was strong with little overlap between adjacent forest types using TRFLP analyses [5]. Similar patterns have been observed along elevation gradients in Borneo, Mexico and Peru [52,53,70]. The turnover of protists (bacteria and Chrenarecheaota) between adjacent forest types was not as strong as the turnover in soil fungi, but the two highest elevation forests had species unique to this zone [5]. Changing the location of organisms along an elevation gradient is often used as a proxy for detecting responses to climate change. Differences in soil microbiota were thought to have contributed to increased carbon loss when soil cores were translocated from both low to high and high to low elevation relative to cores translocated within the same habitat [77].

2.3. Disturbance Gradients

Tropical forests are exposed to an array of disturbance types that vary in intensity, frequency and duration [78,79]. These include events such as tropical cyclones, landslides and droughts [78–80] and anthropogenic disturbances such as timber and charcoal extraction and conversion of forest to plantations, agricultural crops and pastures. The responses of soil biota to these disturbances depends on characteristics of the phylogenetic group or species, the nature and severity of the disturbance, and interactions with other organisms in the below-ground food web.

2.3.1. Invertebrate Responses to Disturbance Gradients

The responses of invertebrates to disturbance in tropical forests vary among phylogenetic groups from phyla to species. For example, native earthworms were abundant in natural second growth forest in Puerto Rico, but absent from adjacent tree plantations, indicating that they are sensitive to disturbance [81]. Further, earthworm dry weight and abundance were twice as high in native second growth forest than in more disturbed tree plantations [81]. Once an introduced species has been established in a new place, the site and species characteristics seem to be key factors determining its spread [82]. In contrast to introduced species, native earthworms are not as tolerant to a shift to dryer grassland microclimate conditions, and are mostly restricted to natural ecosystems [81,83,84]. For example, the introduced *P. corethrurus* can reach an abundance of 1000 individuals per square meter (25 cm deep) in disturbed agricultural pastures [83]. The rapid population growth of this worm may increase competition pressure on food resources on the local earthworm community [85], further leading to changes in N dynamics at the site. It has been shown that *P. corethrurus* enhance nitrogen availability and mineralization in pasture soils [86]. However, Huang et al. [85] showed soil N mineralization by individual *Estherella* spp. and *O. borincana* (native worms) was reduced in the mixed-species treatments containing *P. corethrurus*. Huang et al. [85] proposed that biotic factors, such as competitive exclusion of native earthworms by introduced earthworms, may have considerable effects on retarding their re-colonization and/or causing the disappearance of native earthworm population in disturbed areas. It has been suggested that habitat disturbance, such as fertilizer amendments or vegetation conversion, increase resource availability to anthropochorous earthworms, thus enhancing their ability to invade disturbed sites [87,88]. However, results from the subtropical wet forest (tabonuco) in Puerto Rico support the contention that worm density and biomass can be decreased by fertilization via changes in soil acidity [89]. Barberena-Arias and Aide [90] and Osorio-Pérez et al. [91] studied litter insect diversity and trophic composition during plant secondary succession in Puerto Rico. They found that arthropod species composition was significantly different between early and intermediate/late forests where early successional habitats had few unique species, and intermediate/late habitats had more species specific to woody habitats—suggesting the recovery of arthropod diversity during plant secondary succession is dependent not only on the increase of wood and concomitant resources but also on the recovery of plant diversity [92].

Canopy opening in a simulated hurricane treatment induced shifts in dominance in the litter from macroarthropods such as isopods and millipedes, which are light-averse, to microarthropods, particularly mites [14,24]. Further, González et al. [15] found a negative correlation between the Margalef index of diversity of the litter arthropods and the percent of mass remaining of mixed species of litter, suggesting functional complexity is an important determinant of decay in the LEF. Snail species responded idiosyncratically to the effects of canopy opening and debris deposition. Abundances of all gastropods (combined) as well as abundances of each of three species responded to canopy opening, while abundances of two other species in the same genera responded to debris deposition but not canopy opening [93,94]. Similarly, Torres and González [95] studied the decomposition of *Cyrilla racemiflora* logs over a 13-year period in tropical dry and wet forests in Puerto Rico and found that termites were more abundant in the logs from the tropical dry forest than from the tropical wet forest. High moisture content and low animal diversity seemed to retard wood decay in wet forest, while high diversity of species and functional groups of wood-inhabiting organisms appeared to increase wood decay rates in tropical dry forest.

2.3.2. Microbial Responses to Disturbances

The responses of microbes to disturbance and environmental stress depends on the sensitivity of the organisms and the nature and severity of the disturbance. According to Stephenson et al., Dictyostelid (cellular), protostelid (amoeboid) and myxomycete (plasmodial) slime molds were most abundant in disturbed habitats in mid-elevation wet forest of Puerto Rico [76]. These disturbed sites correspond to the highest functional diversity of the slime mold's prey (primarily bacteria and yeasts) [96]. In contrast, fungi of wet forest in Puerto Rico were found to be surprisingly sensitive to disturbance, especially those associated with a drier environment. Lodge [97] found that fungal biomass in the litter layer sometimes tripled or decreased by half depending on the number of days in the preceding week in which throughfall reached the forest floor. Similarly, reduced litter moisture in plots where the canopy had been removed relative to untrimmed forest was associated with reductions in basidiomycete fungi. Lodge et al. [98,99] found large reductions in ground cover by basidiomycete leaf decomposer mycelial mats in the drier litter of trimmed plots, and abundances of basidiomycete fungal connections between litter layers were also reduced. Soil fungal biomass also varied with soil moisture [97]. Lodge and Ingham [100] found that soil fungal hyphal diameter distributions had almost no overlap between the wetter and drier seasons, indicating a radical change in fungal community dominance between seasons despite the mid-elevation forest where the samples were taken being classified as 'non-seasonal' in the Holdridge Life Zone system. Li and González [44] found significant decreases in total and active fungal and bacterial biomass in the drier season compared to the wetter season working at the same site as Lodge and Ingham [100]. Soil bacteria as well as fungi were found to be highly sensitive to low moisture using a throughfall exclosure experiment, particularly at a low-moisture threshold [101,102]. While canopy trimming in the hurricane simulation experiment decreased litter moisture, soil moisture increased due to reduction of evapotranspiration [14,17,103]. It is therefore not surprising that Cantrell et al. [6] found no effects of canopy trimming and debris deposition treatments in soil microbial communities using FAME and TRFLP analyses, but did find differences attributable to drought in the control plots between years.

2.3.3. Biotic Changes and Interactions in the Detrital Food Web Affect Nutrient and Carbon Cycling

While invertebrates generally have a dominant role in decomposition of organic matter in tropical forests, they also interact with microbial decomposers. For example, freshwater shrimp consumed leaf litter differentially depending on preconditioning by different types of terrestrial fungi. Using paired presentations of leaf discs cut from different parts of the same leaves that had been decomposed by basidiomycete macrofungi versus microfungi, freshwater shrimp selected tough leaves with basidiomycete fungi over microfungi, but had no preferences between rot types in soft leaves [104]. Biotic interactions between plants and detrital communities have also been seen in cross-site leaf decomposition studies. Home-field advantage has been observed where detrital processing and mass loss was faster in the forest type of origin than when translocated to other forest types in the same region or across latitudinal gradients [25,105,106]. Basidiomycete fungi soften decomposing leaves by degrading lignocellulose, so are more important for preconditioning of tough leaves to make them palatable to invertebrates. In a hurricane simulation experiment, a shift in dominance in fungal decomposers from basidiomycete macrofungi to microfungi was associated with increases in fungivore specialist groups (mites, collembola and psocoptera) [14,17,24]. Furthermore, reductions in macroarthropod comminuters of litter and basidiomycete decomposer fungi that degrade lignin together were likely responsible for reduction in rates of leaf decomposition in plots where the canopy was opened [14,17,24,98]. Reduction of basidiomycete fungi was associated with reduced accumulation of phosphorus via translocation by fungal root-like structures, which could have contributed to slowing of leaf decomposition in plots where the canopy was opened [98]. In undisturbed wet forest under closed canopy, drying cycles that kill basidiomycete fungal hyphae followed by rewetting can lead to a pulse of phosphorus released from the litter to soil [107,108]. Such pulsed releases of phosphorus may favor plant root uptake at times when competing soil microbial biomass has been reduced by lower soil moisture [104,105]. Exclusion of macroarthropods from leaf litter via mesh bags confirmed their strong contribution to decomposition rates [14,15]. González et al. [25] showed that soil fauna activity depressed salicylate oxidizers in litter. Methyl salicylate elicits plant defenses and is part of the defense signaling pathway. In the soil, Huang et al. [85] used ^{13}C-labeled litter and showed the exotic earthworm *P. corethrurus* facilitated soil respiration by stimulating microbial activity; however, this effect was suppressed possibly due to the changes in the microbial activities or community when coexisting with the native worm *O. borincana*. Macroarthropod activity in decomposing wood may have especially strong effects on both wood decomposition [35,95] and nutrient cycling in the soil beneath the logs [99,109–111]. Zimmerman et al. [109] found that soil microbial biomass increased beginning 5–7 months after hurricane Hugo, corresponding to the disappearance of soil nitrate via nutrient immobilization and slowing of canopy closure in plots where woody debris was left on the forest floor, whereas soil microbial biomass was lower, soil nitrate levels were higher, and canopy closure was more rapid in plots from which debris was removed. The timing of soil microbial and plant competition for nitrogen (and possibly other limiting nutrients) coincided with deposition of large scolytine (Curculionidae, Subfamily Scolytinae) bark beetle frass piles beneath fallen logs. Lodge et al. [112] found that roots in the upper 10 cm of soil were more abundant away from trunks felled by hurricane Georges seven months earlier, and that carbon to nitrogen ratios in scolytine bark beetle frass were high enough to stimulate microbial nutrient immobilization. Several studies in wet mid-elevation forests in Puerto Rico have shown that tree roots change their foraging patterns depending on relative availability of resources. Zalamea et al. [111] found that soil under decaying wood had fewer roots and lower nitrate and magnesium concentrations than paired samples collected 50 cm away from the logs. Lodge et al. [112] found that root abundance under versus away from logs changed seasonally, likely due to shifts in relative nutrient availability.

3. Summary of Key Findings

Many of the studies summarized here were among the first to examine latitudinal differences in biotic control of litter decomposition, elevation gradients in litter and soil biota in tropical forests, and effects of disturbance on below-ground organisms and the processes they mediate. Studies elsewhere in the tropics or part of the same studies presented here have confirmed the general patterns reported from Puerto Rico.

Salient results from research on soil biota from Puerto Rico along gradients are:

Latitude

- Soil fauna are more diverse and accelerate leaf decomposition more in wet tropical forests than in cold temperate forests, but not warm temperate forest.
- Reciprocal translocation experiments across latitudinal gradients sometimes show that leaf litter decomposes faster at the home-site owing to biotic influences, overriding climatic effects.
- Soil fauna have stronger effects on leaf decomposition than microbes in wet neotropical forests.
- Macroarthropods indirectly affect tropical leaf decomposition via comminution whereas microarthropods have stronger direct effects in temperate forests via fungivory.
- Microfungal diversity is related to rates of leaf decomposition.
- Tropical soil fungi are more diverse and have smaller geographic ranges than temperate fungi.

Elevation

- In the Luquillo Mountains, the number of native and total earthworm species significantly increased as elevation and annual rainfall increased and air temperature decreased.
- Abundance of litter invertebrates and NPP declined with increasing elevation, but species richness and animal biomass peaked at mid-elevation, as in other tropical elevation studies.
- Termites have stronger effects on wood decomposition in tropical low elevation dry forests than in higher elevation wet forests.
- Fungal assemblages turn over more rapidly than protists along tropical elevation gradients.
- Fungi and Mycetozoa decline with elevation or reach a peak at mid-elevation whereas bacteria and Chrenarchaeota are most abundant and diverse at the extreme ends of the gradient.

Disturbance

- Native earthworms of tropical forest are sensitive to disturbance.
- Bark beetle frass from freshly fallen wood apparently stimulates microbial immobilization of nitrogen in the underlying soil, causing roots to initially proliferate away from logs.
- Root abundance changes spatially between seasons based on ephemeral resource hotspots.
- Litter and soil fungi from wet tropical forests are more sensitive to dry cycles and drought than to other types of disturbance.
- Soil microbial communities of wet tropical forest are highly sensitive to drought.
- Small changes and interactions in the detrital food web affect nutrient and carbon cycling.

4. Conclusions

Invertebrates and microbes in litter and soil of tropical forests in Puerto Rico independently and together influence rates of decomposition and availability of nutrients to tree roots. Research from the

Luquillo Mountains of Puerto Rico was among the first to show that litter and soil microbes of wet tropical forests were especially sensitive to drying. Narrow ecological tolerances are consistent with the high turnover of microbial communities along the elevation gradient in the Luquillo Mountains and elsewhere in the tropics, and also the strong response to reciprocal soil transplants across the elevation gradient. Macroinvertebrates have stronger effects than microbes on decomposition of both litter and wood in neotropical forests. Yet, understanding how environmental variation affects the dynamics of different soil microbial and faunal assemblages, and how variation in the composition of such assemblages controls decomposition processes and nutrient cycling is critical for long-term sustainability and management of ecosystems that are subject to global change. Additional future work in the Luquillo Mountains and other tropical forests might focus on (1) the potential deleterious effects of the abundant surface earthworm casting on soil erosion and aeration, and seed germination in pasture lands; (2) whether multi-trophic richness and abundance support ecosystem functioning (i.e., Soliveres et al. [50]); (3) biotic effects of decomposer microorganisms and detritivores in the uppermost soil horizons, where fine roots are concentrated; (4) whether carbon sequestered at greater soil depths contributes to soil fertility and forest productivity; and (5) the potential effects of introduced predators (such as planaria) on earthworms.

Acknowledgments: Much of the research summarized here was performed under grants DEB-0218039 and 1239764 from the National Science Foundation to the Institute of Tropical Ecosystem Studies, University of Puerto Rico, and the United States Department of Agriculture, Forest Service, International Institute of Tropical Forestry as part of the Long-Term Ecological Research Program in the Luquillo Experimental Forest. D.J. Lodge was supported by the USDA Forest Service Northern Research Station and the Forest Products Laboratory. Additional support for G. González was provided by the Luquillo LCZO grant (EAR-1331841). Forest Service research in Puerto Rico is done in collaboration with the University of Puerto Rico. We thank M.F. Barberena-Arias, F.H. Wadsworth, and A.E. Lugo for helpful comments on an earlier version of the manuscript, and very helpful suggestions from two anonymous reviewers.

Author Contributions: G. González presented this summary at the 75th Anniversary celebration of the International Institute of Tropical Forestry. D.J. Lodge drafted the manuscript based on the presentation. D.J. Lodge wrote sections on fungi, microorganisms and microbial interactions with arthropods, and G. González wrote sections on invertebrates and contributed to the sections on interactions between invertebrates and microorganisms, summary of key findings and conclusions.

Conflicts of Interest: The authors declare no conflict of interest.

References

1. Odum, H.T.; Pigeon, R.F. *A tropical Rain Forest: A Study of Irradiation and Ecology at El Verde, Puerto Rico*; Odum, H.T., Pigeon, R.F., Eds.; Division of Technical Information, United States Atomic Energy Commission: Washington, DC, USA, 1970.
2. Lodge, D.J. Microorganisms. In *The Food Web of a Tropical Forest*; Regan, D.P., Waide, R.B., Eds.; University of Chicago Press: Chicago, IL, USA, 1996; pp. 53–108.
3. Holler, J.R. 1966. Microfungi of Soil, Roots and Litter of a Puerto Rican Lower Montane Rain Forest. Ph.D. Thesis, University of South Carolina, Columbia, SC, USA, 1966.
4. Holler, J.R.; Cowley, G.T. Response of soil, root and litter microfungal populations to radiation. In *A Tropical Rain Forest: A Study of Irradiation and Ecology at El Verde, Puerto Rico*; Odum, H.T., Pigeon, R.F., Eds.; United States Atomic Energy Commission: Washington, DC, USA, 1970; pp. F35–F39.
5. Cantrell, S.A.; Lodge, D.J.; Cruz, C.A.; García, L.M.; Pérez-Jiménez, J.R.; Molina, M. Differential abundance of microbial functional groups along the elevation gradient from the coast to the Luquillo Mountains. *Ecol. Bull.* **2013**, *54*, 87–100.
6. Cantrell, S.A.; Molina, M.; Jean Lodge, D.; Rivera-Figueroa, F.J.; Ortiz-Hernández, M.L.; Marchetti, A.A.; Cyterski, M.J.; Pérez-Jiménez, J.R. Effects of a simulated hurricane disturbance on forest floor microbial communities. *For. Ecol. Manag.* **2014**, *332*, 22–31. [CrossRef]
7. Silver, W.L.; Lugo, A.E.; Keller, M. Soil oxygen availability and biogeochemistry along rainfall and topographic gradients in upland wet tropical forest soils. *Biogeochemistry* **1999**, *44*, 301–328. [CrossRef]
8. Silver, W.L.; Liptzin, D.; Almaraz, M. Soil redox dynamics and biogeochemistry along a tropical elevation gradient. *Ecol. Bull.* **2013**, *54*, 195–209.

9. Wood, T.E.; Detto, M.; Silver, W.L. Sensitivity of Soil Respiration to Variability in Soil Moisture and Temperature in a Humid Tropical Forest. *PLoS ONE* **2013**, *8*, e80965. [CrossRef] [PubMed]

10. Wall, D.H.; González, G.; Simmons, B. Seasonally dry forest soil biodiversity and functioning. In *Seasonally Dry Tropical Forests, Ecology and Conservation*; Dirzo, R., Young, H.S., Mooney, H.A., Ceballos, G., Eds.; Island Press: Washington, DC, USA, 2011; pp. 61–70.

11. González, G.; Barberena-Arias, M.F. Ecology of soil arthropod fauna in tropical forests: A review of studies from Puerto Rico. *J. Ag. UPR.* **2017**, in press.

12. Odum, H.T.; Abbott, W.; Selander, R.K.; Golley, F.B.; Wilson, R.F. Estimates of chlorophyll and biomass of the Tabonuco forest of Puerto Rico. In *A Tropical Rain Forest: A Study of Irradiation and Ecology at El Verde, Puerto Rico*; Odum, H.T., Pigeon, R.F., Eds.; Division of Technical Information, United States Atomic Energy Commission: Washington, DC, USA, 1970; pp. I3–I19.

13. Pfeiffer, W.J. Litter invertebrates. In *The Food Web of a Tropical Forest*; Regan, D.P., Waide, R.B., Eds.; University of Chicago Press: Chicago, IL, USA, 1996; pp. 137–181.

14. Richardson, B.A.; Richardson, M.J.; González, G.; Shiels, A.B.; Srivastava, D.S. A Canopy Trimming Experiment in Puerto Rico: The Response of Litter Invertebrate Communities to Canopy Loss and Debris Deposition in a Tropical Forest Subject to Hurricanes. *Ecosystems* **2010**, *13*, 286–301. [CrossRef]

15. González, G.; Lodge, D.J.; Richardson, B.A.; Richardson, M.J. A canopy trimming experiment in Puerto Rico: The response of litter decomposition and nutrient release to canopy opening and debris deposition in a subtropical wet forest. *For. Ecol. Manag.* **2014**, *332*, 32–46. [CrossRef]

16. Shiels, A.B.; González, G.; Willig, M.R. Responses to canopy loss and debris deposition in a tropical forest ecosystem: Synthesis from an experimental manipulation simulating effects of hurricane disturbance. *For. Ecol. Manag.* **2014**, *332*, 124–133. [CrossRef]

17. Shiels, A.B.; González, G.; Lodge, D.J.; Willig, M.R.; Zimmerman, J.K. Cascading Effects of Canopy Opening and Debris Deposition from a Large-Scale Hurricane Experiment in a Tropical Rain Forest. *Bioscience* **2015**, *65*, 871–881. [CrossRef]

18. González, G.; Willig, M.; Waide, R. Ecological gradient analyses in a tropical landscape: Multiples perspectives and emerging themes. *Ecol. Bull.* **2013**, *54*, 13–20.

19. González, G.; Waide, R.B.; Willig, M.R. Advancements in the understanding of spatiotemporal gradients in tropical landscapes: A Luquillo focus and global perspective. *Ecol. Bull.* **2013**, *54*, 245–250.

20. Fragoso, C.; Lavelle, P. Earthworm communities in tropical rain forests. *Soil Biol. Biochem.* **1992**, *24*, 1397–1408. [CrossRef]

21. Richardson, B.A.; Richardson, M.J.; Soto-Adames, F.N. Separating the effects of forest type and elevation on the diversity of litter invertebrate communities in a humid tropical forest in Puerto Rico. *J. Anim. Ecol.* **2005**, *74*, 926–936. [CrossRef]

22. Hansen, R.A. Red oak litter promotes a microarthropod functional group that accelerates its decomposition. *Plant Soil* **1999**, *209*, 37–45. [CrossRef]

23. Heneghan, L.; Coleman, D.C.; Zou, X.; Crossley, D.A.; Haines, B.L. Soil microarthropod contributions to decomposition dynamics: Tropical-temperate comparisons of a single substrate. *Ecology* **1999**, *80*, 1873–1882.

24. Irmler, U. Changes in the fauna and its contribution to mass loss and N release during leaf litter decomposition in two deciduous forests. *Pedobiologia (Jena)* **2000**, *44*, 105–118. [CrossRef]

25. Gonzalez, G.; Seastedt, T. Soil fauna and plant litter decomposition in tropical and subalpine forests. *Ecology* **2001**, *82*, 955–964. [CrossRef]

26. Gonzalez, G. Soil Organisms and LitterDecomposition. In *Modern Trends in Applied Terrestrial Ecology*; Ambasht, R.S., Ambasht, N.K., Eds.; Kluwer Academic/Plenum Publishers: Berlin, Germany, 2002; pp. 315–329.

27. González, G.; Ley, R.E.; Schmidt, S.K.; Zou, X.; Seastedt, T.R. Soil ecological interactions: Comparisons between tropical and subalpine forests. *Oecologia* **2001**, *128*, 549–556. [CrossRef]

28. Seastedt, T.R. The role of microarthropods in decomposition and mineralization processes. *Annu. Rev. Entomol.* **1984**, *29*, 25–46. [CrossRef]

29. Brown, G.G. How do earthworms affect microfloral and faunal community diversity? *Plant Soil* **1995**, *170*, 209–231. [CrossRef]

30. Lavelle, P.; Bignell, D.; Lepage, M.; Wolters, W.; Roger, P.; Ineson, P.; Heal, O.W.; Dhillion, S. Soil function in a changing world: The role of invertebrate ecosystem engineers. *Eur. J. Soil Biol.* **1997**, *33*, 159–193.

31. Heneghan, L.; Coleman, D.; Zou, X.; Crossley, D.; Haines, B. Soil microarthropod community structure and litter decomposition dynamics: A study of tropical and temperate sites. *Appl. Soil Ecol.* **1998**, *9*, 33–38. [CrossRef]

32. Coleman, D.; Crossley, D.A., Jr.; Hendrix, P.F. *Fundamentals of Soil Ecology*, 2nd ed.; Elsevier Academic Press: New York, NY, USA, 2004.

33. González, G.; Murphy, C.M.; Belén, J. Direct and indirect effects of millipedes on the decay of litter of varying lignin Content. *Trop. For.* **2012**, *2*, 37–50.

34. Ruan, H.; Li, Y.; Zou, X. Soil communities and plant litter decomposition as influenced by forest debris: Variation across tropical riparian and upland sites. *Pedobiologia (Jena)* **2005**, *49*, 529–538. [CrossRef]

35. Zalamea-Bustillo, M. 2005. Soil biota, nutrients, and organic matter dynamics under decomposing wood. Master's Thesis, University of Puerto Rico, Puerto Rico, UT, USA, 2005.

36. González, G.; Gould, W.A.; Hudak, A.T.; Hollingsworth, T.N. Decay of aspen (*Populus tremuloides* Michx.) wood in moist and dry boreal, temperate, and tropical forest fragments. *AMBIO* **2008**, *37*, 588–597. [CrossRef] [PubMed]

37. González, G.; Luce, M.M. Woody debris characterization along an elevation gradient in northeastern Puerto Rico. *Ecol. Bull.* **2013**, *54*, 181–193.

38. Mueller, G.M.; Schmit, J.P.; Leacock, P.R.; Buyck, B.; Cifuentes, J.; Desjardin, D.E.; Halling, R.E.; Hjortstam, K.; Iturriaga, T.; Larsson, K.H.; et al. Global diversity and distribution of macrofungi. *Biodivers. Conserv.* **2007**, *16*, 37–48. [CrossRef]

39. Hawksworth, D.L. The magnitude of fungal diversity: The 1.5 million species estimate revisited. *Mycol. Res.* **2001**, *105*, 1422–1432. [CrossRef]

40. Lodge, D.; Baroni, T.; Cantrell, S. Basidiomycetes of the Greater Antilles Project. *Mycologist* **2002**, *15*, 107–112.

41. Rapoport, E. *Aereogeography: Geographical Strategies of Species*; Pergamon Press: Oxford, UK, 1982.

42. Anderson, J. The organization of soil animal communities. In *Organisms as Components of Ecosystems. Proceedings of the VI International Zoology Colloquim of the International Society of Soil Science (SSSA)*; Lohm, U., Persson, T., Eds.; Swedish Natural Science Research Council: Stockholm, Sweden, 1977; pp. 15–23.

43. Swift, M.J.; Heal, O.W.; Anderson, J.W. *Decomposition in Terrestrial Ecosystems*; Blackwell: Oxford, UK, 1979.

44. Li, Y.; González, G. Soil Fungi and Macrofauna in the Neotropics. In *Post-Agricultural Succession in the Neotropics*; Myster, R.W., Ed.; Springer: Berlin, Germany, 2008; pp. 93–114.

45. Tedersoo, L.; Bahram, M.; Polme, S.; Koljalg, U.; Yorou, N.S.; Wijesundera, R.; Ruiz, L.V.; Vasco-Palacios, A.M.; Thu, P.Q.; Suija, A.; et al. Global diversity and geography of soil fungi. *Science* **2014**, *346*, 1256688. [CrossRef] [PubMed]

46. Polishook, J.D.; Bills, G.F.; Lodge, D.J. Microfungi from decaying leaves of two rain forest trees in Puerto Rico. *J. Ind. Microbiol. Biotechnol.* **1996**, *17*, 284–294. [CrossRef]

47. Santana, M.E.; Lodge, D.J.; Lebow, P. Relationship of host recurrence in fungi to rates of tropical leaf decomposition. *Pedobiologia* **2005**, *49*, 549–564. [CrossRef]

48. Eisenhauer, N. Aboveground—Belowground interactions as a source of complementarity effects in biodiversity experiments. *Plant Soil* **2012**, *351*, 1–22. [CrossRef]

49. Bardgett, R.D.; van der Putten, W.H. Belowground biodiversity and ecosystem functioning. *Nature* **2014**, *515*, 505–511. [CrossRef] [PubMed]

50. Soliveres, S.; van der Plas, F.; Manning, P.; Prati, D.; Gossner, M.M.; Renner, S.C.; Alt, F.; Arndt, H.; Baumgartner, V.; Binkenstein, J.; et al. Biodiversity at multiple levels is needed for ecosystem multifunctionality. *Nature* **2017**, *536*, 456–459. [CrossRef] [PubMed]

51. Lodge, D.; Læssøe, T.; Aime, M.; Henkel, T. Montane and cloud forest specialists among neotropical Xylaria species. *N. Am. Fungi* **2008**, *3*, 193–213. [CrossRef]

52. Merckx, V.S.F.T.; Hendriks, K.P.; Beentjes, K.K.; Mennes, C.B.; Becking, L.E.; Peijnenburg, K.T.C.A.; Afendy, A.; Arumugam, N.; de Boer, H.; Biun, A.; et al. Evolution of endemism on a young tropical mountain. *Nature* **2015**, *524*, 347–350. [CrossRef] [PubMed]

53. Geml, J.; Pastor, N.; Fernandez, L.; Pacheco, S.; Semenova, T.A.; Becerra, A.G.; Wicaksono, C.Y.; Nouhra, E.R. Large-scale fungal diversity assessment in the Andean Yungas forests reveals strong community turnover among forest types along an altitudinal gradient. *Mol. Ecol.* **2014**, *23*, 2452–2472. [CrossRef] [PubMed]

54. Gomez-Hernandez, M.; Williams-Linera, G.; Lodge, D.J. Phylogenetic diversity of macromycetes and woody plants along an elevational gradient in Eastern Mexico. *Biotropica* **2016**, *48*, 577–585. [CrossRef]

55. Wiens, J.J.; Graham, C.H. Niche Conservatism: Integrating Evolution, Ecology, and Conservation Biology. *Annu. Rev. Ecol. Evol. Syst.* **2005**, *36*, 519–539. [CrossRef]

56. Crisp, M.D.; Arroyo, M.T.K.; Cook, L.G.; Gandolfo, M.A.; Jordan, G.J.; McGlone, M.S.; Weston, P.H.; Westoby, M.; Wilf, P.; Linder, H.P. Phylogenetic biome conservatism on a global scale. *Nature* **2009**, *458*, 754–756. [CrossRef] [PubMed]

57. Geml, J.; Morgado, L.N.; Semenova-Nelsen, T.A.; Schilthuizen, M. Changes in richness and community composition of ectomycorrhizal fungi among altitudinal vegetation types on Mount Kinabalu in Borneo. *New Phytol.* **2017**. [CrossRef] [PubMed]

58. Dalling, J.W.; Heineman, K.; González, G.; Ostertag, R. Geographic, environmental and biotic sources of variation in the nutrient relations of tropical montane forests. *J. Trop. Ecol.* **2016**, *32*, 368–383. [CrossRef]

59. Lugo, A.E. Up, down, and across the mountains: A new look at the Luquillo Mountains. *Ecol. Bull.* **2013**, *54*, 9–11.

60. Weaver, P.; Gould, W. Forest vegetation along environmental gradients in northeastern Puerto Rico. *Ecol. Bull.* **2013**, *54*, 43–65.

61. Richardson, B.A.; Richardson, M.J. Litter-based invertebrate communities in forest floor and bromeliad microcosms along an elevational gradient in Puerto Rico. *Ecol. Bull.* **2013**, *54*, 101–115.

62. Willig, M.; Presley, S.; Bloch, C.P.; Alvarez, J. Population, community, and metacommunity dynamics of terrestrial gastropods in the Luquillo Mountains: A gradient perspective. *Ecol. Bull.* **2013**, *54*, 117–140.

63. González, G.; García, E.; Cruz, V.; Borges, S.; Zalamea, M.; Rivera, M.M. Earthworm communities along an elevation gradient in Northeastern Puerto Rico. *Eur. J. Soil Biol.* **2007**, *43*, S24–S32. [CrossRef]

64. Richardson, B.A.; Richardson, M.J.; Scatena, F.N.; Mcdowell, W.H.; Richardson, B.A.; Richardson, M.J.; Scatena, F.N. Effects of nutrient availability and other elevational changes on bromeliad populations and their Effects of nutrient availability and other elevational changes on bromeliad populations and their invertebrate communities in a humid tropical forest in Puer. *J. Trop. Ecol.* **2000**, *16*, 167–188. [CrossRef]

65. Harmon, M.E.; Hua, C. Coarse Woody Debris Dynamics in Two Old-Growth Ecosystems. *Bioscience* **1991**, *41*, 604–610. [CrossRef]

66. Torres, J.A. Wood Decomposition of *Cyrilla racemiflora* in a Tropical Montane Forest. *Biotropica* **1994**, *26*, 124. [CrossRef]

67. Creed, I.F.; Morrison, D.L.; Nicholas, N.S. Is coarse woody debris a net sink or source of nitrogen in the red spruce—Fraser fir forest of the southern Appalachians, U.S.A.? *Can. J. For. Res.* **2004**, *34*, 716–727. [CrossRef]

68. Delaney, M.; Brown, S.; Lugo, A.E.; Torres-Lezama, A.; Quintero, N.B. The Quantity and Turnover of Dead Wood in Permanent Forest Plots in Six Life Zones of Venezuela1. *Biotropica* **1998**, *30*, 2–11. [CrossRef]

69. Nascimento, H.E.M.; Laurance, W.F. Total aboveground biomass in central Amazonian rainforests: A landscape-scale study. *For. Ecol. Manag.* **2002**, *168*, 311–321. [CrossRef]

70. Gómez-Hernández, M.; Williams-Linera, G.; Guevara, R.; Lodge, D.J. Patterns of macromycete community assemblage along an elevation gradient: Options for fungal gradient and metacommunity analyse. *Biodivers. Conserv.* **2012**, *21*, 2247–2268. [CrossRef]

71. Zalamea, M.; González, G. Substrate-Induced Respiration in Puerto Rican Soils: Minimum glucose amendment. *Acta Científica* **2007**, *21*, 11–17.

72. Lodge, D.J.; McDowell, W.H.; Macy, J.; Ward, S.K.; Leisso, R.; Claudio Campos, K.; Kuhnert, K. Distribution and role of mat-forming saprobic basidiomycetes in a tropical forest. In *Ecology of Saprobic Basidiomycetes*; Boddy, L., Frankland, J.C., Eds.; Academic Press, Elsevier LTD: Amsterdam, The Netherland, 2008; pp. 197–209.

73. Moore, D.L.; Spiegel, F.W. Microhabitat Distribution of Protostelids in Tropical Forests of the Caribbean National Forest, Puerto Rico. *Mycologia* **2000**, *92*, 616. [CrossRef]

74. Novozhilov, Y.K.; Schnittler, M.; Rollins, A.W.; Stephenson, S. L. Myxomycetes from different forest types in Puerto Rico. *Mycotaxon* **2001**, *77*, 285–299.

75. Schnittler, M.; Stephenson, S.L. Inflorescences of Neotropical herbs as a newly discovered microhabitat for myxomycetes. *Mycologia* **2002**, *94*, 6–20. [CrossRef] [PubMed]

76. Stephenson, S.L.; Landolt, J.C.; Moore, D.L. Protostelids, dictyostelids, and myxomycetes in the litter microhabitat of the Luquillo Experimental Forest, Puerto Rico. *Mycol. Res.* **1999**, *103*, 209–214. [CrossRef]

77. Chen, D.; Yu, M.; González, G.; Zou, X.; Gao, Q. Climate Impacts on Soil Carbon Processes along an Elevation Gradient in the Tropical Luquillo Experimental Forest. *Forests* **2017**, *8*, 90. [CrossRef]

78. Scatena, F.N.; Blanco, J.F.; Beard, K.H.; Waide, R.B.; Lugo, A.E.; Brokaw, N.V.L.; Silver, W.L.; Haines, B.L.; Zimmerman, J.K. Disturbance regime. In *A Caribbean Forest Tapestry: The Multidimensional Nature of Disturbance and Response*; Brokaw, N.V.L., Crowl, A.T., Lugo, A.E., McDowell, W.H., Waide, R.B., Willig, M., Eds.; Oxford University Press: Oxford, UK, 2012; pp. 42–71.

79. Waide, R.B.; Willig, M.R. Conceptual overview: disturbance, gradients, and response. In *A Caribbean Forest Tapestry: The Multidimensional Nature of Disturbance and Response*; Brokaw, N.V.L., Crowl, A.T., Lugo, A.E., McDowell, W.H., Waide, R.B., Willig, M., Eds.; Oxford University Press: Oxford, UK, 2012; pp. 42–71.

80. Lundquist, J.E.; Camp, A.E.; Tyrrell, M.L.; Seybold, S.J.; Cannon, P.; Lodge, D.J. *Earth, Wind, and Fire: Abiotic Factors and the Impacts of Global Environmental Change on Forest Health*; Cambridge University Press: Cambridge, UK, 2011.

81. González, G.; Zou, X.; Borges, S. Earthworm abundance and species composition in abandoned tropical crop lands comparisons of tree plantations and secondary forests. *Pedobiologia* **1996**, *40*, 385–391.

82. González, G.; Huang, C.Y.; Zou, X.; Rodríguez, C. Earthworm invasions in the tropics. *Biol. Invasions* **2006**, *8*, 1247–1256. [CrossRef]

83. Zou, X.; González, G. Changes in earthworm density and community structure during secondary succession in abandoned tropical pastures. *Soil Biol. Biochem.* **1997**, *29*, 627–629. [CrossRef]

84. Leon, Y.S.-D.; Zou, X.; Borges, S.; Ruan, H. Recovery of Native Earthworms in Abandoned Tropical Pastures. *Conserv. Biol.* **2003**, *17*, 999–1006. [CrossRef]

85. Huang, C.Y.; González, G.; Hendrix, P. Resource Utilization by Native and Invasive Earthworms and Their Effects on Soil Carbon and Nitrogen Dynamics in Puerto Rican Soils. *Forests* **2016**, *7*, 277. [CrossRef]

86. González, G.; Zou, X. Earthworm influence on N availability and the growth of Cecropia schreberiana in tropical pasture and forest soils. *Pedobiologia* **1999**, *43*, 824–829.

87. Fragoso, C.; Lavelle, P.; Blanchart, E.; Senapati, B.K.; Jimenez, J.J.; Angeles Martinez, M.; de los Decaëns, T.; Tondoh, J. Earthworm communities of tropical agroecosystems: Origin, structure and influence of management practices. In *Earthworm Management in Tropical Agroecosystems*; CABI: Wallingford, UK, 1999; pp. 27–55.

88. Winsome, T.; Epstein, L.; Hendrix, P.F.; Horwath, W.R. Habitat quality and interspecific competition between native and exotic earthworm species in a California grassland. *Appl. Soil Ecol.* **2006**, *32*, 38–53. [CrossRef]

89. González, G.; Li, Y.; Zou, X. Effects of post-hurricane fertilization and debris removal on earthworm abundance and biomass in subtropical forests in Puerto Rico. In *Minhocas na America Latina: Biodiversidade e Ecologia*; Brown, G.G., Fragoso, C., Eds.; EMBRAPA: London, UK, 2007; pp. 99–108.

90. Barberena-Arias, M.F.; Aide, T.M. Species diversity and trophic composition of litter insects during plant secondary succession. *Caribb. J. Sci.* **2003**, *39*, 161–169.

91. Osorio-Pérez, K.; Barberena-Arias, M.F.; Aide, T.M. Changes in Ant Species Richness and Composition During Plant Secondary Succession in Puerto Rico. *Caribb. J. Sci.* **2007**, *43*, 244–253. [CrossRef]

92. Barberena-Arias, M.F.; Aide, T.M. Variation in species and trophic composition of insect communities in Puerto Rico. *Biotropica* **2002**, *34*, 357–367. [CrossRef]

93. Secrest, M.F. 1995. The Impacts of Hurricane Hugo on Two Common Tree Snails in the Luquillo Experimental Forest of Puerto Rico. Master's Thesis, Texas Tech University, Lubbock, TX, USA, 1995.

94. Willig, M.R.; Bloch, C.P.; Presley, S.J. Experimental decoupling of canopy opening and debris addition on tropical gastropod populations and communities. *For. Ecol. Manag.* **2014**, *332*, 103–117. [CrossRef]

95. Torres, J.A.; Gonzalez, G. Wood Decomposition of *Cyrilla racemiflora* (Cyrillaceae) in Puerto Rican Dry and Wet Forests: A 13-year Case Study1. *Biotropica* **2005**, *37*, 452–456. [CrossRef]

96. Willig, M.R.; Willig, M.R.; Moorhead, D.L.; Moorhead, D.L.; Cox, S.B.; Cox, S.B.; Zak, J.C. Functional diversity of soil bacterial communities in the tabonuco forest: Interaction of anthropogenic and natural disturbance. *Biotropica* **1996**, *28*, 471–483. [CrossRef]

97. Lodge, D.J. Nutrient cycling by fungi in wet tropical forests. In *Aspects of Tropical Mycology*; Isaac, S., Frankland, J.C., Watling, R., Whalley, A.J.S., Eds.; Cambridge University Press: Cambridge, UK, 1993; pp. 37–57.

98. Lodge, D.J.; Cantrell, S.A.; González, G. Effects of canopy opening and debris deposition on fungal connectivity, phosphorus movement between litter cohorts and mass loss. *For. Ecol. Manag.* **2014**, *332*, 11–21. [CrossRef]

99. Lodge, D.J.; Cantrell, S.A.; González, G.; Stankavich, S.; Shaffer, A.; Stock, M.; Colón Hernández, V.N. Simulated Hurricane Treatment Reduces Basidiomycete Litter Mat Cover in Subtropical Wet Forest. Available online: http://2015.botanyconference.org/engine/search/ind2015 (accessed on 20 October 2015).

100. Lodge, D.J.; Ingham, E.R. A comparison of agar film techniques for estimating fungal biovolumes in litter and soil. *Agric. Ecosyst. Environ.* **1991**, *34*, 131–144. [CrossRef]

101. Bouskill, N.J.; Lim, H.C.; Borglin, S.; Salve, R.; Wood, T.E.; Silver, W.L.; Brodie, E.L. Pre-exposure to drought increases the resistance of tropical forest soil bacterial communities to extended drought. *ISME J.* **2013**, *7*, 384–394. [CrossRef] [PubMed]

102. Bouskill, N.J.; Wood, T.E.; Baran, R.; Ye, Z.; Bowen, B.P.; Lim, H.C.; Zhou, J.; van Nostrand, J.D.; Nico, P.; Northen, T.R.; et al. Belowground response to drought in a tropical forest soil. I. Changes in microbial functional potential and metabolism. *Front. Microbiol.* **2016**, *7*, 1–11. [CrossRef] [PubMed]

103. Shiels, A.B.; González, G. Understanding the key mechanisms of tropical forest responses to canopy loss and biomass deposition from experimental hurricane effects. *For. Ecol. Manag.* **2014**, *332*, 1–10. [CrossRef]

104. De Jesús, M.; Lodge, D.; Crowl, T. Palatability of leaf litter conditioned by white-rot vs. non-white-rot fungi to leaf shredders in a freshwater stream. In Proceedings of the North American Benthalogical Society (NABS) 56th Annual Meeting, Salt Lake City, UT, USA, 25–28 May 2008; p. P3233.

105. González, G.; Seastedt, T.R.; Donato, Z. Earthworms, arthropods and plant litter decomposition in aspen (Populus tremuloides) and lodgepole pine (Pinus contorta) forests in Colorado, USA. *Pedobiologia* **2003**, *47*, 863–869. [CrossRef]

106. Harmon, M.E.; Silver, W.L.; Fasth, B.; Chen, H.; Burke, I.C.; Parton, W.J.; Hart, S.C.; Currie, W.S. LIDET Long-term patterns of mass loss during the decomposition of leaf and fine root litter: An intersite comparison. *Glob. Chang. Biol.* **2009**, *15*, 1320–1338. [CrossRef]

107. Lodge, D.J.; McDowell, W.H.; McSwiney, C.P. The importance of nutrient pulses in tropical forests. *Trends Ecol. Evol.* **1994**, *9*, 384–387. [CrossRef]

108. Miller, R.M.; Lodge, D.J. Fungal responses to disturbance—Agriculture and forestry. In *The Mycota, Second ed., IV, Environmental and Microbial Relationships*; Esser, K., Kubicek, P., Druzhinina, I.S., Eds.; Springer-Verlag: Berlin, Germany, 2007; pp. 44–67.

109. Zimmerman, J.K.; Pulliam, W.M.; Lodge, D.J.; Quiñones-Orfila, V.; Fetcher, N.; Guzmán-Grajales, S.; Parrotta, J.A.; Asbury, C.E.; Walker, L.R.; Waide, R.B. Nitrogen Immobilization by Decomposing Woody Debris and the Recovery of Tropical Wet Forest from Hurricane Damage. *Oikos* **1995**, *72*, 314–322. [CrossRef]

110. Zalamea, M.; González, G.; Ping, C.L.; Michaelson, G. Soil organic matter dynamics under decaying wood in a subtropical wet forest: effect of tree species and decay stage. *Plant Soil* **2007**, *296*, 173–185. [CrossRef]

111. Zalamea, M.; González, G.; Lodge, D. Physical, Chemical, and Biological Properties of Soil under Decaying Wood in a Tropical Wet Forest in Puerto Rico. *Forests* **2016**, *7*, 168. [CrossRef]

112. Lodge, D.; Winter, D.; González, G.; Clum, N. Effects of Hurricane-Felled Tree Trunks on Soil Carbon, Nitrogen, Microbial Biomass, and Root Length in a Wet Tropical Forest. *Forests* **2016**, *7*, 264. [CrossRef]

forests

MDPI

Review

Trailblazing the Carbon Cycle of Tropical Forests from Puerto Rico

Sandra Brown [1],[†] and Ariel E. Lugo [2],*

[1]   Winrock International, Little Rock, AR 72202, USA; dr.sbrown44@btinternet.com
[2]   USDA Forest Service International Institute of Tropical Forestry, Río Piedras, Puerto Rico 00926, USA
*   Correspondence: alugo@fs.fed.us; Tel.: +1-787-764-7743
†   Deceased on 13 February 2017.

Academic Editor: Timothy A. Martin
Received: 14 February 2017; Accepted: 23 March 2017; Published: 29 March 2017

Abstract: We review the literature that led to clarifying the role of tropical forests in the global carbon cycle from a time when they were considered sources of atmospheric carbon to the time when they were found to be atmospheric carbon sinks. This literature originates from work conducted by US Forest Service scientists in Puerto Rico and their collaborators. It involves the classification of forests by life zones, estimation of carbon density by forest type, assessing carbon storage changes with ecological succession and land use/land cover type, describing the details of the carbon cycle of forests at stand and landscape levels, assessing global land cover by forest type and the complexity of land use change in tropical regions, and assessing the ecological fluxes and storages that contribute to net carbon accumulation in tropical forests. We also review recent work that couples field inventory data, remote sensing technology such as LIDAR, and GIS analysis in order to more accurately determine the role of tropical forests in the global carbon cycle and point out new avenues of carbon research that address the responses of tropical forests to environmental change.

Keywords: biomass; allometry; volume expansion factors; soil organic carbon; tropical forest area; forest inventory data; novel forests; tree plantations; secondary forests; mature forests

1. Introduction

When Leslie Holdridge became the first scientist at the Tropical Forest Experiment Station of the USDA Forest Service in Río Piedras, Puerto Rico in 1939, no one knew that the International Institute of Tropical Forestry (as it became known in 1993) would become heavily engaged in helping unravel the role of tropical forests in the carbon cycle of the world. Based on his experience in the Caribbean, Holdridge developed the Life Zone System for identifying vegetation formations from climatic data [1,2]. Decades later, we showed that Holdridge life zones correlated with carbon storages and fluxes of mature tropical forests [3]. As a contribution to the 75th anniversary of the Institute, we review the contributions of its scientists and collaborators to the understanding of the carbon cycle of tropical forests. We also make observations about the evolving role of tropical forests in storing carbon in the context of the Anthropocene Epoch.

2. Foundational Research

Carbon-related research requires considerable background information about tropical forests such as knowledge about dendrology including wood properties and stand volume stocks. This type of information facilitates estimates of stand biomass and conversion of stand structure to biomass. Institute scientists developed some of that information during the 1950s and 1960s [4,5]. This included tree identification [6,7], the wood properties of tree species [8,9], regressions for tree volume and biomass estimation [10–13], and volume tables for timber species [14,15]. A system of long-term observation plots was also established in the early 1940s to assess tree growth and stand volume yields [14,16–18]. Some of these plots in the Luquillo Mountains, the oldest under continuous measurement in the Neotropics, are described in [19] (pp. 91–92). The tree plantation program of the Institute also yielded information about plantation wood yields under different conditions and plantation species [20].

3. Carbon Flux and Storage Studies at the Institute

During the late 1950s and early 1960s, Institute collaborators Howard T. Odum and Frank B. Golley estimated the biomass of a *Rhizophora mangle* L. forest [21] and a tabonuco (*Dacryodes excelsa* Vahl) subtropical wet forest [22–24]. Both studies were pioneering in the field of ecology and both led to ecosystem-level carbon budgets for these two types of tropical forests. The Odum studies at the Luquillo Mountains developed into the Radiation Experiment at El Verde [25], which explored in depth the carbon dynamics of a tropical forest [10]. In Figure 1 we reproduce the resulting carbon budget (expressed in energy units) for a stand of tabonuco forest at El Verde. This analysis stimulated similar work in other tropical forests in Puerto Rico, such as the carbon budgets for a dry forest [26–28] (see Figure 7.8 in [28]), a palm floodplain forest (see Figure 8 in [29]), various tree plantation species [30,31], and for tree plantations and secondary forests of similar age [32]. In Brazil, Institute scientist Michael Keller and collaborators addressed the carbon cycle of Amazonian forests through allometry [33,34], remote sensing and LIDAR [35], and detailed studies of the effects of logging on the carbon cycle and coarse woody debris (necromass) dynamics [36–38]. More recently, Institute scientists addressed the carbon dynamics of novel forests. Novel forests are secondary forests growing on deforested and degraded lands, and are dominated by naturalized tree species [39–44]. These forests behaved as net carbon sinks.

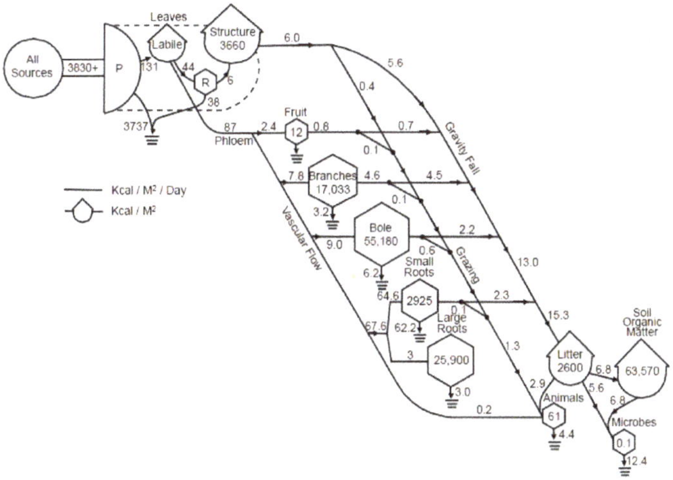

Figure 1. Carbon budget of a subtropical wet forest at El Verde, Puerto Rico [10] expressed in energy units (about 4 kcalories per g of vegetation material, because Odum used caloric equivalents of the various materials). The budget was constructed assuming steady state for the whole system. However, plants do exhibit greater production than respiration. Photosynthesis is P, R is respiration, structure refers to leaf biomass, and the symbols are those of the energy language of Odum [10]: bullet symbol represents photosynthesis, hexagons are consumer compartments, tanks are storages, the circle represents external energy inputs, the heat sinks are energy losses as a result of energy transformations, and lines represent the fluxes.

The above studies resulted in a number of insights about the carbon dynamics of tropical forests, many of which still hold or apply under comparable conditions. For example:

➢ A large fraction of the carbon uptake in mature forests is consumed by stand respiration, with low net yields [10]. This result confirms early research that showed slow tree growth and low volume yields in these forests [4,15].

➢ Large pools of carbon and nutrients in these forests tend to be belowground (mostly in soil), contrary to early beliefs that tropical forests stored most of their nutrients and mass aboveground [45].

➢ Stand biomass of mature forest varies as much as five-fold depending on topographic position on the landscape [11].

➢ Root biomass tends to be higher in mature forests compared to successional ones and higher in native forests compared to timber tree plantations of similar age [32].

➢ Forested watershed export of carbon is proportional to runoff [29,46,47]. The export of organic matter from a floodplain forest at high elevation is high (35 g/m^2·year; [29]) compared to forested tropical watersheds at lower elevations (2.2 to 15.5 g/m^2·year). Forested watersheds in turn exhibit higher organic matter exports than intensively used watersheds (weighted average of 3.7 g/m^2·year (range of 1.5 to 10.5 g/m^2·year) for 14 intensively used watersheds) [47].

➢ Secondary and novel forests accumulate aboveground biomass and nutrients at high rates, and circulate a large fraction of their net primary productivity to the forest floor [41,48,49]. Novel forests have a higher rate of litterfall and nutrient return to the forest floor than native forests in similar climates and soils [39–43].

➢ Wood density increases with age and maturity of forest stands. For example, stand-weighted specific wood density increased by 3.9% (from 304 Kg C/m^3 to 316 Kg C/m^3) among dichotyledon trees in mature *Cyrilla racemiflora* L. forests over a 35-year period in the Luquillo Mountains [50].

➢ Logging can be an atmospheric carbon source and its carbon effects can be mitigated, but not eliminated, through management such as reduced impact logging [36].

4. Estimating the Global Role of Tropical Forests in the Carbon Cycle

The carbon budget of tropical forests gained increased scientific attention when Woodwell et al. asserted that "analysis shows through convergent lines of evidence that the biota is not a sink and may be a source of CO$_2$ as large as or larger than the fossil fuel [51] (p. 141)." They estimated that the biota was a global source of carbon of up to 8 Pg/year. This statement and analysis was controversial because it undermined the prevailing understanding of the global carbon cycle, as scientists could not anticipate how the atmosphere and oceans could absorb such a high level of carbon input [52] (Table 1). Research was needed to improve the understanding of the carbon budgets of tropical forests and decrease the uncertainty of global estimates of carbon fluxes.

The summary of forest biome biomass density of Whitaker and Likens [53] was the state of understanding of biomass distribution around the world, later updated by Ajtay et al. [54]. We argued that the types of tropical forests, and therefore their biomass density, were more diverse than the two entries used by those authors (see Table 1 in [53] and Table 5.2 in [54]). To test our hypothesis, we used a Holdridge life zone approach for assessing the biomass of tropical forests [3]. We did an intensive search of the literature on tropical forests and collected all the biomass data we located and then used the ratio of mean annual temperature to mean annual precipitation (T/P) as a surrogate of the potential evapotranspiration to precipitation ratio in the life zone chart to determine if any patterns existed between these variables. These ratios are indicators of the potential water availability to forests. We found that the T/P ratio correlated with forest and soil carbon storage (Figure 2). Using estimates of global distributions of life zones, we developed a new estimate of 185 Mg C/ha for moist tropical forests. Our estimate was lower than the values used in carbon models based on Whittaker and Likens [53] and Ajtay et al. [54]. A lower forest biomass implied lower carbon release to the atmosphere as a result of deforestation, fire, and decomposition.

Table 1. Consensus on the magnitude of atmospheric global carbon source and sink fluxes (Pg/year) at four historic moments. The perceived role of vegetation as a carbon sink and the level of uncertainty in the budget estimate are highlighted (rows in bold). Values are based on [55–57].

Global Process and Sinks	1980	1980–1989	1990–1999	2000–2007
	Sources			
Fossil fuel burning and cement manufacture	4.5–5.9	5.4 ± 0.5	6.5 ± 0.4	7.6 ± 0.4
Change in land use *	1.8–3.3	1.6 ± 1.0	1.5 ± 0.7	1.1 ± 0.7
Total	6.3–9.2	7.0	8.0 ± 0.8	8.7 ± 0.8
	Sinks			
Atmosphere	2.3–2.7	3.4 ± 0.2	3.2 ± 0.1	4.1 ± 0.1
Oceans	1.5–2.5	2.0 ± 0.8	2.2 ± 0.4	2.3 ± 0.4
Terrestrial vegetation			**2.5 ± 0.4**	**2.3 ± 0.5**
Total	3.8–5.2	5.4	7.9 ± 0.6	8.7 ± 0.7
Residual (uncertainty)	**1.1–6.8**	**1.6 ± 1.4**	**0.1 ± 1.0**	**0.0 ± 1.0**

* Estimated at 8 Pg/year by [51].

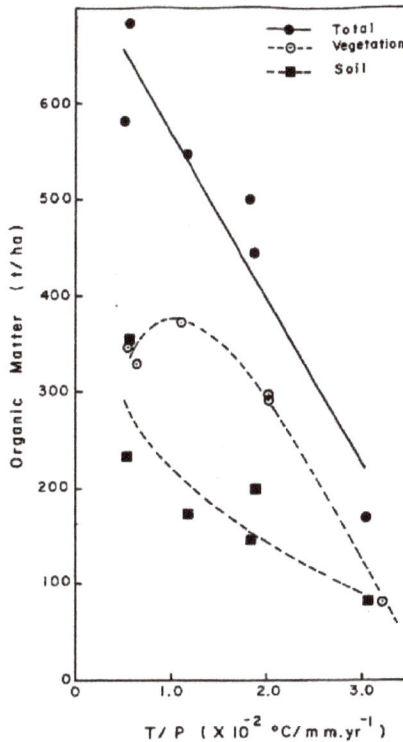

Figure 2. Relationship between aboveground and soil organic matter of mature tropical forests and the ratio of mean annual temperature to mean annual precipitation (T/P) [3].

However, we soon realized that we were synthetizing ecological information that was poorly representative of tropical forests worldwide. The ecological literature usually focused on mature forests on a few sites. For example, the database that we used for that first biomass density estimate was based on field measurements of plots covering less than 30 ha of forest cover, mostly of moist tropical forests. A larger and more representative sample size was required for a more accurate and realistic estimate of the biomass density of tropical forests. We visited the Headquarters of the Food and Agriculture Organization (FAO) in Rome, and found unpublished reports of extensive timber cruises throughout the tropical world, including the Amazon. Using knowledge from volume and biomass studies from Puerto Rico and elsewhere, we converted volume data to biomass [58]. This analysis required developing volume expansion factors for tropical forests that turned out to be different from those used in temperate regions [59,60] (Figure 3). We also developed regression equations and procedures for estimating tropical tree biomass under different life zone conditions and with different starting tree inventory information [60–62], and procedures for converting truncated volume tables to biomass estimates [63]. For root biomass, we conducted a comparative global analysis and found that root biomass density was best correlated with aboveground biomass density regardless of latitudinal location [64].

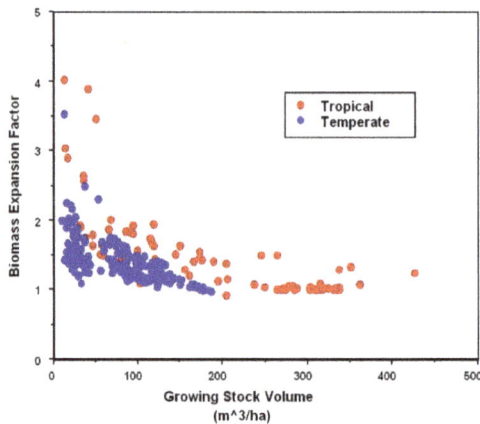

Figure 3. Relationship between the biomass expansion factor and growing stock volume of tropical and temperate forests [59,60].

Our best estimate of carbon density for closed tropical forests around the globe (99 Mg C/ha) was lower than the one based on mature forest data [59]. We deemed these forestry data more indicative of the actual biomass of tropical forests than ecological data because of the larger sample area of the timber cruises and because these volume studies were designed to estimate the actual wood volume on the landscape for economic harvests. Ecological data focused mostly on mature forest stands of limited areal extent. Decades later, when the biomass of island-wide forests in Puerto Rico was estimated from inventory data by Brandeis and Suárez Rozo [13], aboveground biomass was found to be a function of forest structure and age of stands, with the highest values in mature forests, which had the lowest area coverage.

During this stage of our work, we also focused on the carbon storage in tropical timber tree plantations and soils. For plantations, we determined both carbon storage and production rates [65,66]. For soils, our research included soil carbon content of forests and timber tree plantations [67–70], soil carbon storage in agricultural soils [71,72], and soil carbon under a variety of land covers and through succession [73–75].

Some of the insights gained about the carbon dynamics of tropical forests from the studies described in this section are listed below. Subsequent research has expanded the database, the range of values for carbon budget parameters, and addressed the carbon budget in greater detail. However, the general tendencies continue to apply:

➢ In mature tropical forests, carbon storage above and belowground was negatively correlated with the T/P ratio, (i.e., lower at drier locations (Figure 2)). Aboveground carbon storage peaked in moist tropical and subtropical forests and decreased towards the wet, rain, and dry forests, with dry forests exhibiting the lowest carbon density. Soil carbon storage was higher in wet and rain forests and declined towards dry forests.

➢ While deforestation and subsequent agricultural land use caused reductions in soil organic carbon, land abandonment and plant succession accumulated soil organic carbon at rates of 0.3 to 0.5 Mg/ha·year over 40 to 100 years [71,76]. Past land use, life zone, and stage of succession influenced soil organic carbon accumulation with higher values in older and more structurally developed moist secondary forests [67].

➢ Conversion of forests to pasture caused very small or negligible changes in the carbon content of the soil to a meter depth [73–75].

➢ Soil carbon and nutrient retention was resilient following agricultural activity [71,72].

➢ Succession from pasture to forests was associated with reductions in pasture derived soil organic carbon (−0.4 Mg/ha·year) and increases in forest derived soil organic carbon of 0.9 Mg/ha·year with a net increase in soil carbon of 33 Mg/ha over 61 years [70].

➢ Litterfall rate peaked in the moist forests and declined towards the wet, rain, and dry forests [3].

➢ Rate of carbon storage in secondary forests is a function of age, peaking at about 20 years depending on the life zone [74].

➢ Functional attributes related to organic matter production and circulation such as net primary productivity and leaf litterfall were up to ten times faster in secondary forests compared to mature forests [48].

➢ Large trees, defined as those with a diameter at breast height equal or greater than 70 cm, contributed few stems, generally no more than 3% of stand tree density, but can account for more than 40% of the aboveground biomass. Total aboveground biomass of stands generally increased with increasing number of large trees. These results were first reported for the Brazilian Amazon [77] but were also found to be true for Southeast Asia [78], and southeastern United States [79]. Because of their longevity, big trees collectively operate as a large and slow carbon sink in forests.

➢ The removal of large trees by legal or illicit felling lowers the biomass of stands and stimulates the growth of remaining trees for a limited period. This degradation process transforms mature forests into short-term successional forests with changes in rates of carbon sequestration without canopy opening. Any increase in carbon sequestration of residual trees, usually of a limited time period [80], does not make up for the loss of the felling of the large diameter trees. The assumption that closed forests are mature forests in carbon steady state is invalidated when those stands are experiencing net growth as a result of recovery from past disturbances [78].

➢ As stands age and larger trees develop, their rate of carbon accumulation through succession can be larger than when they were younger [70].

➢ Secondary forests accelerate the carbon cycle of tropical forests by turning over as much as 100 Mg/ha of biomass in a decade [48].

➢ Tree plantations also function as strong carbon sinks both above and belowground [81]. Their global role has changed as a result of the dramatic increase in their area since 1980 (from about 10 million hectares to 81.6 million hectares in 2015 [82]).

➢ Disturbances such as hurricanes accelerate the carbon cycle even more in secondary forests [83]. Hurricanes can also generate carbon sinks because initial biomass regeneration after the event can be faster than the decomposition of downed woody debris and burial of organic matter associated with landslides [84].

➤ Although aboveground biomass of forests can quickly recover from mechanical disturbances such as hurricanes and tree harvesting [85,86] recovery after physiological stressors such as ionizing radiation or soil degradation slow down biomass recovery rates [86–88].

➤ Coarse woody debris (necromass) production in Amazon intact and logged forests can account for 14% to 19% of the forests' annual carbon flux. The residence time of this necromass is 4.2 year. However, the amount of necromass in the carbon cycle of a forest cannot be accurately estimated from tree mortality data [38].

5. Tropical Forests: Carbon Sources or Sinks?

As we advanced our understanding of the carbon cycle of tropical forests, we focused attention on the question of whether tropical forests were sources or sinks of atmospheric carbon. We had already found that the biomass density of tropical forests was lower than initially thought and that maturing secondary forests dramatically increased the rate of carbon sequestration. Although with newer studies we are finding that many forests under timber concessions in Indonesia, Central Africa, and Guyana have very high biomass density before logging commences, up to 250 Mg C/ha or more [80]. We had also found that tropical soils could be important carbon sinks, particularly after abandonment of agricultural lands and through pasture or forest succession. We found that even at maturity tropical forests continued to accumulate carbon in the growth of developing large trees, increasing the weighted wood density, accumulating soil carbon and aboveground biomass (including necromass), and exporting organic carbon. The leaching of organic carbon from mature forests is a slow atmospheric carbon sink that is exported downstream from the terrestrial sector of the biosphere. Thus, to address the question of carbon balance, we needed information on land cover and land cover change to expand our findings on carbon dynamics of stands to larger scales.

While browsing in the stacks of the Commerce Department Library in Washington, DC, we ran across the two-volume analysis of the global forest resource by Zon and Sparhawk [89]. This work, commissioned for the first global Forestry Congress in the United States, gave us an insight about the perceived area of tropical forest in the world (Figure 4). It appeared that estimates of the area of tropical forests since the time of Zon and Sparhawk had shown an increase. Moreover, recent estimates are similar to estimates in the 1970s, which considering the inaccuracies of such global estimates, suggested little change in actual forest area. It was possible that global carbon models were using rates of tropical forest deforestation that were much higher than perception from a number of estimates conducted by different organizations. At the time of the Woodwell et al. paper [60], estimates of annual tropical deforestation ranged from 2% to 4%. Higher deforestation rates, multiplied by higher estimates of biomass density of tropical forests would yield higher carbon emissions to the atmosphere than would the use of our lower estimates of biomass density and lower rates of deforestation (1% per year or less).

Moreover, the dynamics of land use in the tropics are more complex than the dynamics used by global carbon models of the time [48]. Global models typically included three states for tropical forests: mature, undergoing deforestation, and recovering from deforestation [55]. For convenience, mature forests were assumed to be in carbon steady state, and thus neutral with respect to their effects on the carbon content of the atmosphere. This fraction of the tropical forest "biome" had the largest land area assigned in the model. Thus, the global role of tropical forests in the carbon cycle was limited to those forestlands in transition since the pre-industrial atmosphere in 1860 (i.e., forestlands were either being deforested or recovering from deforestation since 1860). Models allowed 120 years for forests to reach maturity or carbon steady state with the atmosphere.

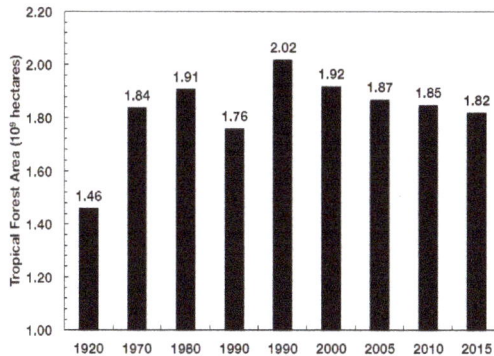

Figure 4. Area of tropical forests between the 1920s [89] and 2015 [3,82,90,91]. The 1990 estimates are both from the Food and Agriculture Organization (FAO), but estimated at different times; the last five estimates are from 2015. We selected these estimates as the most credible for their time because they contained supporting empiric information. Nevertheless, each estimate has unique assumptions and definitions that preclude precise comparisons among them.

Our analysis of the role of tropical forests on the global carbon cycle revealed that the area of forests contributing to net carbon exchange with the atmosphere was much greater than in the carbon models because the role of secondary forests was underrepresented and the carbon budget of presumed mature tropical forests was not necessarily in steady state as assumed. Moreover, we inferred from tropical forest succession literature that 120 years was too short a span of time for achieving carbon steady state. More time is needed to develop a forest structure with large trees and soil carbon at steady state [92]. Our empirical support for our arguments was summarized above and as early as 1980, when we proposed that tropical forests were sinks of atmospheric carbon [93–95]. As more information became available, we updated our case for considering tropical forests as net carbon sinks [92,96–100]. By 1992 and 1993, a shift in scientific consensus was signaled by two international conferences that were dedicated to the identification of carbon sinks in the terrestrial biota [101,102]

Today, there is general agreement that tropical forests are sinks of atmospheric carbon due to secondary forests that are predominant in the tropical landscape [103], as we had suggested many years ago [48]. Moreover, intact old-growth forests have also been shown to be carbon sinks [104], as was demonstrated in Africa [105] and confirmed in the Amazon [106,107]. However, Clark [108], questioned the analysis of plot data, and disputed this conclusion that was based on Phillips et al. [109]. Subsequently, Espírito Santo et al. [106] accounted for the effects of disturbances at various scales, and found that these sources of atmospheric carbon were smaller than the atmospheric sink functions in those mature forests, thus supporting the notion that mature forests in the Amazon were sinks of atmospheric carbon.

Table 1 shows that the 2007 magnitude of carbon sinks in the atmosphere and oceans remain within the range recognized in the 1970s but that the role of vegetation (particularly in the tropics) changed from a source to a sink of carbon, suggesting that the early models underestimated the functioning of tropical forests. The result has been that while more fossil fuel carbon is added to the global carbon cycle, the uncertainty of the global carbon budget has declined, although one can argue about the precision of those estimates. Nevertheless, increased scientific attention to the global carbon cycle has yielded better estimates of carbon fluxes and storages and highlighted additional smaller carbon fluxes that contribute to the source-sink question. An example is carbon burial [110]. According to McLeod et al., carbon burial in tropical forests is 0.04 Mg C·ha^{-1} year^{-1}, that when added to other smaller fluxes such as carbon exports from the terrestrial biota, accumulation of dead woody debris (necromass), increases in wood carbon density, carbon accumulation in large trees, soil carbon sequestration through succession, or carbon in wood products, collectively make a global difference and which we estimated as 3.1 Pg C/year in the 1990s [99].

With the onset of the Anthropocene Epoch, many worry that the sink function of tropical forests might come to an end due to changes in the atmosphere, climate change, and land cover changes [111–118]. Recent studies focus on the potential effects on the carbon balance of land degradation [119], lianas [120] residence time of woody biomass [121], stem mortality [122], selective logging [123] as well as the ecophysiological responses of tropical trees and forests to environmental change [124–129]. Clearly, there are many factors that potentially can change the carbon balance of tropical forests and convert them from sinks to sources of atmospheric carbon. However, just as forests adapted to human activity during the Holocene, they are likely to adapt to Anthropocene conditions through novelty [130]. The emergence of novel forests [131] and Anthromes [132] suggests that a major restructuring of the biota is underway and that these changes are adaptive and unlikely to fundamentally change the functioning of ecosystems. We expect novel ecosystems to adapt to Anthropocene conditions and continue to function as carbon sinks in a new world order where the speed of ecological processes is accelerated.

6. Outlook

Most aspects of our early work have been addressed recently and improved considerably in terms of geographic and ecological coverage and detail of analyses. Some examples include tree allometry [133–136], measurements of net primary productivity [137,138], the relationship between above and belowground biomass allocation [139], and the effects of hyperdominance in carbon cycling [140]. The use of permanent plots to assess carbon dynamics is now common in the literature (e.g., [141–143]). Comprehensive estimates of carbon stocks within countries (e.g., [144–146]) or across various landscape gradients (e.g., [147–150]) are also common in the literature.

New technology allows for a stronger empirical basis for estimating the global role of tropical forests. For example, our life zone approach has been expanded to encompass larger data sets and diverse controls on biomass accumulation other than climatic. These new efforts lead to cross latitudinal comparisons of the carbon cycle of forests [151,152]. We applied remote sensing and GIS technology to detect changes of carbon density of southeast Asian forests even in the absence of changes of forest cover [153–155]. The changes in biomass were due to either maturation of forests (gains of biomass) or degradation of forest stands (loss of biomass). Losses in carbon density were correlated with the perimeter to area (P/A) ratio of forest fragments, suggesting that human access was a causal factor of forest degradation. Fragmented forests (P/A > 0) had net biomass decreases while non-fragmented forests (P/A < 0) had net biomass increases. We also applied remote sensing technologies to estimate forest biomass using regressions of tree canopy area to tree carbon storage obtained from intensive fieldwork [156].

Recent field approaches for assessing carbon stocks of extensive tropical forest landscapes were also developed to assure the estimation of error of field carbon stock determinations (e.g., [157]) and expand the area covered by estimates (e.g., [158–161]). These inventory techniques using thousands of plots, coupled with high-resolution remote sensing images, LIDAR, and GIS analysis were used to conduct carbon accounting procedures for producing maps of carbon stocks and carbon emissions at various geographical scales [162,163]. Producing continental [153,164] and global [165,166] carbon density maps based on site-specific empirical information is now possible (Figure 5) in spite of the challenges involved in the development of these maps [167]. Airborne observatories such as the Carnegie Airborne Observatory [168] are shedding new light on the complexities of tropical forest cover, logging, deforestation, and climate change. These complexities of land use and cover change due to human activities can shift the carbon balance of whole landscapes and lead to new cycles of trailblazing research activity much like what happened between the 1940s and 1990s when we and other Institute scientists and collaborators had the opportunity to address similar questions but starting from a different perspective.

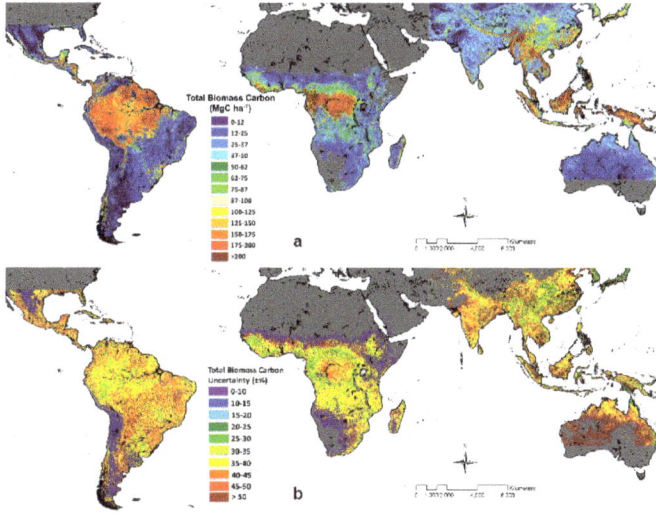

Figure 5. A global map of carbon density of tropical and subtropical forests (**a**) and the geographic distribution of the uncertainty of the estimate (**b**) [165].

Acknowledgments: This study was done in collaboration with the University of Puerto Rico. We acknowledge financial support from the USA Department of Energy Office of Environment, especially through Program Manager Roger Dahlman, and from the USDA Forest Service. Colleagues Charles Hall, H.S. Louise Iverson, Gisel Reyes, Nancy Harris, and the staffs of the International Institute of Tropical Forestry and Winrock International were also instrumental in the success of this research. We also thank Benjamin Branoff and Michael Keller for improving the manuscript.

Author Contributions: The authors contributed equally to this manuscript.

Conflicts of Interest: The authors declare no conflicts of interest.

References

1. Holdridge, L.R. Determination of the world plant formations from simple climatic data. *Science* **1947**, *105*, 367–368. [CrossRef] [PubMed]
2. Holdridge, L.R. *Life Zone Ecology*; Tropical Science Center: San José, Costa Rica, 1967.
3. Brown, S.; Lugo, A.E. The storage and production of organic matter in tropical forests and their role in the global carbon cycle. *Biotropica* **1982**, *14*, 161–187. [CrossRef]
4. Wadsworth, F.H. A review of past research in the Luquillo Mountains. In *A tropical Rain Forest*; Odum, H.T., Pigeon, R.F., Eds.; United States Atomic Energy Commission (AEC), Division of Technical Information: Oak Ridge, TN, USA, 1970; pp. B33–B46.
5. Wadsworth, F.H. A forest research institution in the West Indies: The first 50 years. In *Tropical forests: Management and Ecology*; Lugo, A.E., Lowe, C., Eds.; Springer: New York, NY, USA, 1995; pp. 33–56.
6. Little, E.L.; Wadsworth, F.H. *Common Trees of Puerto Rico and the Virgin Islands*; Agriculture Handbook 249; USDA Forest Service: Washington, DC, USA, 1964.
7. Little, E.L.; Woodbury, R.O.; Wadsworth, F.H. *Trees of Puerto Rico and the Virgin Islands*; USDA Forest Service, Agriculture Handbook 449; United States Department of Agriculture (USDA): Washington, DC, USA, 1974; Volume 2.
8. Longwood, F.R. *Puerto Rican Woods: Their Machining, Seasoning, and Related Characteristics*; United States Department of Agriculture Forest Service, Agricultural Handbook 205; United States Department of Agriculture (USDA): Washington, DC, USA, 1961.
9. Reyes, G.; Brown, S.; Chapman, J.; Lugo, A.E. *Wood Densities of Tropical Tree Species*; USDA Forest Service, General Technical Report SO-88; Southern Forest Experiment Station: New Orleans, LA, USA, 1992.

10. Odum, H.T. Summary: An emerging view of the ecological systems at El Verde. In *A Tropical Rain Forest*; Odum, H.T., Pigeon, R.F., Eds.; National Technical Information Service: Springfield, VA, USA, 1970; pp. I191–I289.

11. Ovington, J.D.; Olson, J.S. Biomass and chemical content of El Verde lower montane rain forest plants. In *A Tropical Rain Forest. A Study of Irradiation and Ecology at El Verde, Puerto Rico*; Odum, H.T., Pigeon, R.F., Eds.; National Technical Information Service: Springfield, VA, USA, 1970; pp. H53–H77.

12. Weaver, P.L.; Gillespie, A.J.R. Tree biomass equations for the forests of the Luquillo Mountains, Puerto Rico. *Common. For. Rev.* **1992**, *71*, 35–39.

13. Brandeis, T.J.; del Rocio, M.R.; Rozo, S. Effects of model choice and forest structure on inventory-based estimations of Puerto Rican forest biomass. *Caribb. J. Sci.* **2005**, *41*, 250–268.

14. Wadsworth, F.H. The development of the forest land resources of the Luquillo Mountains of Puerto Rico. Ph.D. Thesis, University of Michigan, Ann Arbor, MI, USA, 1949.

15. Briscoe, C.B.; Wadsworth, F.H. Stand structure and yield in the tabonuco forest of Puerto Rico. In *A Tropical Rain Forest: A Study of Irradiation and Ecology at El Verde, Puerto Rico*; Odum, H.T., Pigeon, R.F., Eds.; National Technical Information Service: Springfield, VA, USA, 1970.

16. Wadsworth, F.H. Growth in the lower montane rain forest of Puerto Rico. *Caribb. For.* **1947**, *8*, 27–43.

17. Weaver, P.L. *Tree Growth in Several Tropical Forests of Puerto Rico*; USDA Forest Service Research Paper SO-152; Southern Forest Experiment Station: New Orleans, LA, USA, 1979.

18. Wadsworth, F.H.; Parresol, B.R.; Figueroa Colón, J.C. Tree increment indicators in a subtropical wet forest. In *Proceedings of Seminar on Growth and Yield in Tropical Mixed/Moist Forests*; Wan Razali, W.M., Chan, H.T., Appanah, S., Eds.; Forest Research Institute: Kuala Lumpur, Malaysia, 1989; pp. 205–212.

19. Brown, S.; Lugo, A.E.; Silander, S.; Liegel, L. *Research History and Opportunities in the Luquillo Experimental Forest*; USDA Forest Service, Southern Forest Experiment Station, General Technical Report SO-44; United States Department of Agriculture (USDA): New Orleans, LA, USA, 1983.

20. Francis, J.K. Forest plantations in Puerto Rico. In *Tropical Forests: Management and Ecology*; Lugo, A.E., Lowe, C.A., Eds.; Springer: New York, NY, USA, 1995; pp. 210–223.

21. Golley, F.B.; Odum, H.T.; Wilson, R.F. The structure and metabolism of a Puerto Rican red mangrove forest in May. *Ecology* **1962**, *43*, 9–19. [CrossRef]

22. Odum, H.T. Man and the ecosystem. In *Lockwood Conference on the Suburban Forest and Ecology*; The Connecticut Agricultural Experiment Station: New Haven, CN, USA, 1962; pp. 57–75.

23. Odum, H.T.; Copeland, B.J.; Brown, R.Z. Direct and optical assay of leaf mass of the lower montane rain forest of Puerto Rico. *Proc. Natl. Acad. Sci. USA* **1963**, *49*, 429–434. [CrossRef] [PubMed]

24. Odum, H.T.; Abbott, W.; Selander, R.K.; Golley, F.B.; Wilson, R.F. Estimates of chlorophyll and biomass of the tabonuco forest of Puerto Rico. In *A Tropical Rain Forest a Study of Irradiation and Ecology at El Verde, Puerto Rico*; Odum, H.T., Pigeon, R.F., Eds.; National Technical Information Service: Springfield, VA, USA, 1970; pp. I3–I19.

25. Odum, H.T.; Pigeon, R.F. *A Tropical Rain Forest*; National Technical Information Service: Springfield, VA, USA, 1970.

26. Lugo, A.E.; González Liboy, J.A.; Cintrón, B.; Dugger, K. Structure, productivity, and transpiration of a subtropical dry forest in Puerto Rico. *Biotropica* **1978**, *10*, 278–291. [CrossRef]

27. Murphy, P.G.; Lugo, A.E. Structure and biomass of a subtropical dry forest in Puerto Rico. *Biotropica* **1986**, *18*, 89–96. [CrossRef]

28. Murphy, P.G.; Lugo, A.E.; Murphy, A.J.; Nepstad, D.C. The dry forests of Puerto Rico's south coast. In *Tropical Forests: Management and Ecology*; Lugo, A.E., Lowe, C., Eds.; Springer: New York, NY, USA, 1995; pp. 178–209.

29. Frangi, J.L.; Lugo, A.E. Ecosystem dynamics of a subtropical floodplain forest. *Ecol. Monogr.* **1985**, *55*, 351–369. [CrossRef]

30. Wang, D.; Bormann, F.H.; Lugo, A.E.; Bowden, R.D. Comparison of nutrient-use efficiency and biomass production in five tropical tree taxa. *For. Ecol. Manag.* **1991**, *46*, 1–21. [CrossRef]

31. Lugo, A.E.; Wang, D.; Bormann, F.H. A comparative analysis of biomass production in five tropical tree species. *For. Ecol. Manag.* **1990**, *31*, 153–166. [CrossRef]

32. Lugo, A.E. Comparison of tropical tree plantations with secondary forests of similar age. *Ecol. Monogr.* **1992**, *62*, 1–41. [CrossRef]

33. Keller, M.; Palace, M.; Hurtt, G.E. Biomass estimation in the Tapajos National Forest, Brazil: Examination of sampling and allometric uncertainties. *For. Ecol. Manag.* **2001**, *154*, 371–382. [CrossRef]
34. Lefsky, M.A.; Harding, D.J.; Keller, M.; Cohen, W.B.; Carabajal, C.C.; Del Espirito-Santo, F.B.; Hunter, M.O.; de Oliveira, R. Estimates of forest canopy height and aboveground biomass using ICESat. *Geophys. Res. Lett.* **2005**, *33*. [CrossRef]
35. Chen, Q.; Lu, D.; Keller, M.; dos-Santos, M.; Bolfe, E.; Feng, Y.; Wang, C. Modeling and mapping agroforestry aboveground biomass in the Brazilian Amazon using airborne LIDAR data. *Remote Sens.* **2015**, *8*, 21. [CrossRef]
36. Keller, M.; Asner, G.P.; Silva, N.; Palace, M. Sustainability of selective logging of upland forests in the Brazilian Amazon: Carbon budgets and remote sensing as tools for evaluating logging effects. In *Working Forests in the Neotropics: Conservation Through Sustainable Management?* Zarin, D.J., Alavalapati, J.R.R., Putz, F.E., Schmink, M., Eds.; Columbia University Press: New York, NY, USA, 2004; pp. 41–63.
37. Palace, M.; Keller, M.; Asner, G.P.; Silva, J.N.M.; Passos, C. Necromass in undisturbed and logged forests in the Brazilian Amazon. *For. Ecol. Manag.* **2007**, *238*, 309–318. [CrossRef]
38. Palace, M.; Keller, M.; Silva, H. Necromass production: Studies in undisturbed and logged Amazon forests. *Ecol. Appl.* **2008**, *18*, 873–884. [CrossRef] [PubMed]
39. Lugo, A.E.; da-Silva, J.F.; Sáez-Uribe, A. Balance de carbono en un bosque de *Castilla elastica*: Resultados preliminares. *Acta Cient.* **2008**, *22*, 13–28.
40. Lugo, A.E.; Abelleira, O.J.; Collado, A.; Viera, C.A.; Santiago, C.; Vélez, D.O.; Soto, E.; Amaro, G.; Charón, G.; Colón, H.; et al. Allometry, biomass, and chemical content of novel African tulip tree (*Spathodea campanulata*) forests in Puerto Rico. *New For.* **2011**, *42*, 267–283. [CrossRef]
41. Lugo, A.E.; Martínez, O.A.; da Silva, J.F. Aboveground biomass, wood volume, nutrient stocks and leaf litter in novel forests compared to native forests and tree plantations in Puerto Rico. *Bois For. Trop.* **2012**, *314*, 7–16.
42. Abelleira Martínez, O.J. Flooding and profuse flowering result in high litterfall in novel *Spathodea campanulata* forests in northern Puerto Rico. *Ecosphere* **2011**, *2*, 105.
43. Da Silva, J.F. Ecophysiology and Productivity of Castilla Elastica, an Introduced Tropical Tree Species. Master's Thesis, University of Puerto Rico, Rio Piedras, Puerto Rico, 2011.
44. Del Arroyo, G.; Santiago, O. Soil Respiration of a Novel Subtropical Moist Forest: From Diel to Seasonal Patterns. Master's Thesis, University of Puerto Rico, San Juan, Puerto Rico, 2014.
45. Lugo, A.E.; Brown, S. Tropical lands: Popular mis2014 conceptions. *Mazingira* **1981**, *5*, 10–19.
46. Lugo, A.E. Organic carbon export by riverine waters of Spain. In *Transport of Carbon and Minerals in Major World Rivers*; Degens, E.T., Kempe, S., Soliman, H., Eds.; University of Hamburg: Hamburg, Germany, 1983; pp. 267–279.
47. Lugo, A.E.; Quiñones, F. Organic carbon export from intensively used watersheds in Puerto Rico. In *Transport of Carbon and Minerals in Major World Rivers*; Degens, E.T., Kempe, S., Soliman, H., Eds.; University of Hamburg: Hamburg, Germany, 1983; pp. 237–242.
48. Brown, S.; Lugo, A.E. Tropical secondary forests. *J. Trop. Ecol.* **1990**, *6*, 1–32. [CrossRef]
49. Lugo, A.E.; Domínguez Cristóbal, C.; Santos, A.; Torres Morales, E. Nutrient return and accumulation in litter of a secondary forest in the coffee region of Puerto Rico. *Acta Cient.* **1999**, *13*, 43–74.
50. Weaver, P.L. The colorado and dwarf forests of Puerto Rico's Luquillo Mountainsin. In *Tropical Forests: Management and Ecology*; Lugo, A.E., Lowe, C., Eds.; Springer: New York, NY, USA, 1995; pp. 109–141.
51. Woodwell, G.M.; Whittaker, R.H.; Reiners, W.A.; Likens, G.E.; Delwiche, C.S.; Botkin, D.B. The biota and the world carbon budget. *Science* **1978**, *199*, 141–146. [CrossRef] [PubMed]
52. Broecker, W.S.; Takahashi, T.; Simpson, H.J.; Peng, T.P. Fate of fossil fuel carbon dioxide and the global carbon budget. *Science* **1979**, *206*, 409–418. [CrossRef] [PubMed]
53. Whittaker, R.H.; Likens, G.E. Carbon in the biota. In *Carbon and the Biosphere*; Technical Information Center: Springfield, VA, USA, 1973; pp. 281–302.
54. Ajtay, G.L.; Ketner, P.; Duvigneaud, P. Terrestrial primary production and phytomass. In *The Global Carbon Cycle*; Bolin, B., Degens, E.T., Kempe, S., Ketner, P., Eds.; John Wiley & Sons: Chichester, UK, 1979; pp. 129–181.
55. Woodwell, G.M.; Hobbie, J.E.; Houghton, R.A.; Melillo, J.M.; Moore, B.; Peterson, B.J.; Shaver, G.R. Global deforestation: Contribution to atmospheric carbon dioxide. *Science* **1983**, *222*, 1081–1086. [CrossRef] [PubMed]

56. Pan, Y.; Birdsey, R.A.; Fang, J.; Houghton, R.; Kauppi, P.E.; Kurz, W.A.; Phillips, O.L.; Shvidenko, A.; Lewis, S.L.; Canadell, J.G.; et al. A large and persistent carbon sink in the world's forests. *Science* **2011**, *333*, 988–993. [CrossRef] [PubMed]

57. Watson, R.T.; Rodhe, H.; Oeschger, H.; Siegenthaler, U. Greenhouse gases and aerosols. In *Cimate Change. The IPCC Scientific Assessment*; Houghton, J.T., Jenkins, G.J., Ephraums, J.J., Eds.; Cambridge University Press: Cambridge, UK, 1990; pp. 1–40.

58. Brown, S.; Lugo, A.E. Biomass of tropical forests: A new estimate based on forest volumes. *Science* **1984**, *223*, 1290–1293. [CrossRef] [PubMed]

59. Brown, S.; Gillespie, A.J.R.; Lugo, A.E. Biomass estimation methods for tropical forests with applications to forest inventory data. *For. Sci.* **1989**, *35*, 881–902.

60. Brown, S. *Estimating Biomass and Biomass Change of Tropical Forests: A Primer*; FAO Forestry Paper 134; Food and Agriculture Organization of the United Nations: Rome, Italy, 1997.

61. Brown, S.; Gillespie, A.J.R.; Lugo, A.E. Use of forest inventory data for biomass estimation of tropical forests. In *Global Natural Resource Monitoring and Assessments: Preparing for the 21st Century*; Lund, H.G., Preto, G., Eds.; American Society for Photogrammetry and Remote Sensing: Bethesda, MD, USA, 1990; pp. 1046–1055.

62. Gillespie, A.J.R.; Brown, S.; Lugo, A.E. Biomass estimates for tropical forests based on existing inventory data. In *State-of-the-Art Methodology of Forest Inventory: A Symposium Proceedings*; Bau, V.J.L., Cunia, T., Eds.; USDA Forest Service Pacific Northwest Research Station General Technical Report PNW-GTR-263; United States Department of Agriculture (USDA): Portland, OR, USA, 1990; pp. 246–253.

63. Gillespie, A.J.R.; Brown, S.; Lugo, A.E. Tropical forest biomass estimation from truncated stand tables. *For. Ecol. Manag.* **1992**, *48*, 69–87. [CrossRef]

64. Cairns, M.A.; Brown, S.; Helmer, E.H.; Baumgardner, G.A. Root biomass allocation in the world's upland forests. *Oecologia* **1997**, *111*, 1–11. [CrossRef] [PubMed]

65. Lugo, A.E.; Schmidt, R.; Brown, S. Preliminary estimates of storage and production of stemwood and organic matter in tropical tree plantations. In *Wood Production in the Neotropics via Plantations*; Whitmore, J.L., Ed.; IUFRO/MAB/Forest Service Symposium: Washington, DC, USA, 1981; pp. 8–17.

66. Lugo, A.E.; Brown, S.; Chapman, J. An analytical review of production rates and stemwood biomass of tropical forest plantations. *For. Ecol. Manag.* **1988**, *23*, 179–200. [CrossRef]

67. Weaver, P.L.; Birdsey, R.A.; Lugo, A.E. Soil organic matter in secondary forests of Puerto Rico. *Biotropica* **1987**, *19*, 17–23. [CrossRef]

68. Lugo, A.E.; Cuevas, E.; Sanchez, M.J. Nutrients and mass in litter and top soil of ten tropical tree plantations. *Plant Soil* **1990**, *125*, 263–280. [CrossRef]

69. Cuevas, E.; Brown, S.; Lugo, A.E. Above and belowground organic matter storage and production in a tropical pine plantation and a paired broadleaf secondary forest. *Plant Soil* **1991**, *135*, 257–268. [CrossRef]

70. Silver, W.L.; Kueppers, L.M.; Lugo, A.E.; Ostertag, R.; Matzek, V. Carbon sequestration and plant community dynamics following reforestation of tropical pasture. *Ecol. Appl.* **2004**, *14*, 1115–1127. [CrossRef]

71. Lugo, A.E.; Sánchez, M.J.; Brown, S. Land use and organic carbon content of some subtropical soils. *Plant Soil* **1986**, *96*, 185–196. [CrossRef]

72. Beinroth, F.H.; Vázquez, M.A.; Snyder, V.A.; Reich, P.F.; Pérez Alegría, L.R. *Factors Controlling Carbon Sequestration in Tropical Soils: A Case Study of Puerto Rico*; University of Puerto Rico at Mayagüez and USDA Natural Resources Conservation Service: Mayagüez, Puerto Rico, 1996.

73. Brown, S.; Glubczynski, A.; Lugo, A.E. Effects of land use and climate on the organic carbon content of tropical forest soils in Puerto Rico. In *Proceedings of the Convention of the Society of American Foresters*; Society of American Foresters: Washington, DC, USA; Portland, Oregon, OR, USA, 1984; pp. 204–209.

74. Brown, S.; Lugo, A.E. Effects of forest clearing and succession on the carbon and nitrogen content of soils in Puerto Rico and US Virgin Islands. *Plant Soil* **1990**, *124*, 53–64. [CrossRef]

75. Lugo, A.E.; Brown, S. Management of tropical soils as sinks or sources of atmospheric carbon. *Plant Soil* **1993**, *149*, 27–41. [CrossRef]

76. Silver, W.L.; Ostertag, R.; Lugo, A.E. The potential for carbon sequestration through reforestation of abandoned tropical agricultural and pasture lands. *Restor. Ecol.* **2000**, *8*, 394–407. [CrossRef]

77. Brown, S.; Lugo, A.E. Above ground biomass estimates for tropical moist forests of the Brazilian Amazon. *Interciencia* **1992**, *17*, 8–18.

78. Brown, S.; Gillespie, A.J.R.; Lugo, A.E. Biomass of tropical forests of southeast Asia. *Can. J. For. Res.* **1991**, *21*, 111–117. [CrossRef]

79. Brown, S.; Schroeder, P.; Birdsey, R. Aboveground biomass distribution of US Eastern hardwood forests and the use of large trees as an indicator of forest development. *For. Ecol. Manag.* **1997**, *96*, 37–47. [CrossRef]

80. Pearson, T.R.H.; Brown, S.; Casarim, F.M. Carbon emissions from tropical forest degradation caused by logging. *Environ. Res. Lett.* **2014**, *9*, 034017. [CrossRef]

81. Brown, S.; Lugo, A.E.; Chapman, J. Biomass of tropical tree plantations and its implications for the global carbon budget. *Can. J. For. Res.* **1986**, *16*, 390–394. [CrossRef]

82. Keenan, R.J.; Reams, G.A.; Achard, F.; de-Freitas, J.V.; Grainger, A.; Lindquist, E. Dynamics of global forest area: Results from the FAO Global Forest Resources Assessment 2015. *For. Ecol. Manag.* **2015**, *352*, 9–20. [CrossRef]

83. Lugo, A.E.; Domínguez Cristóbal, C.; Méndez, N. Hurricane Georges accelerated litterfall fluxes of a 26-year-old novel secondary forest in Puerto Rico. In *Recent Hurricane Research: Climate, Dynamics, and Societal Impacts*; Lupo, A.R., Ed.; InTech: Rijeka, Croatia, 2011; pp. 535–554.

84. Lugo, A.E. Visible and invisible effects of hurricanes on forest ecosystems: an international review. *Austral Ecol.* **2008**, *33*, 368–398. [CrossRef]

85. Scatena, F.N.; Moya, S.; Estrada, C.; Chinea, J.D. The first five years in the reorganization of aboveground biomass and nutrient use following Hurricane Hugo in the Bisley Experimental Watersheds, Luquillo Experimental Forest, Puerto Rico. *Biotropica* **1996**, *28*, 424–440. [CrossRef]

86. Molina Colón, S.; Lugo, A.E. Recovery of a subtropical dry forest after abandonment of different land uses. *Biotropica* **2006**, *38*, 354–364. [CrossRef]

87. Lugo, A.E.; Heartsill Scalley, T. Research in the Luquillo Experimental Forest has advanced understanding of tropical forests and resolved management issues. In *USDA Forest Service Experimental Forests and Ranges: Research for the Long-Term*; Hayes, D.C., Stout, S.L., Crawford, R.H., Hoover, A.P., Eds.; Springer: New York, NY, USA, 2014; pp. 435–461.

88. Aide, T.M.; Zimmerman, J.K.; Pascarella, J.B.; Rivera, L.; Marcano-Vega, H. Forest regeneration in a chronosequence of tropical abandoned pastures: Implications for restoration. *Restor. Ecol.* **2000**, *8*, 328–338. [CrossRef]

89. Zon, R.; Sparhawk, W.N. *Forest Resources of the World*; McGraw-Hill Book Co.: New York, NY, USA, 1923.

90. Food and Agriculture Organization (FAO). *Forest Resources Assessment 1990: Tropical Countries*; Food and Agriculture Organization Forestry Paper 112; Food and Agriculture Organization (FAO): Rome, Italy, 1993.

91. Food and Agriculture Organization (FAO). *Global Forest Resources Assessment 2015: How Are the World's Forests Changing?* Food and Agriculture Organization: Rome, Italy, 2016.

92. Lugo, A.E.; Brown, S. Steady state ecosystems and the global carbon cycle. *Vegetatio* **1986**, *68*, 83–90.

93. Lugo, A.E. Are tropical forest ecosystems sources or sinks of carbon? In *The Role of Tropical Forests on the World Carbon Cycle*; Brown, S., Lugo, A.E., Liegel, B., Eds.; CONF-800350 UC-11; U.S. Department of Energy, National Technical Information Service: Springfield, VA, USA, 1980; pp. 1–18.

94. Lugo, A.E.; Brown, S. Tropical forest ecosystems: Sources or sinks of atmospheric carbon? *Unasylva* **1980**, *32*, 8–13.

95. Lugo, A.E.; Brown, S. Ecological issues associated with the interpretation of atmospheric CO_2 data. In *The Role of Tropical Forests on the World Carbon Cycle*; Brown, S., Lugo, A.E., Liegel, B., Eds.; CONF-800350 UC-11; U.S. Department of Energy, National Technical Information Service: Springfield, VA, USA, 1980; pp. 30–43.

96. Brown, S.; Lugo, A.E. *The Role of Terrestrial Biota in the Global CO_2 Cycle*; American Chemical Society (ACS) Division of Petroleum Inc. Preprints: San Diego, CA, USA, 1981; Volume 26, pp. 1019–1025.

97. Brown, S.; Gertner, G.Z.; Lugo, A.E.; Novak, J.M. Carbon dioxide dynamics of the biosphere. In *Energy and Ecological Modelling*; Mitsch, W.J., Bosserman, R.W., Klopatek, J.M., Eds.; Elsevier Scientific Publishing Company: Amsterdam, The Netherlands, 1981; pp. 19–28.

98. Lugo, A.E. Influence of green plants on the world carbon budget. In *Alternative Energy Sources V. Part E: Nuclear/Conservation/Environment*; Veziroglu, T.N., Ed.; Elsevier Science Publishers B.V. Hemisphere Publishing Corporation: Amsterdam, The Netherlands, 1983; pp. 391–398.

99. Lugo, A.E.; Brown, S. Tropical forests as sinks of atmospheric carbon. *For. Ecol. Manag.* **1992**, *54*, 239–255. [CrossRef]

100. Brown, S.; Lugo, A.E.; Wisniewski, J. Missing carbon dioxide. *Science* **1992**, *257*, 11.

101. Wisniewski, J.; Lugo, A.E. Natural sinks of CO_2. *Water Air Soil Pollut.* **1992**, *64*, 1–463.

102. Wisniewski, J.; Sampson, R.N. Terrestrial biospheric carbon fluxes: Quantification of sinks and sources of CO_2. *Water Air Soil Pollut.* **1993**, *70*, 1–696.

103. Chazdon, R.L. *Second Growth: The Promise of Tropical Forest Regeneration in an Age of Deforestation*; The University of Chicago Press: Chicago, IL, USA, 2014.

104. Luyssaert, S.; Schulze, E.-D.; Börner, A.; Knohl, A.; Hessesmöler, D.; Law, B.E.; Ciais, P.; Grace, J. Old-growth forests as global carbon sinks. *Nature* **2008**, *455*, 213–215. [CrossRef] [PubMed]

105. Lewis, S.L.; Lopez Gonzalez, G.; Sonké, B.; Affum-Baffoe, K.; Baker, T.R.; Ojo, L.O.; Phillips, O.L.; Reitsma, J.M.; White, L.; Comiskey, J.A.; et al. Increasing carbon storage in intact African tropical forests. *Nat. Lett.* **2009**, *457*, 1003–1006. [CrossRef] [PubMed]

106. Espírito Santo, F.D.B.; Gloor, M.; Keller, M.; Malhi, Y.; Saatchi, S.; Nelson, B.; Junior, R.C.; Pereira, C.; Lloyd, J.; Frolking, S.; Palace, M.; et al. Size and frequency of natural forest disturbances and the Amazon forest carbon balance. *Nat. Commun.* **2014**, *5*, 3434. [PubMed]

107. Phillips, O.L.; Brienen, R.J.W. Carbon uptake by mature Amazon forests has mitigated Amazon nation's carbon emissions. *Carbon Balance Manag.* **2017**, *12*, 1–9. [CrossRef] [PubMed]

108. Clark, D.A. Are tropical forests an important carbon sink? Reanalysis of the long-term plot data. *Ecol. Appl.* **2002**, *12*, 3–7. [CrossRef]

109. Phillips, O.L.; Mahli, Y.; Higuchi, N.; Laurance, W.F.; Nunez, P.V.; Vazquez, R.M.; Laurance, S.G.; Ferreira, L.V.; Stern, M.; Brown, S.; et al. Changes in the carbon balance of tropical forests: Evidence from long-term plots. *Science* **1998**, *282*, 439–442. [CrossRef] [PubMed]

110. Mcleod, E.; Chmura, G.L.; Bouillon, S.; Salm, R.; Björk, M.; Duarte, C.M.; Lovelock, C.E.; Schlesinger, W.H.; Silliman, B.R. A blueprint for blue carbon: Toward an improved understanding of the role of vegetated coastal habitats in sequestering CO_2. *Front. Ecol. Environ.* **2011**, *9*, 552–560. [CrossRef]

111. Brienen, R.J.W.; Phillips, O.L.; Feldpausch, T.R.; Gloor, E.; Baker, T.R.; Lloyd, J.; Lopez-Gonzalez, G.; Monteagudo-Mendoza, A.; Malhi, Y.; Lewis, S.L.; et al. Long-term decline of the Amazon carbon sink. *Nature* **2015**, *519*, 344–348. [CrossRef] [PubMed]

112. Clark, D.A. Sources or sinks? The responses of tropical forests to current and future climate and atmospheric composition. *Philos. Trans. R. Soc.* **2004**, *359*, 477–491. [CrossRef] [PubMed]

113. Clark, D.A. Tropical forests and global warming: Slowing it down or speeding it up? *Front. Ecol. Environ.* **2004**, *2*, 73–80. [CrossRef]

114. Cramer, W.; Bondeau, A.; Schaohoff, S.; Lucht, W.; Smith, B.; Sitch, S. Tropical forests and the global carbon cycle: Impacts of atmospheric carbon dioxide, climate change and rate of deforestation. *Philos. Trans. R. Soc. B* **2004**, *359*, 331–343. [CrossRef] [PubMed]

115. Dutra Aguilar, A.P.; Guimarães Vieira, I.C.; Oliveira Assis, T.; Dalla-Nora, E.L.; Toledo, P.M.; Santos-Junior, R.A.O.; Batistella, M.; Coelho, A.S.; Savaget, E.K.; Nobre, C.A.; et al. Land use change emission scenarios: Anticipating a forest transition process in the Brazilian Amazon. *Glob. Chang. Biol.* **2016**, *22*, 1821–1840.

116. Gloor, M.; Phillips, O.L.; Lloyd, J.J.; Lewis, S.L.; Malhi, Y.; Baker, T.R.; Lopez-Gonzalez, G.; Peacock, J.; Almeida, S.; Alves de Oliveiraet, A.C.; et al. Does the disturbance hypothesis explain the biomass increase in basin-wide Amazon forest plot data? *Glob. Chang. Biol.* **2009**, *15*, 2418–2430. [CrossRef]

117. Phillips, O.L.; Lewis, S.L.; Baker, T.R.; Chao, K.J.; Higuchi, N. The changing Amazon forest. *Philos. Trans. R. Soc. B* **2008**, *363*, 1819–1827. [CrossRef] [PubMed]

118. Willcock, S.; Phillips, O.L.; Platts, P.J.; Swetnam, R.D.; Balmford, A.; Burgess, N.D.; Ahrends, A.; Bayliss, J.; Doggart, N.; Doody, K.; et al. Land cover change and carbon emissions over 100 years in an African biodiversity hotspot. *Glob. Chang. Biol.* **2016**, *22*, 2787–2800. [CrossRef] [PubMed]

119. Pearson, T.R.H.; Brown, S.; Murray, L.; Sidman, G. Greenhouse gas emissions from tropical forest degradation: An underestimated source. *Carbon Balance Manag.* **2017**, *12*, 1–11. [CrossRef] [PubMed]

120. Phillips, O.L.; Martínez, R.V.; Arroyo, L.; Baker, T.R.; Killeen, T.; Lewis, S.L.; Malhi, Y.; Monteagudo Mendoza, A.; Neill, D.; Núñez Vargas, P.; et al. Increasing dominance of large lianas in Amazonian forests. *Nature* **2002**, *418*, 770–773. [CrossRef] [PubMed]

121. Galbraith, D.; Malhi, Y.; Affum-Baffoe, K.; Castanho, A.D.A.; Doughty, C.E.; Fisher, R.A.; Lewis, S.L.; Peh, K.S.-H.; Phillips, O.L.; Quesada, C.A.; et al. Residence times of woody biomass in tropical forests. *Plant Ecol. Divers.* **2013**, *6*, 139–157. [CrossRef]

122. Johnson, M.O.; Galbraith, D.; Gloor, M.; De Deurwaerder, H.; Guimberteau, M.; Rammig, A.; Thonicke, K.; Verbeeck, H.; von Randow, C.; Monteagudo, A.; et al. Variation in stem mortality rates determines patterns of above-ground biomass in Amazonian forests: Implications for dynamic global vegetation models. *Glob. Chang. Biol.* **2016**, *22*, 3996–4013. [CrossRef] [PubMed]

123. Blanc, L.; Echard, M.; Herault, B.; Bonal, D.; Marcon, E.; Chave, J.; Baraloto, C. Dynamics of aboveground carbon stocks in a selectively logged tropical forest. *Ecol. Appl.* **2009**, *19*, 1397–1404. [CrossRef] [PubMed]

124. Clark, D.A. Detecting tropical forests responses to global climatic and atmospheric change: Current challenges and a way forward. *Biotropica* **2007**, *39*, 4–19. [CrossRef]

125. Clark, D.B.; Clark, D.A.; Oberbauer, S.F. Annual wood production in a tropical rain forest in NE Costa Rica linked to climatic variation but not to increasing CO_2. *Glob. Chang Biol.* **2010**, *16*, 747–759. [CrossRef]

126. Clark, D.A.; Clark, D.B.; Oberbauer, S.F. Field-quantified responses of tropical rainforest aboveground productivity to increasing CO_2 and climate stress, 1997–2009. *J. Geophys. Res. Biogeosci.* **2013**, *118*, 783–794. [CrossRef]

127. Clark, D.A.; Piper, S.C.; Keeling, C.D.; Clark, D.B. Tropical rain forest tree growth and atmospheric carbon dynamics linked to interannual temperature variation during 1984–2000. *Proc. Natl. Acad. Sci. USA* **2003**, *100*, 5852–5857. [CrossRef] [PubMed]

128. Malhi, Y.; Doughty, C.E.; Goldsmith, G.R.; Metcalfe, D.B.; Girardin, C.A.; Marthews, T.R.; del Aguila-Pasquel, J.; Aragão, L.E.; Araujo-Murakami, A.; Brando, P.; et al. The linkages between photosynthesis, productivity, growth and biomass in lowland Amazonian forests. *Glob. Chang Biol.* **2015**, *21*, 2283–2295. [CrossRef] [PubMed]

129. Wagner, F.H.; Hérault, B.; Bonal, D.; Stahl, C.; Anderson, L.O.; Baker, T.R.; Becker, G.S.; Beeckman, H.; Boanerges Souza, D.; Botosso, P.C.; et al. Climate seasonality limits leaf carbon assimilation and wood productivity in tropical forests. *Biogeosciences* **2016**, *13*, 2537–2562. [CrossRef]

130. Lugo, A.E. Novel tropical forests: Nature's response to global change. *Trop. Conserv. Sci.* **2013**, *6*, 325–337. [CrossRef]

131. Hobbs, R.J.; Higgs, E.S.; Hall, C.M. *Novel Ecosystems: Intervening in the New Ecological World Order*; Willey-Blackwell: West Sussex, UK, 2013.

132. Ellis, E.C. Ecology in an anthropogenic biosphere. *Ecol. Monogr.* **2015**, *85*, 287–331. [CrossRef]

133. Chave, J.; Andalo, C.; Brown, S.; Cairns, M.A.; Chambers, J.Q.; Eamus, D.; Fölster, H.; Fromard, F.; Higuchi, N.; Kira, T.; et al. Tree allometry and improved estimation of carbon stocks and balance in tropical forests. *Oecologia* **2005**, *145*, 87–99. [CrossRef] [PubMed]

134. Chave, J.; Réjou-Méchain, M.; Búrquez, A.; Chidumayo, E.; Colgan, M.S.; Delitti, W.B.; Duque, A.; Eid, T.; Fearnside, P.M.; Goodman, R.C.; et al. Improved allometric models to estimate the aboveground biomass of tropical trees. *Glob. Change Biol.* **2015**, *20*, 3177–3190. [CrossRef] [PubMed]

135. Feldpausch, T.R.; Banin, L.; Phillips, O.L.; Baker, T.R.; Lewis, S.L.; Quesada, C.A.; Affum-Baffoe, K.; Arets, E.G.M.M.; Berry, N.G.; Bird, M.; et al. Height-diameter allometry of tropical forest trees. *Biogeosciences* **2011**, *8*, 1081–1106. [CrossRef]

136. Goodman, R.C.; Phillips, O.L.; Baker, T.R. The importance of crown dimensions to improve tropical tree biomass estimates. *Ecol. Appl.* **2014**, *24*, 680–698. [CrossRef] [PubMed]

137. Clark, D.A.; Brown, S.; Kicklighter, D.W.; Chambers, J.Q.; Thomlinson, J.R.; Ni, J. Measuring net primary production in forests: Concepts and field methods. *Ecol. Appl.* **2001**, *11*, 356–370. [CrossRef]

138. Clark, D.A.; Brown, S.; Kicklighter, D.W.; Chambers, J.Q.; Thomlinson, J.R.; Ni, J.; Holland, E.A. Net primary production in tropical forests: An evaluation and synthesis of existing field data. *Ecol. Appl.* **2001**, *11*, 371–384. [CrossRef]

139. Raich, J.W.; Clark, D.A.; Schwendenmann, L.; Wood, T.E. Aboveground tree growth varies with belowground carbon allocation in a tropical rainforest environment. *PLoS ONE* **2014**, *9*, e100275. [CrossRef] [PubMed]

140. Fauset, S.; Johnson, M.O.; Gloor, M.; Baker, T.R.; Monteagudo, M.A.; Brienen, R.J.W.; Feldpausch, T.R.; Lopez-Gonzalez, G.; Malhi, Y.; ter Steege, H.; et al. Hyperdominance in Amazonian forest carbon cycling. *Nat. Commun.* **2015**, *6*, 6857. [CrossRef] [PubMed]

141. Báez, S.; Malizia, A.; Carilla, J.; Blundo, C.; Aguilar, M.; Aguirre, N.; Aquirre, Z.; Álvarez, E.; Cuesta, F.; Duque, Á.; et al. Large-scale patterns of turnover and basal area change in Andean forests. *PLoS ONE* **2015**, *10*, e0126594. [CrossRef] [PubMed]

142. Chave, J.; Condit, R.; Lao, Z.; Caspersen, J.P.; Foster, R.B.; Hubbell, S.P. Spatial and temporal variation of biomass in a tropical forest: Results from a large census plot in Panama. *J. Ecol.* **2003**, *91*, 240–252. [CrossRef]

143. Lewis, S.L.; Phillips, O.L.; Baker, T.R.; Lloyd, J.; Malhi, Y.; Almeida, S.; Higuchi, N.; Laurance, W.F.; Neill, D.A.; Silva, J.N.; et al. Concerted changes in tropical forest structure and dynamics: Evidence from 50 South American long-term plots. *Philos. Trans. R. Soc.* **2004**, *359*, 421–436. [CrossRef] [PubMed]

144. Phillips, J.; Duque, A.; Scott, C.; Wayson, C.; Galindo, G.; Cabrera, E.; Chave, J.; Peña, M.; Álvarez, E.; Cárdenas, D.; et al. Live aboveground carbon stocks in natural forests of Colombia. *For. Ecol. Manag.* **2016**, *374*, 119–128. [CrossRef]

145. Vieira, S.A.; Alves, L.F.; Aidar, M.; Araújo, L.S.; Baker, T.; Batista, J.L.F.; Campos, M.C.; Camargo, P.B.; Chave, J.; Delitti, W.B.C.; et al. Estimation of biomass and carbon stocks: The case of the Atlantic Forest. *Biota Neotrop* **2008**, *8*, 21–29. [CrossRef]

146. Yepes, A.; Herrera, J.; Phillips, J.; Cabrera, E.; Galindo, G.; Granados, E.; Duque, A.; Barbosa, A.; Olarte, C.; Cardona, M. Contribución de los bosques tropicales de montaña en el almacenamiento de carbono en Colombia. *Rev. Biol. Trop.* **2015**, *63*, 69–82. [CrossRef] [PubMed]

147. Clark, D.B.; Clark, D.A. Landscape-scale variation in forest structure and biomass in a tropical rain forest. *For. Ecol. Manag.* **2000**, *137*, 185–198. [CrossRef]

148. Laurance, W.F.; Fearnside, P.M.; Laurance, S.G.; Delamonica, P.; Lovejoy, T.E.; Rankin-de-Merona, J.M.; Chambers, J.Q.; Gascon, C. Relationship between soils and Amazon forest biomass: A landscape-scale study. *For. Ecol. Manag.* **1999**, *118*, 127–138. [CrossRef]

149. Longo, M.; Keller, M.; Dos-Santos, M.N.; Leitold, V.; Pinagé, E.R.; Baccini, A.; Saatchi, S.; Nogueira, E.M.; Batistella, M.; Morton, D.C. Aboveground biomass variability across intact and degraded forests in the Brazilian Amazon. *Glob. Biogeochem. Cycles* **2016**, *30*, 1639–1660. [CrossRef]

150. Sullivan, M.J.P.; Talbot, J.; Lewis, S.L.; Phillips, O.L.; Qie, L.; Begne, S.K.; Chave, J.; Cuni-Sanchez, A.; Hubau, W.; Lopez-Gonzalez, G.; et al. Diversity and carbon storage across the tropical forest biome. *Sci. Rep.* **2017**, *7*, 39102. [CrossRef] [PubMed]

151. Fernández Martínez, M.; Vicca, S.; Janssens, I.A.; Luyssaert, S.; Campioli, M.; Sardans, J.; Estiarte, M.; Peñuelas, J. Spatial variability and controls over biomass stocks, carbon fluxes, and resourse-use efficiencies across forest ecosystems. *Trees* **2014**, *28*, 597–611. [CrossRef]

152. Luyssaert, S.; Inglima, I.; Jung, M.; Richardson, A.D.; Reichstein, M.; Papale, D.; Piao, S.L.; Schulze, E.D.; Wingate, L.; Matteucci, G.; et al. CO_2 balance of boreal, temperate, and tropical forests derived from a global database. *Glob. Change Biol.* **2007**, *13*, 2509–2537. [CrossRef]

153. Brown, S.; Iverson, L.R.; Prasad, A.; Liu, D. Geographic distribution of carbon in biomass and soils of tropical Asian forests. *Geocarto Int.* **1993**, *8*, 45–59. [CrossRef]

154. Brown, S.; Iverson, L.R.; Lugo, A.E. Land-use and biomass changes of forests in Peninsular Malaysia from 1972 to 1982: A GIS approach. In *Effects of Land Use Change on Atmospheric CO_2 Concentrations. Southeast Asia as a Case Study*; Dale, V.H., Ed.; Springer: New York, NY, USA, 1994; pp. 117–143.

155. Iverson, L.R.; Brown, S.; Prasad, A.; Mitasova, H.; Gillespie, A.J.R.; Lugo, A.E. Use of GIS for estimating potential and actual forest biomass for continental south and southeast Asia. In *Effects of Land Use Change on Atmospheric CO_2 Concentrations. Southeast Asia as a Case Study*; Dale, V.H., Ed.; Springer: New York, NY, USA, 1994; pp. 67–116.

156. Brown, S.; Pearson, T.; Slaymaker, D.; Ambagis, S.; Moore, N.; Novelo, D.; Sabido, W. Creating a virtual tropical forest from three-dimensional aerial imagery: Application for estimating carbon stocks. *Ecol. Appl.* **2005**, *15*, 1083–1095. [CrossRef]

157. Chave, J.; Condit, R.; Aguilar, S.; Hernandez, A.; Lao, S.; Perez, R. Error propagation and scaling for tropical forest biomass estimates. *Philos. Trans. R. Soc.* **2004**, *359*, 409–420. [CrossRef] [PubMed]

158. Campioli, M.; Malhi, Y.; Vicca, S.; Luyssaert, S.; Papale, D.; Peñuelas, J.; Reichstein, M.; Migliavacca, M.; Arain, M.A.; Janssens, I.A. Evaluating the convergence between eddy-covariance and biometric methods for assessing carbon budgets of forests. *Nat. Commun.* **2016**, *7*, 13717. [CrossRef] [PubMed]

159. Morel, A.C.; Fisher, J.B.; Malhi, Y. Evaluating the potential to monitor aboveground biomass in forest and oil palm in Sabah, Malaysia, for 2000–2008 with Landsat ETM+ and ALOS-PALSAR. *Int. J. Remote Sens.* **2012**, *33*, 3614–3639. [CrossRef]

160. Réjou-Méchain, M.; Tymen, B.; Blanc, L.; Fauset, S.; Feldpausch, T.R.; Monteagudo, A.; Phillips, O.L.; Richard, H.; Chave, J. Using repeated small-footprint LIDAR acquisitions to infer spatial and temporal variations of a high-biomass Neotropical forest. *Remote Sens. Environ.* **2015**, *169*, 93–101. [CrossRef]

161. Tong Minh, D.H.; Toan, T.L.; Rocca, F.; Tebaldini, S.; Villard, L.; Réjou-Méchain, M.; Phillips, O.L.; Feldpausch, T.R.; Dubois-Fernandez, P.; Scipal, K.; et al. SAR tomography for the retrieval of forest biomass and height: Cross-validation at two tropical forest sites in French Guiana. *Remote Sens. Environ.* **2016**, *175*, 138–147. [CrossRef]

162. Espírito Santo, F.D.B.; Keller, M.M.; Linder, E.; Junior, R.C.O.; Pereira, C.; Oliveira, C.G. Gap formation and carbon cycling in the Brazilian Amazon: Measurement using high-resolution optical remote sensing and studies of large forest plots. *Plant Ecol. Divers.* **2013**, *7*, 305–318. [CrossRef]

163. Harris, N.L.; Brown, S.; Hagen, S.C.; Saatchi, S.S.; Petrova, S.; Salas, W.; Hansen, M.C.; Potapov, P.V.; Lotsch, A. Baseline map of carbon emissions from deforestation in tropical regions. *Science* **2012**, *336*, 1573–1576. [CrossRef] [PubMed]

164. Gaston, G.; Brown, S.; Lorenzini, M.; Singh, K.D. State and change in carbon pools in the forests of tropical Africa. *Glob. Change Biol.* **1998**, *4*, 97–114. [CrossRef]

165. Saatchi, S.S.; Harris, N.L.; Brown, S.; Lefsky, M.; Mitchard, E.T.A.; Salas, W.; Zutta, B.R.; Buermann, W.; Lewis, S.L.; Hagen, S.; et al. Benchmark map of forest carbon stocks in tropical regions across three continents. *Proc. Natl. Acad. Sci. USA* **2011**, *108*, 9899–9904. [CrossRef] [PubMed]

166. Avitabile, V.; Herold, M.; Heuvelink, G.B.M.; Lewis, S.L.; Phillips, O.L.; Asner, G.P.; Armston, J.; Ashton, P.S.; Banin, L.; Bayol, N.; et al. An integrated pan-tropical biomass map using multiple reference datasets. *Glob. Chang. Biol.* **2016**, *22*, 1406–1420. [CrossRef] [PubMed]

167. Mitchard, E.T.A.; Feldpausch, T.R.; Brienen, R.J.; Lopez Gonzalez, G.; Monteagudo, A.; Baker, T.R.; Lewis, S.L.; Lloyd, J.; Quesada, C.A.; Gloor, M.; et al. Markedly divergent estimates of Amazon forest carbon density from ground plots and satellites. *Glob. Ecol. Biogeogr.* **2014**, *23*, 935–946. [CrossRef] [PubMed]

168. Asner, G. Carnegie Airborne Observatory. Available online: https://cao.carnegiescience.edu (accessed on 27 March 2017).

Communication

Insights on Forest Structure and Composition from Long-Term Research in the Luquillo Mountains

Tamara Heartsill Scalley

USDA Forest Service, International Institute of Tropical Forestry, Jardín Botánico Sur, 1201 Calle Ceiba, Río Piedras, PR 00926, USA; theartsill@fs.fed.us

Academic Editors: Grizelle González and Ariel E. Lugo
Received: 3 April 2017; Accepted: 5 June 2017; Published: 10 June 2017

Abstract: The science of ecology fundamentally aims to understand species and their relation to the environment. At sites where hurricane disturbance is part of the environmental context, permanent forest plots are critical to understand ecological vegetation dynamics through time. An overview of forest structure and species composition from two of the longest continuously measured tropical forest plots is presented. Long-term measurements, 72 years at the leeward site, and 25 years at windward site, of stem density are similar to initial and pre-hurricane values at both sites. For 10 years post-hurricane Hugo (1989), stem density increased at both sites. Following that increase period, stem density has remained at 1400 to 1600 stems/ha in the leeward site, and at 1200 stems/ha in the windward site. The forests had similar basal area values before hurricane Hugo in 1989, but these sites are following different patterns of basal area accumulation. The leeward forest site continues to accumulate and increase basal area with each successive measurement, currently above 50 m^2/ha. The windward forest site maintains its basal area values close to an asymptote of 35 m^2/ha. Currently, the most abundant species at both sites is the sierra palm. Ordinations to explore variation in tree species composition through time present the leeward site with a trajectory of directional change, while at the windward site, the composition of species seems to be converging to pre-hurricane conditions. The observed differences in forest structure and composition from sites differently affected by hurricane disturbance provide insight into how particular forest characteristics respond at shorter or longer time scales in relation to previous site conditions and intensity of disturbance effects.

Keywords: disturbance; hurricane; succession; long-term; basal area; species composition; trees; tropical; Luquillo Experimental Forest

1. Introduction

Long-term forest monitoring provides vegetation succession measurements needed to understand resiliency and recovery of forest systems from disturbance events. Two frequently evaluated components of forest succession in response to disturbance are stand level structural properties and the dynamics of species composition through time [1]. Understanding forest structure and composition during vegetation succession in tropical forests may mean following a high number of species, each with their own particular life-history traits [2,3]. The environment of the Caribbean, with its natural disturbance regime and predicted climate change scenarios serves as a template to understand forest dynamics responses to disturbance events such as hurricanes [4–9].

Our understanding of forest dynamics has benefited immensely from the ability to observe changes through time, thus chronicling disturbance events and vegetation responses in an adequate and increasing time span [10–12]. This short communication aims to present an overview of succession dynamics in two continuously measured sites in the same forest type [12,13], located in the Luquillo Experimental Forest (LEF), Puerto Rico. Because of differences in location relative to prevailing winds, hurricane disturbance effects and the measured responses at these two sites were different [14], but these observations have not been analyzed together to understand overall response patterns at the level of the forest type. A comparison of responses in forest composition and structure during hurricane-induced succession in mature secondary forests is presented. This includes contrasting patterns of stem density, stem structure, and basal area responses at each site, and changes in abundance of dominant species and those with contrasting life histories through time. Brief notes on some of the species that have been lost and gained to the plots throughout the measurement period are also presented. An exploratory ordination analysis is used to highlight the differences in trajectories of species composition dynamics through time and in response to hurricane disturbance at each forest site. The role of hurricane disturbance at these sites is discussed in relation to observed and predicted forest structure.

2. Methods

2.1. Study Site

The LEF is located in northeastern Puerto Rico, with the El Verde Research Area (El Verde) on the west and the Bisley Experimental Watersheds (Bisley) on the east (Figure 1). The forests of the LEF grow in the context of environmental events such as hurricanes, volcanoes, Saharan-dust, and earthquakes [15–17]. The forest vegetation communities in the LEF are associated with different intensities and scales of disturbances such as treefalls, landslides, and hurricanes [18–21]. The lower montane forest contains the *Dacryodes-Sloanea* forest association, commonly known as tabonuco forest. This forest type occurs in the subtropical wet forest life zone, *sensu* Holdridge, at 200 to 600 m elevation with an average 3482 mm/yr of rainfall [17,22,23]. This forest type occupies the greatest extent of area in the LEF and is also found throughout Caribbean islands [13,14,24]. This forest type has trees with canopies that range from 25 to 30 m in height and lianas (i.e., woody vines) that form a common structural element in these forest communities (Figure 2a).

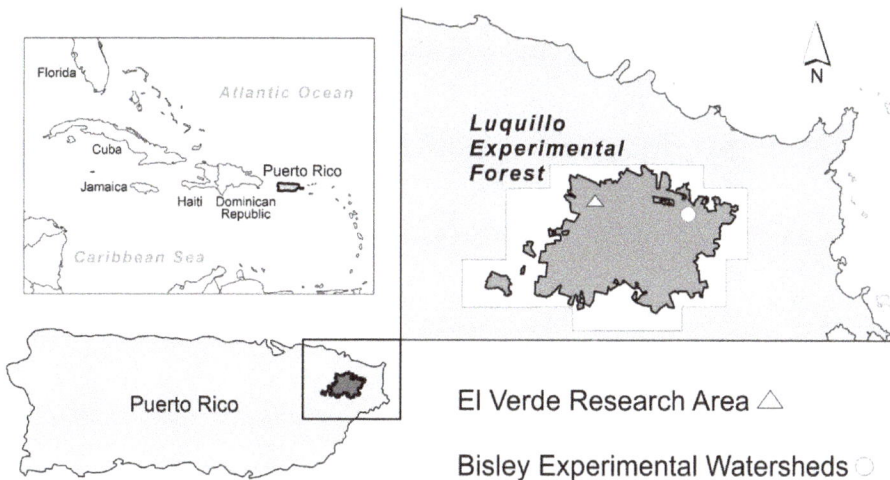

Figure 1. Location of Puerto Rico in the context of the Caribbean, and inset of the El Verde Research Area and Bisley Experimental Watersheds, both long-term sites in the Luquillo Experimental Forest.

(a)

(b)

Figure 2. Views of vegetation from the Luquillo Experimental Forest, Puerto Rico. Top panel (**a**) Tabonuco, *Dacryodes excelsa*, the dominant tree species in forests at lower montane elevations; Bottom panel (**b**) Sierra palm, *Prestoea montana*, with stems found at all elevations and able to form mono-specific forest stands. Photos by Jerry Bauer.

Trees in this forest type are distributed in relation to terrain geomorphology, with the dominant tree species *Dacryodes excelsa* Vahl (tabonuco) occupying ridges and ridge tops (Figure 2a) [25,26]. The sierra palm, *Prestoea montana* (synonymous with *P. acuminata*, var. *montana*) (R. Graham) Nichols, tends to occupy riparian valleys, concave areas and slopes, and can form mono-specific forest stands (Figure 2b) [27]. Other canopy level trees include *Sloanea berteriana* Choisy (motillo), *Manilkara bidentata* (A. DC.) Chev. (ausubo), *Guarea guidonia* L. Sleumer (guaraguao), *Buchenavia capitata* (Aubl.) Howard (granadillo), and *Ocotea leucoxylon* (Sw.) De Laness. (laurel geo) [23]. Younger tabonuco forest stands often include the tree species *Tabebuia heterophylla* (DC.) Britton, (roble blanco) *Cecropia schreberiana* Miq. (yagrumo), and *Schefflera morototoni* (Aubl.) Maguire, Steyerm. & Frodin (yagrumo hembra). In the past, some lower elevation tabonuco forests were selectively harvested for trees with commercially valuable wood from the species that are characteristically dominant in this forest type such as *D. excelsa*, *M. bidentata*, *S. berteriana*, and *Magnolia splendens* Urban (laurel sabino) [10,28–30]. The LEF has had a suite of climatic disturbance events that have been described, including hurricanes San Nicolás in 1931, San Ciprián in 1932, Santa Clara (Betsy) in 1956, Hugo in 1989, and Georges in 1998 [10,12,13]. Hurricane Hugo, the largest storm to affect the area since 1932, passed over the LEF on 18 September 1989. The hurricane defoliated entire areas, and although it reduced the aboveground biomass by 50 percent in windward Bisley [31], there was minimal biomass reduction in leeward El Verde [12]. After hurricane Hugo, eight other storms passed near the LEF [18,21]. Of those, hurricane Georges in 1998 was the largest and resulted in localized defoliation and uprooting [32]. At least six meteorological droughts of varying intensity have also been recorded during a recent 25-year study period [17,33,34].

2.1.1. Permanent Plots

The El Verde 3 plot (El Verde, 18°19′ N, 65°49′ W) is a 0.72 hectare plot that was established in 1943 by Frank Wadsworth to assess stem growth of trees of commercially valuable species [10,12]. In 1937 and in 1958, a timber-stand treatment was conducted, cutting and removing stems. The plot site has a leeward, northwest aspect and is situated on a ridge on the south side of the Quebrada Sonadora. The El Verde plot is ~0.5 km from the 16 ha Center for Tropical Forest Science (CTFS) Luquillo Forest Dynamics Plot, in what is now the El Verde Research Management Area. The El Verde Plot was measured upon establishment and then at intervals between 3 to 12 years, with recent measurement intervals set at 5 years (measurements in 1943, 1946, 1951, 1976, 1988, 1993, 1998, 2005, 2010, and 2015). In censuses of El Verde, all stems of all tree species ≥4.0 cm diameter at 1.3 m from the ground (dbh) were measured. Palms were measured for dbh and were included in the census when their internode reached 1.37 m. As new trees and palms reached the minimum diameter class, they were numbered and marked with aluminum tags. A detailed narrative of each El Verde plot assessment can be found in [12]. The Bisley Experimental Watersheds (Bisley; 18°20′ N, 65°50′ W) study area spans 13 ha of 3 monitored watersheds, tributaries to the Río Mameyes, established in 1987 as a research area with gaged streams, canopy towers for meteorological data, and sampling of forest vegetation. At this site, selective logging is presumed to have minimally occurred [29,30]. In 1989, a series of 78.54 m^2 permanent forest plots, with a total cumulative sampled area of 0.71 ha, were established in Bisley [13,30,31]. These plots are measured at 5 year intervals (measurements in pre-hurricane 1989, post-hurricane 1989, 1994, 1999, 2004, 2009, and 2014). In censuses of Bisley, all stems ≥2.5 cm diameter at 1.3 m from the ground (dbh) were measured, but for comparative purposes in this communication only stems ≥4 cm dbh are included. Palms were measured for dbh and were included in the census when their internode reached 1.5 m. As new trees and palms reached the minimum diameter class, they were numbered and marked with aluminum tags.

2.1.2. Data Analyses

Data spanning 1943 to 2015 (10 censuses) for El Verde and 1989 to 2014 (7 censuses) for Bisley were used for analyses. Previous studies with data from these sites were limited to 5 censuses from Bisley and 8 from El Verde [12,13]. Analyses include data from both permanent plot sites with all stems ≥4 cm in diameter. Data on stem density and basal area are presented per hectare for all identified and living stems per census. To illustrate individual species stem density through time, six species that were dominant in terms of basal area and or stem density in the plots were selected. These species represent primary forest species (late successional) and secondary forest species (early successional), as classified by studies that combine the species-specific characteristics of seedling regeneration under different light conditions, and relative densities of various life history stages in this forest type [34,35]. The early successional species are *C. schreberiana*, and *Psychotria berteriana* DC. (cachimbo). The late successional species are *D. excelsa*, *M. bidentata*, and *S. berteriana*. The sierra palm, *P. montana*, is classified as a mature forest species although it has previously also been associated with early succession due to its ability to form slope and floodplain stands [10,14]. These six species have also been used in other studies of this forest type to explore responses to experimental hurricane disturbance [36] or to model simulations [37]. Examples of species that have been lost and gained to the plot sites throughout the measurement period are presented. To assess tree community species composition dynamics through time, two non-metric multidimensional scaling (NMS) ordination analyses were conducted with number of stems per species per census year. An NMS ordination was conducted because of its efficiency when reducing high dimensional multivariate species space (high number of species) to two dimensions, which provides ease when plotting simple 2 axis graphs. Another benefit is that NMS has minimal assumptions about relationships among variables [38,39]. The multivariate analyses contained only species that occurred in at least two censuses per study site. The NMS ordination was made using species abundance, with which a matrix was generated for each site using Bray–Curtis distance in PC Ord-6 (PC-ORD, Gleneden Beach, OR, USA; [40]). For El Verde, ordination data contained 75 species from 10 census years, and for Bisley, 62 species from 7 census years. Differences in community composition among census year (i.e., census year as the grouping variable) were also compared via simple standardized chi-square distances for species associations per census year on a matrix of species presence and absence. A value of −1 is a perfect negative association, or no similarity in composition. A value of 1 is a perfect positive association, or great similarity in composition. This is a complementary procedure to explore further comparisons of community species composition (presence and absence) among census year per site.

3. Results

3.1. Observations on Structural Characteristics

Stem density values in El Verde were between 1400 and 1800 per hectare during the first eight years of plot censuses (Figure 3). From 1976 to 2015, stem values have been constrained from 1400 to 1600 stems per hectare. Although stem density seems to have fluctuated during the 72-year study period at El Verde, these fluctuations have been minimal compared to dynamics observed at Bisley. Stem density values at Bisley were 1200 stems per hectare before hurricane Hugo, and these stem density values returned to that level five years after the hurricane. Maximum stem density values at Bisley were observed ten years after the hurricane, in 1999. The current trend in Bisley stem density is for a steady decrease in values, becoming closer to 1200 stems per hectare as observed in the last three censuses.

In terms of basal area, El Verde has an increasing trend with values from the last three plot censuses 2005, 2010, and 2015 almost doubled from initial plot measurements in 1943 and 1951 (Figure 4). Basal area at Bisley has maintained values between 35 and 40 m^2 per hectare, which is similar to the basal area recorded at the site before hurricane Hugo (Figure 4). Although basal area decreased almost by half after hurricane Hugo, ten years later, basal area had consistent values at Bisley. Both El Verde and Bisley were at similar values of basal area (35 m^2/ha) before hurricane Hugo in 1989.

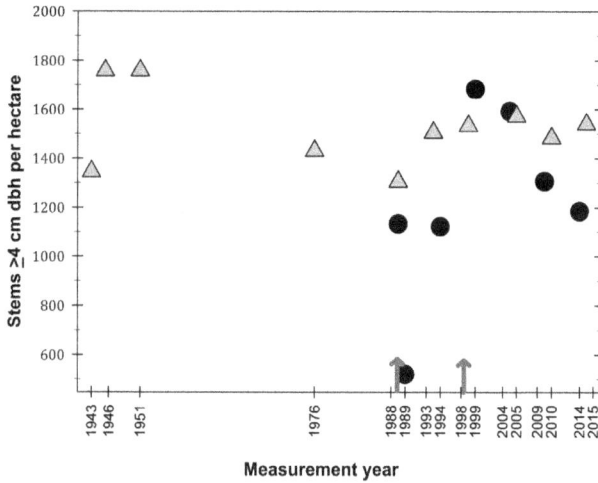

Figure 3. Density of stems (≥4 cm dbh) per hectare in the El Verde permanent plot (grey triangles) and in Bisley (filled circles), Luquillo Experimental Forest, Puerto Rico. Arrows represent hurricane Hugo (1989) and Georges (1998).

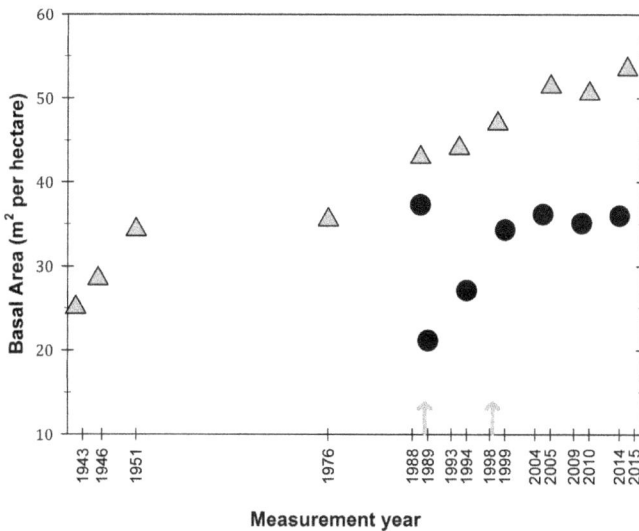

Figure 4. Basal area (m^2/ha) values in the El Verde permanent plot site (grey triangles) and the Bisley site (filled circles), Luquillo Experimental Forest, Puerto Rico. Arrows represent hurricane Hugo (1989) and Georges (1998).

Distribution of stem diameter size classes through time reveal that at El Verde, the stem category of 10 to 20 cm consistently increased in plot censuses from 1943 to 2005 and has contained the greatest number of stems since 1998 (Figure 5). Stems in the smaller size category of 5 to 10 cm accounted for the greatest proportion of the plot in the early 1943, 1946 and 1951 censuses, but this size class decreased consistently since 1993. The categories of 30 to 40 cm and of 40 to 50 cm have both maintained an increasing trend at El Verde. Since plot establishment, El Verde stems in the >60 cm category have gradually increased. In Bisley, however, trees >60 cm reached pre-hurricane Hugo values in 2015. The greatest proportion of stems in Bisley since 1994 originate from the 10 to 20 cm size category. Similarly, in El Verde and Bisley, the greatest number of stems is in the category of 10 to 20 cm; however, Bisley currently has a greater quantity of stems in that size category than El Verde. At the Bisley study site, the 30 to 40 cm and the 40 to 50 cm categories account for a smaller component of the stems in contrast to El Verde.

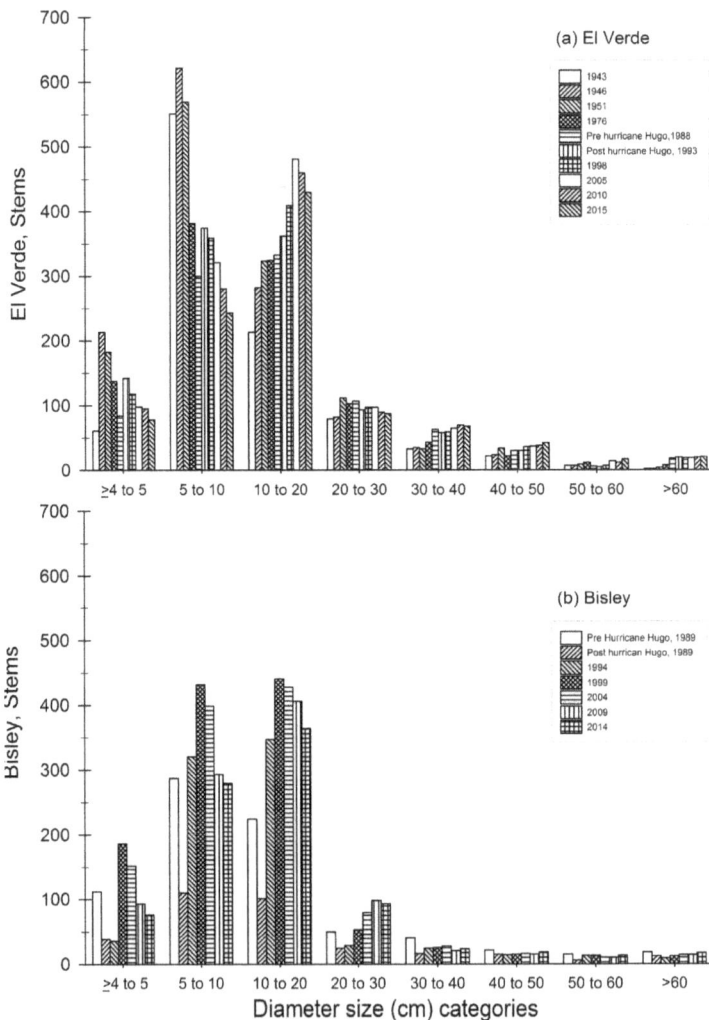

Figure 5. Diameter size category distribution for each census year for El Verde (**a**) top panel and (**b**) Bisley (bottom panel). Each bar represents a census year.

3.2. Observations on Species Abundances and Plot Species Composition

The long-lived and shade tolerant primary forest species, *D. excelsa*, had slightly increasing stem densities at both El Verde and Bisley (Figure 6a,b). This trend was observed even with the effect of hurricane Hugo decreasing basal area at Bisley. However, this was not the case with *S. berteriana*, which presented a slightly decreasing, perhaps stabilizing, trend in stem density at Bisley ten years after hurricane Hugo. At the El Verde site, *S. berteriana* presented a decrease in stem density after the 1951 census (Figure 6a). This decreasing trend started to stabilize in the 2005, 2010 censuses, but at much lower stem density values than in the initial 1943 census. The slow growing, long lived and dense hardwood *M. bidentata* in El Verde follows a very similar stem density pattern with a consistent and slight increase during the past 72 years as observed with *D. excelsa*. At Bisley, *M. bidentata* has remained at low stem density and has not changed during the study period (Figure 6b). In the case of *M. bidentata*, although the stem density remained constant at Bisley, it was one order of magnitude lower than at El Verde.

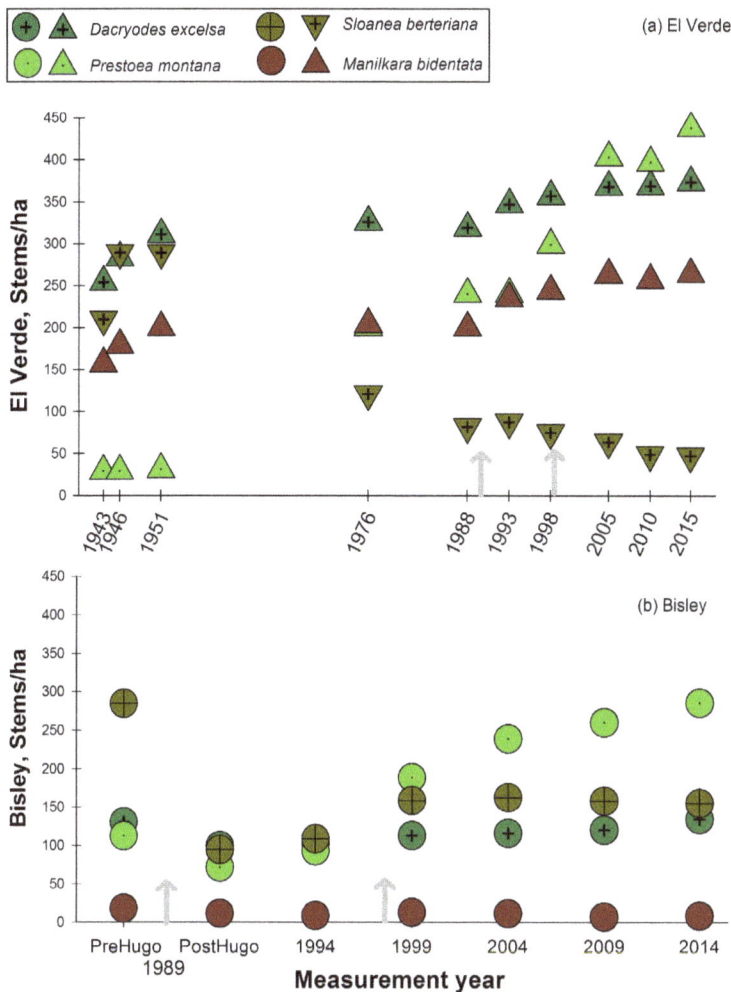

Figure 6. Stem density per each census year for selected shade tolerant/late successional species in two long-term research plots, Luquillo Experimental Forest, Puerto Rico. Panel (**a**) El Verde, with triangle symbols; panel (**b**) Bisley with circle symbols.

The sierra palm, *P. montana*, pattern of stem density has consistently increased through the study period at both Bisley and El Verde (Figure 6a,b). At El Verde, the lowest stem densities were observed in the initial censuses 1943, 1946 and 1951, and values have continued to increase since 1976 in all following censuses. At Bisley, *P. montana* decreased in the census after hurricane Hugo, but ten years later, values were higher than previous to hurricane Hugo (Figure 6b). The patterns that contrast the most are the stem density fluctuations observed in two of the most abundant species, *P. montana* and *S. berteriana*. The palm *P. montana* has increased steadily at both sites while *S. berteriana* has presented a consistently decreasing trend at El Verde and a relatively constant stem density 10 years post-hurricane Hugo at Bisley. At Bisley, stem densities of the shade intolerant *C. schreberiana* have not returned to the lower pre-hurricane Hugo densities, while at El Verde, the values are lower than the initial 1943 census (Figure 7a). In the case of *P. berteriana*, both Bisley and El Verde stem densities have returned to values similar to initial census conditions at each site (Figure 7a,b). Both of these species were in low abundance initially and had a dramatic increase after hurricane Hugo. Although stem densities for *C. schreberiana* are still higher than pre-hurricane Hugo values, these are much lower at El Verde (35–40 stem/ha) compared to Bisley (100–200 stems/ha). In the last 3 censuses at both study sites, there has been a consistently decreasing trend of early successional *C. schreberiana* and *P. berteriana* stem density (Figure 7a,b). Other changes in species observed include the tree fern *Cyathea arborea*, documented for the first time inside the El Verde plot in 2005, after 62 years. This is in contrast to Bisley plots where this species was present since before hurricane Hugo and has remained in constant stem abundance values. The opposite case is observed with the wind dispersed and high-light associated *T. heterophylla*, which has remained present in El Verde plots during the 72 years of censuses, while it disappeared from Bisley plots 5 years after hurricane Hugo. The species *Magnolia splendens* was rare in the initial plot measurements at Bisley and El Verde and is now absent from the plots at both sites.

In the El Verde plot, species composition of each census through time occupies a distinct part of ordination space that continues to change through time (Figure 8a). The last three censuses at El Verde, 2005, 2010 and 2015, have high association in their species composition as evidenced by their very close scatter in the multivariate species space. This is also represented with standardized chi-square distance pairwise comparisons between the 2005 and 2010 census (0.60), and the 2010 and 2015 census (0.91). In contrast, the initial census dates, 1943, 1946, and 1951, are very distantly scattered in El Verde species ordination space in relation to the most recent census dates 2005, 2010, and 2015. This is also evidenced by the pairwise distance comparisons between 1943 and 2005 (0.11), 1946 and 2005 (0.01), 1951 and 2005 (−0.07), 1951 and 2010 (0.05), 1951 and 2015 (0.07), i.e., all very low values for similarity in species composition. The three initial censuses, 1943, 1946 and 1951, share great similarity in species composition, evidenced by the pairwise comparison distance values between 1943 and 1951 (0.74), 1946 and 1951 (0.77). These initial census dates occupy a distinctly separate area of ordination space to that of the most recent El Verde censuses. In contrast, community species composition seems to be converging in Bisley, with the most recent censuses, 2009 and 2014, occupying a closely scattered area of species ordination space as the pre-hurricane 1989 census (Figure 8b). The 2009 and 2014 census were the closest in the Bisley multivariate species space (pairwise distance 0.84), and the 2014 census was marginally similar to the species composition of the 1989 census after hurricane Hugo (0.43). The 1994 census species composition, five years after Hugo, had the greatest difference in species composition with respect to the pre-hurricane census (0.26). The species *C. schreberiana* and *P. berteriana* were at their highest stem densities in Bisley during the 1994 census and in El Verde during the 1993 census.

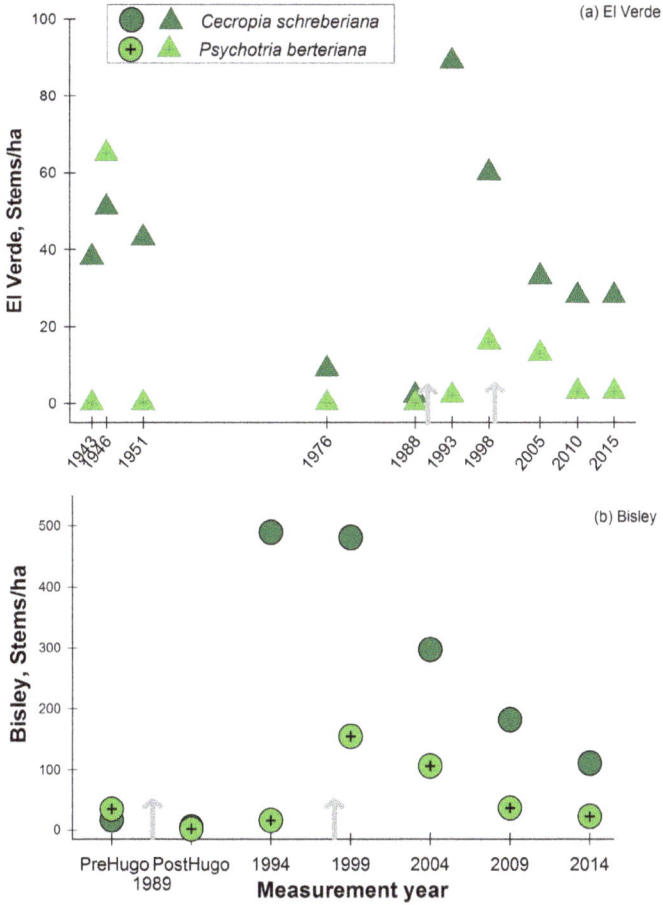

Figure 7. Stem density per each census year for selected shade intolerant/early successional species in two long-term research plots, Luquillo Experimental Forest, Puerto Rico. Panel (**a**) El Verde, with triangle symbols; panel (**b**) Bisley with circle symbols.

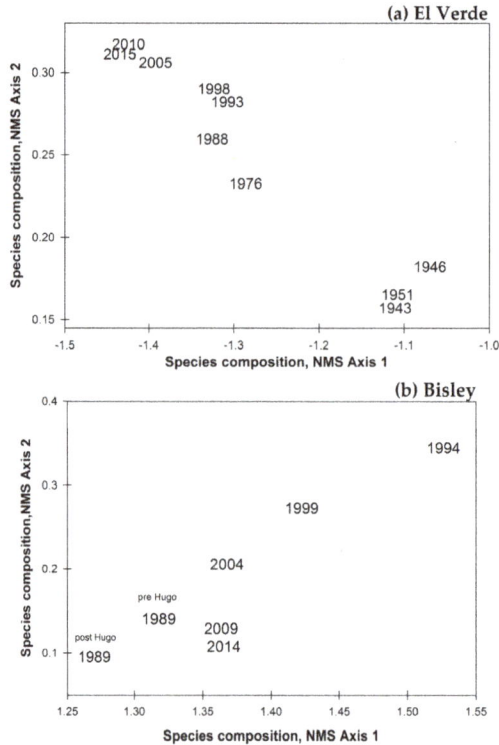

Figure 8. Non-metric multidimensional scaling (NMS) ordination based on species composition and abundance per census of two tabonuco forest permanent plot sites at Luquillo Experimental Forest, Puerto Rico. Top panel (**a**) El Verde and bottom panel (**b**) Bisley.

4. Discussion

The patterns of vegetation dynamics observed have components that are broadly representative of tabonuco forests and also display site-specific responses associated with location in relation to hurricane disturbance events. Site-specific responses such as accumulated basal area depend on the magnitude of effects and frequency of disturbance events [7,14,24,41,42]. Tabonuco forest's structural attributes such as basal area are able to recover from disturbances that remove as much as half of the above-ground biomass [13] and also have the capacity to continue accumulating basal area through time when cumulative disturbance effects are minimal [12].

The factors that contribute to these responses are various and include the life-history patterns of dominant species in this forest type [9,37,43]. The patterns presented in common at both El Verde and Bisley sites were those of the dominant species in these long-term plots, which had stem densities with stable and mostly increasing trends in all censuses after hurricane Hugo in 1989. In this forest type, the attributes of the dominant species, *D. excelsa*, *M. bidentata* and *P. montana*, drive the combination of responses of stem density and basal area at these long-term plot sites [14,24,35]. The resistance and increases observed from dominant species drive forest structure responses. This was also found in modeled simulations of tabonuco forest with individual-based species responses to disturbance, where low temporal variation was observed as the response of dominant species [37]. At both long-term study sites, a rapid response of increased stem density from the early successional species *C. schreberiana* and *P. berteriana* is a proxy for canopy formation and light availability, which starts the process of facilitation and changing conditions in the understory while increasing basal area [2,20,36,37].

Once the canopy closes and light availability changes, these early successional species not only respond negatively to canopy closure, but they are also susceptible to pathogens due to their high intra-species post-hurricane densities [44,45]. However, the patterns that contrast the most are those between stem density fluctuations for two of the most abundant species, *P. montana* and *S. berteriana*. The palm *P. montana* has increased steadily at both sites while *S. berteriana* has presented a trend towards decrease. An increase in palm density in tabonuco forest stands had been noted at El Verde since the 1976 measurements [10] and discussed in a long-term analysis of palm forests compared with tabonuco forests stands [14]. The decreasing trend observed in the El Verde plot with *S. berteriana* could be due to initial stem density values being relatively high, as these could have been still responding to the effects of the previous San Ciprián hurricane [10]. Because decreases in stem density of *S. berteriana* at El Verde, and to a lesser degree at Bisley, occur simultaneously with increases in *P. montana*, another possibility could be that perhaps *P. montana* could be occupying sites for *S. berteriana* recruitment as both of these species are found in wet and concave locations [13,26,27,30,34].

These two tabonuco forest sites experienced different disturbance conditions and are at different points in their response to disturbance or successional trajectories. Due to a difference in their location and geographical aspect, the plots seem to be responding to effects from different disturbance events. Analyses of the effect of the passage of hurricanes over Caribbean, Philippines and Australian forests highlighted the importance of site location and aspect relative to wind direction in the interpretation of hurricane effects [9,14,36,42,43,46–48,51]. These are fundamental observations when assessing forest responses to these events. The increases in basal area at El Verde seem primarily due to increases in already established stems of the 30 to 40 cm and 40 to 50 cm diameter categories, which are trees that have been long-time residents of the plot and are resisting and growing through time. This is not the case in the Bisley plot, where trees of the 30 to 40 cm and 40 to 50 cm dbh categories are less abundant, contribute proportionally less to basal area, and have not had a net increase in stem density during the study period. Similar windward and leeward site responses were observed in palm forest stands in the LEF. Greater hurricane Hugo effects were measured on the windward slopes where a long-term palm forest study plot is located, while much fewer effects were observed in a palm forest study plot on a leeward slope [14].

The effects from hurricane disturbances maintain these forests in a constant state of structural and compositional change in response to intensity of the cumulative effects. At Bisley, there was a greater successional trajectory dynamic due to the stronger hurricane effect on stem density and basal area. At El Verde, the hurricane defoliation had minimal immediate effects on stem density and basal area, while increasing canopy opening and light to the forest floor [2,20,36].

Thus, El Verde has been maintained in a directional successional transition, as effects from hurricane Hugo were minimal, and it is still in a response trajectory from the effects of San Ciprián in 1932 [10,12]. When hurricane Georges reached the Bisley site, it was within the first ten years of recovery from the changes in structure that resulted following hurricane Hugo. Although no significant effects were observed in terms of stem density and basal area at Bisley following hurricane George in 1998 [13], the forest site benefited from the canopy defoliation and associated increases of light to the forest floor by continuing or extending on the particular succession trajectory set by hurricane Hugo. At the El Verde site, it seems that due to the time since hurricane San Ciprián and the minimal effects from Hugo, the forest has continued on a succession path defined by increased growth, accrued structure and basal area that continues surpassing initial measurement conditions [10,12]. The apparent asymptote at Bisley for basal area values is likely due to the steady loss of the accumulated large stems of early successional species being replaced by smaller shade tolerant stems of the dominant species. If the next hurricane at Bisley has moderate or minimal effects on forest structure, limited to defoliation type effects, then basal area should be expected to increase, as has been observed at El Verde.

In response to disturbance events such as hurricanes, forest community dynamics depend on the interaction of processes that occur at different time scales. Tree structure can recover to pre-disturbance values within a decade in the tabonuco forest of the LEF. This is due to the tree species community being able to shift composition in such a way that only the dominant species are maintained in association with a flux of changes from other species. During the 72 years of vegetation censuses at El Verde and the 25 years at Bisley, concomitant changes in species composition have allowed for forest structure components such as basal area to either recover and maintain pre-disturbance values (such as observed in Bisley), or to progressively increase (as observed in El Verde). If there is now a period of hurricane-free conditions for another 10–15 years, perhaps the tabonuco forests at the LEF will slow down their pattern of accretion of basal area under undisturbed and closed canopy conditions. Before hurricane Hugo in 1989, both El Verde and Bisley were at similar values of basal area, but because of the different disturbance effects, their responses diverged. The defoliation and trimming effect seems to be what has allowed El Verde to increase in basal area, due to a combination of the temporary increase in *C. schreberiana*, and the increases in basal area contributed by the stems of already established and resistant dominant species that are now at the canopy level. It is proposed that another storm or hurricane disturbance in Bisley, similar to hurricane Georges, will produce the canopy defoliation needed to accumulate greater basal area at this forest site.

Following hurricanes, recovery and resilience of different components of the forest ecosystem have their own response intervals [20]. Tree structural characteristics and processes such as litterfall rates are the fastest to recover [31,34]. On the other hand, tree and understory species composition shift in more dramatic ways and are not able to completely return to pre-disturbance conditions at the same time scales [13,49]. Species composition occurs in the context of both changes in environmental conditions set after the particular disturbance event, plus the pool of species present at the site. The range of species at a site is limited by the constraints of their individual life history dynamics and responses to disturbance [37,41,50]. The insights into forest structure and composition point to the nuances of changes in species dynamics during succession, with individual species population dynamics and interactions among species still in need of further exploration.

Acknowledgments: This USDA Forest Service (USFS) International Institute of Tropical Forestry (IITF) research was conducted in collaboration with the University of Puerto Rico (UPR) as part of IITF contributions to the Luquillo Long-Term Ecological Research Program (LTER) in the LEF. Assistance with measurements of the El Verde plot was provided by USFS SRS-FIA and SUNY-ESF collaborators via Cooperative Grants and Agreements (N IITF-98-CA-008 and 05-DG-11120101-016) from the USFS IITF and Edna Bailey Sussman and Farnsworth Funds-SUNY-ESF. This is a contribution part of IITF's 75th Anniversary Research Symposium. The revision and suggestions from F.H. Wadsworth, D.J. Lodge, A.E. Lugo, S.A. Sloan and three anonymous reviewers improved this manuscript. The findings, conclusions, and views expressed in this manuscript are those of the author and do not necessarily represent the views of the USDA Forest Service.

Conflicts of Interest: The author declares no conflict of interest.

References

1. Chazdon, R.L. Tropical forest recovery: Legacies of human impact and natural disturbances. *Perspect. Plant Ecol. Evol. Syst.* **2003**, *6*, 51–71. [CrossRef]
2. Brokaw, N.V.L.; Walker, L.R. Summary of the effects of Caribbean hurricanes on vegetation. *Biotropica* **1991**, *23*, 442–447. [CrossRef]
3. Zimmerman, J.K.; Willig, M.R.; Walker, L.R.; Silver, W.L. Introduction: Disturbance and Caribbean ecosystems. *Biotropica* **1996**, *28*, 414–423. [CrossRef]
4. Myers, N.; Mittermeier, R.A.; Mittermeier, C.G.; Da Fonseca, G.A.; Kent, J. Biodiversity hotspots for conservation priorities. *Nature* **2000**, *403*, 853–858. [CrossRef] [PubMed]
5. López-Marrero, T.; Heartsill-Scalley, T. Get up, stand up: Environmental situation, threats, and opportunities in the insular Caribbean. *Caribb. Stud.* **2012**, *40*, 3–14. [CrossRef]
6. Lugo, A.E.; Helmer, E.H.; Santiago Valentín, E. Caribbean landscapes and their biodiversity. *Interciencia* **2012**, *37*, 705–710.
7. Lugo, A.E. Visible and invisible effects of hurricanes on forest ecosystems: An international review. *Austral Ecol.* **2012**, *33*, 368–398. [CrossRef]
8. Chai, S.L.; Healey, J.R.; Tanner, E.V. Evaluation of forest recovery over time and space using permanent plots monitored over 30 years in a Jamaican montane rain forest. *PLoS ONE* **2012**, *7*, e48859. [CrossRef] [PubMed]

9. Luke, D.; McLaren, K.; Wilson, B. Modeling hurricane exposure in a Caribbean lower montane tropical wet forest: The Effects of frequent, intermediate disturbances and topography on forest structural dynamics and composition. *Ecosystems* **2016**, *19*, 1178–1195. [CrossRef]

10. Crow, T.R. A rainforest chronicle: A 30-year record of change in structure and composition at El Verde, Puerto Rico. *Biotropica* **1980**, *12*, 42–55. [CrossRef]

11. Johnston, M.H. Successional Change and Species/Site Relationships in a Puerto Rican Tropical Forest. Ph.D. Thesis, State University of New York, New York, NY, USA, 1991; p. 253.

12. Drew, A.P.; Boley, J.D.; Zhao, Y.; Johnston, M.H.; Wadsworth, F.H. Sixty-two years of change in subtropical wet forest structure and composition at El Verde, Puerto Rico. *Interciencia* **2009**, *34*, 34–40.

13. Heartsill, S.; Tamara, S.; Frederick, N.; Lugo, A.E.; Moya, S.; Estrada Ruiz, C.R. Changes in structure, composition, and nutrients during 15 years of hurricane-induced succession in a subtropical wet forest in Puerto Rico. *Biotropica* **2010**, *42*, 455–463. [CrossRef]

14. Lugo, A.E.; Frangi, J.L. Long-term response of Caribbean palm forests to hurricanes. *Caribb. Nat.* **2016**, 157–175.

15. Walker, L.R.; Zarin, D.J.; Fetcher, N.; Myster, R.W.; Johnson, A.H. Ecosystem development and plant succession on landslides in the Caribbean. *Biotropica* **1996**, *28*, 566–576. [CrossRef]

16. Coll, M.; Fonseca, A.C.; Cortés, J. The mangrove and others vegetation associations in de Gandoca lagoon, Limón, Costa Rica. *Rev. Biol. Trop.* **2001**, *49*, 321–329. [PubMed]

17. Heartsill Scalley, T.; Scatena, F.N.; Estrada, C.; McDowell, W.H.; Lugo, A.E. Disturbance and long-term patterns of rainfall and throughfall nutrient fluxes in a subtropical wet forest in Puerto Rico. *J. Hydrol.* **2007**, *333*, 472–485. [CrossRef]

18. Gonzalez, G.; Waide, R.B.; Willig, M.R. Advancements in the understanding of spatiotemporal gradients in tropical landscapes: A Luquillo focus and global perspective. *Ecol. Bull.* **2013**, *54*, 245–250.

19. Thomas, S.C.; Martin, A.R.; Mycroft, E.E. Tropical trees in a wind-exposed island ecosystem: Height-diameter allometry and size at onset of maturity. *J. Ecol.* **2015**, *103*, 594–605. [CrossRef]

20. Brokaw, N.; Zimmerman, J.K.; Willig, M.; Camilo, G.; Covich, A.; Crowl, T.; Fetcher, N.; Haines, B.; Lodge, J.; Lugo, A.E.; et al. *A Caribbean Forest Tapestry: The Multidimensional Nature of Disturbance and Response*; Brokaw, N., Crowl, T.A., Lugo, A.E., McDowell, W.H., Scatena, F.N., Waide, R.B., Willig, M.R., Eds.; Oxford University Press: New York, NY, USA, 2012; pp. 201–271.

21. Taylor, M.A.; Stephenson, T.S.; Chen, A.A.; Stephenson, K.A. Climate change and the Caribbean: Review and response. *Caribb. Stud.* **2012**, *40*, 169–200. [CrossRef]

22. Beard, J.S. *The Natural Vegetation of the Windward and Leeward Islands*; Clarendon Press: Oxford, UK, 1949; p. 192.

23. Alvarez Ruiz, M. *Effects of Human Activities on Stand Structure and Composition, and Genetic Diversity of Dacryodes Excelsa Vahl (Tabunuco)*; University of Puerto Rico: San Juan, Puerto Rico, 2002; p. 266.

24. Lugo, A.E.; Scatena, F.N. Ecosystem-level properties of the Luquillo Experimental Forest with emphasis on the Tabonuco Forest. In *Tropical Forests: Management and Ecology*; Springer: New York, NY, USA, 1995; pp. 59–108.

25. Basnet, K.; Scatena, F.N.; Likens, G.E.; Lugo, A.E. Ecological consequences of root grafting in Tabonuco (*Dacryodes excelsa*) trees in the Luquillo Experimental Forest, Puerto Rico. *Biotropica* **1993**, *25*, 28–35. [CrossRef]

26. Weaver, P.L. *Sloanea Berteriana Choisy Motillo. Elaeocarpaceae. Elaeocarpus Family*; USDA Forest Service, International Institute of Tropical Forestry: Rio Piedras, Puerto Rico, 1997; p. 7.

27. Lugo, A.E.; Francis, J.K.; Frangi, J.L. *Prestoea Montana (R. Graham) Nichols. Sierra Palm. Palmaceae. Palm Family*; US Department of Agriculture, Forest Service, International Institute of Tropical Forestry: Rio Piedras, Puerto Rico, 1998; p. 9.

28. Weaver, P.L. Ecological observations on *Magnolia splendens* urban in the Luquillo mountains of Puerto Rico. *Caribb. J. Sci.* **1987**, *23*, 340–351.

29. Garcia-Montiel, D.; Scatena, F.N. The effect of human activity on the structure and composition of a tropical forest in Puerto Rico. *Forest Ecol. Manag.* **1994**, *63*, 57–78. [CrossRef]

30. Scatena, F.N. *An Introduction to the Physiography and History of the Bisley Experimental Watersheds in the Luquillo Mountains of Puerto Rico*; Gen. Tech. Rep. SO-72. U.S.D.A; Forest Service, Southern Forest Experiment Station: New Orleans, LA, USA, 1989; p. 22.

31. Scatena, F.N.; Moya, S.; Estrada, C.; Chinea, J.D. The first five year in the reorganization of aboveground biomass and nutrient use following Hurricane Hugo in the Bisley Experimental Watersheds, Luquillo Experimental Forest, Puerto Rico. *Biotropica* **1996**, *28*, 424–440. [CrossRef]
32. Ostertag, R.; Silver, W.L.; Lugo, A.E. Factors affecting mortality and resistance to damage following hurricanes in a rehabilitated subtropical moist forest. *Biotropica* **2005**, *37*, 16–24. [CrossRef]
33. Larsen, M.C. Analysis of 20th century rainfall and streamflow to characterize drought and water resources in Puerto Rico. *Phys. Geogr.* **2000**, *21*, 494–521.
34. Beard, K.H.; Vogt, K.A.; Vogt, D.J.; Scatena, F.N.; Covich, A.P.; Sigurdardottir, R.; Siccama, T.G.; Crowl, T.A. Structural and functional responses of a subtropical forest to 10 year of hurricanes and droughts. *Ecol. Monogr.* **2005**, *75*, 345–361. [CrossRef]
35. Lugo, A.E.; Zimmerman, J.K. Ecological Life Histories. In *Tropical Tree Seed Manual*; Vozzo, J.A., Ed.; USDA Forest Service: Washington, DC, USA, 2002; pp. 191–213.
36. Shiels, A.B.; Zimmerman, J.K.; García-Montiel, D.C.; Jonckheere, I.; Holm, J.; Horton, D.; Brokaw, N. Plant responses to simulated hurricane impacts in a subtropical wet forest, Puerto Rico. *J. Ecol.* **2010**, *98*, 659–673. [CrossRef]
37. Uriarte, M.; Canham, C.D.; Thompson, J.; Zimmerman, J.K.; Murphy, L.; Sabat, A.M.; Fetcher, N.; Haines, B.L. Natural disturbance and human land use as determinants of tropical forest dynamics: Results from a forest simulator. *Ecol. Monogr.* **2009**, *79*, 423–443. [CrossRef]
38. Johnson, D.E. *Applied Multivariate Methods for Data Analysts*; Duxbury Press: Pacific Grove, CA, USA, 1998; p. 567.
39. McCune, B.; Grace, J.B. *Analysis of Ecological Communities*; MJM Software Design: Gleneden Beach, OR, USA, 2002; p. 300.
40. McCune, B.; Mefford, M.J. PC-ORD. In *Multivariate Analysis of Ecological Data*; Version 5.1; Mjm Software: Gleneden Beach, OR, USA, 2006.
41. Lugo, A.E. Effects and outcomes of Caribbean hurricanes in a climate change scenario. *Sci. Total Environ.* **2000**, *262*, 243–251. [CrossRef]
42. Boose, E.R.; Foster, D.R.; Fluet, M. Hurricane impacts to tropical and temperate forest landscapes. *Ecol. Monogr.* **1994**, *64*, 369–400. [CrossRef]
43. Bellingham, P.J.; Tanner, E.V.J.; Healey, J.R. Damage and responsiveness of Jamaican montane forest tree species after disturbance by a hurricane. *Ecology* **1995**, *76*, 2562–2580. [CrossRef]
44. Lodge, D.J.; Cantrell, S. Fungal communities in wet tropical forests: Variation in time and space. *Can. J. Bot.* **1995**, *73*, 1391–1398. [CrossRef]
45. Lodge, D.J. Microorganisms. In *The Food Wed of a Tropical Forest*; Reagan, D.P., Waide, R.B., Eds.; University of Chicago press: Chicago, IL, USA, 1996; pp. 53–108.
46. Yap, S.L.; Davies, S.J.; Condit, R. Dynamic response of a Philippine dipterocarp forest to typhoon disturbance. *J. Veg. Sci.* **2016**, *27*, 133–143. [CrossRef]
47. Murphy, H.T.; Metcalfe, D.J.; Bradford, M.G.; Ford, A.J. Community divergence in a tropical forest following a severe cyclone. *Austral Ecol.* **2014**, *39*, 696–709. [CrossRef]
48. Foster, D.R.; Knight, D.H.; Franklin, J.F. Landscape patterns and legacies resulting from large, infrequent forest disturbances. *Ecosystems* **1988**, *1*, 497–510. [CrossRef]
49. Willig, M.R.; Bloch, C.P.; Covich, A.P.; Hall, C.A.S.; Lodge, D.J.; Lugo, A.E.; Silver, W.L.; Waide, R.B.; Walker, L.R.; Zimmerman, J.K. Long-term research in the Luquillo Mountains, synthesis and foundations for the future. In *A Caribbean Forest Tapestry: The Multidimensional Nature of Disturbance and Response*; Brokaw, N., Crowl, T.A., Lugo, A.E., Mcdowell, W.H., Scatena, F.N., Waide, R.B., Willig, M.R., Eds.; Oxford University Press: New York, NY, USA, 2012; pp. 361–441.
50. Imbert, D.; Labbe, P.; Rousteau, A. Hurricane damage and forest structure in Guadeloupe, French West Indies. *J. Trop. Ecol.* **1996**, *12*, 663–680. [CrossRef]
51. Royo, A.A.; Scalley, T.H.; Moya, S.; Scatena, F.N. Non-arborescent vegetation trajectories following repeated hurricane disturbance: Ephemeral versus enduring responses. *Ecosphere* **2011**, *2*, 77. [CrossRef]

forests

MDPI

Article

The Plight of Migrant Birds Wintering in the Caribbean: Rainfall Effects in the Annual Cycle

Joseph M. Wunderle Jr. * and Wayne J. Arendt

International Institute of Tropical Forestry, United States Department of Agriculture Forest Service, Sabana Field Research Station, HC 02 Box 6205, Luquillo, Puerto Rico, USA; warendt@fs.fed.us
* Correspondence: jwunderlejr@fs.fed.us or jmwunderle@gmail.com; Tel.: +1-787-888-3673

Academic Editors: Grizelle González and Ariel E. Lugo
Received: 27 January 2017; Accepted: 1 April 2017; Published: 8 April 2017

Abstract: Here, we summarize results of migrant bird research in the Caribbean as part of a 75th Anniversary Symposium on research of the United States Department of Agriculture Forest Service, International Institute of Tropical Forestry (IITF). The fate of migratory birds has been a concern stimulating research over the past 40 years in response to population declines documented in long-term studies including those of the IITF and collaborators in Puerto Rico's Guánica dry forest. Various studies indicate that in addition to forest loss or fragmentation, some migrant declines may be due to rainfall variation, the consequences of which may carry over from one stage of a migrant's annual cycle to another. For example, the Guánica studies indicate that rainfall extremes on either the temperate breeding or tropical wintering grounds affect migrant abundance and survival differently depending on the species. In contrast, IITF's collaborative studies of the migrant Kirtland's Warbler (*Setophaga kirtlandii*) in the Bahamas found that late winter droughts affect its annual survival and breeding success in Michigan. We review these IITF migrant studies and relate them to other studies, which have improved our understanding of migrant ecology of relevance to conservation. Particularly important is the advent of the full annual cycle (FAC) approach. The FAC will facilitate future identification and mitigation of limiting factors contributing to migrant population declines, which for some species, may be exacerbated by global climate change.

Keywords: annual cycle; carry over effects; climate change; Guánica; Nearctic-Neotropical

1. Introduction

Since the 50th anniversary of the International Institute of Tropical Forestry (IITF), the institute's wildlife program has increased its research on migrant birds in the Caribbean basin in response to evidence that migrant populations were decreasing in an effort to better understand their winter ecology to help guide conservation efforts. The expansion of IITF's migrant studies coincided with a substantial increase in migrant research by others working on both the breeding and wintering grounds as well as during the migration period (reviewed in Greenberg and Marra [1]). By studying all stages of the migrant annual cycle—breeding, nonbreeding, and migratory stages—in many cases it is now possible to identify when and where in the annual cycle populations are limited or the stage in which mortality is highest. Our migrant studies have taken advantage of recent technological and conceptual advances to better understand how events throughout a migrant's annual cycle are interrelated and how multiple limiting factors affect wintering migrant populations in the Caribbean. Continued population declines of migrants in North America [2] and Puerto Rico [3] provide the impetus for our research. In this essay, we summarize some results of our studies, particularly on the effects of rainfall on different stages of the annual cycle of migrants that winter in the Caribbean. We use a historical approach to relate our avian migrant research over the past 25 years with the advancements in understanding of migrant biology of relevance to their conservation.

1.1. Historical Background

The fate of migratory birds has been a focus of concern stimulating research and conservation over the past 40 years. This concern was due to population declines in migrant birds, which breed in North America and overwinter in the tropics (i.e., Nearctic-Neotropical migrants). Some of this early apprehension arose from long-term population studies in small parks and woodlots of eastern United States, which indicated major declines of forest-dwelling birds, most of which overwinter in the Neotropics [4,5]. Not only was evidence accumulating for migrant declines in the North American breeding grounds, but evidence also came from the tropical wintering grounds where declines were detected in long-term monitoring studies of winter resident migrants in Puerto Rico [6,7]. These migrant declines were also consistent with the impressions of population declines held by local ornithologists, bird watchers, and conservation stewards on various Caribbean islands [8]. Additional analyses suggested that the migrant declines covered major regions of the continent [9,10]. These declines were not just reducing avian abundance and diversity in eastern forests of the U.S. and Canada, where migrants frequently comprise 65%–85% of the breeding birds [11], but these declines were expected to diminish ecosystem services, such as insect pest control and seed dispersal [12]. Concerns arising from these studies provided support for the initiation of the Neotropical Migratory Bird Conservation Program, currently known as Partners in Flight.

At the time of the institute's 50th anniversary in 1989, the two most commonly cited causes for migrant bird population declines were attributed to deforestation and fragmentation on the temperate breeding grounds and deforestation on the subtropical and tropical wintering grounds. Breeding-ground forest loss and fragmentation were implicated in declines because fragmentation increased the relative amount of edge habitat through which edge-dwelling predators and brood parasites (i.e., cowbirds) obtained access to nests of forest interior species, most of which are Nearctic-Neotropical migrants [5,13,14]. Evidence for tropical deforestation as a driver of migrant declines came from analyses of population trends in long-term Breeding Bird Surveys conducted throughout the continental U.S. by Robbins et al. [15]. Their analyses indicated that 75% of the breeding migrant bird species that wintered in tropical forests declined between 1978 and 1987, a period of rapid deforestation in the Neotropics. In contrast, during the same period migrant species wintering in tropical scrub habitats showed no declines, nor were consistent patterns found in short-distance migrants and resident temperate zone species. The Robbins et al. [15] analyses were expanded and subsequently verified by Askins et al. [4]. Thus, by the time of IITF's 50th anniversary in 1989, the major factors or stresses hypothesized to cause migrant population declines were viewed as either occurring on the breeding or the wintering grounds, with most declines in the recent past attributed primarily to the negative consequences of breeding ground forest loss and fragmentation [5]. It was recognized, however, that identifying the factor or factors contributing to migrant declines was challenging given their complex life cycle, which often involves several habitats distributed across large spatial scales and ignorance of where specific populations bred, migrated, or overwintered [16].

Improvements in our understanding of the migrant annual cycle are attributable to recent technological advances. For example, increased computational power has facilitated use of more sophisticated statistical and population models to estimate migrant survival at different stages of the annual cycle, and refinements in stable isotope usage and modeling enable remotely sensed determination of habitat use and distribution [17]. In addition, improvements in battery and memory storage have facilitated miniaturization of devices that allow studies of habitat use and distribution of small birds at both the scale of a home range during sedentary periods (with radio transmitters) [18], as well as tracking long-distance movements of birds (with light-sensitive geolocators [19]) during migration and the full annual cycle (henceforth, FAC).

Not only have recent technological advances aided identification of when and where in the annual cycle mortality is greatest for some species, but the FAC focus has enabled biologists to determine how events in one stage of the annual cycle can affect a migrant in some other stage(s). This was nicely illustrated in the pioneering work of Marra et al. [20] who demonstrated a relationship between the quality of a tropical habitat (i.e., food rich vs. food poor) occupied by a migrant during the winter and its subsequent reproductive success on its North American breeding grounds. Studying male American Redstarts (*Setophaga ruticilla*) wintering in Jamaica on high quality habitat (i.e., insect-rich, mangrove forest) and nearby poor quality habitat (i.e., insect-poor, dry second-growth scrub), Marra and his team discovered that males in high-quality winter habitat had better body condition and departed earlier on spring migration and subsequently arrived earlier on the temperate breeding grounds than males that occupied poorer quality habitat. Therefore, males arriving early on the breeding grounds were coming from high quality moist winter habitat as ascertained by stable isotopes, and had high reproductive success. In contrast, males arriving later on the breeding grounds came from low quality winter habitats and had low reproductive success. This contribution, subsequently substantiated by others [21–25], indicates that events occurring during one stage of a migrant's annual cycle can "carry over" to other stages of the annual cycle. We now recognize that wintering ground habitat quality, often governed by moisture availability, can affect migrant reproductive success on the breeding grounds, but also that conditions on the breeding ground can carry over to affect overwinter survival.

1.2. The Caribbean as a Wintering Site

The Caribbean is important for Nearctic-Neotropical migrants because many winter near the continental U.S., where migrants constitute half or more of all terrestrial birds present during the winter in parts of Mexico, the Bahamas, Hispaniola, Cuba, Jamaica, Puerto Rico, and the Virgin Islands [12,26]. Although early analyses suggested that migrant abundance in the islands decreased with distance from North America and increased with island size [27], subsequent analyses found that the distance/size relationship with abundance held only when migrants in the same habitat type were compared among islands [26]. Controlling for habitat type in the latter analyses was required because different habitats on the same island can vary substantially in migrant abundance and species richness. Of the estimated 44 terrestrial migrants that winter in the Caribbean (52% in family Parulidae), small gleaning insectivores are disproportionately overrepresented in contrast to hawking, aerial, and large gleaning insectivores and frugivores and granivores, which are rare or absent [28]. Two migrant species winter exclusively in the Caribbean and have populations considered threatened or endangered (Kirtland's Warbler, *Setophaga kirtlandii*; Bicknell's Thrush *Catharus bicknelli*). Presumed extinct, the Bachman's Warbler (*Vermivora bachmanii*) wintered primarily in Cuba but has not been observed since the early 1970s [29]. Habitat loss, either clearing for agricultural crops on its breeding grounds in the Southeastern U.S. and/or conversion of its winter habitat for sugar cane in Cuba, are suspected to have contributed to its disappearance [29,30].

Forested habitats in the Caribbean have been viewed as precarious given the small land areas and high human population densities on many islands resulting in substantial forest loss and degradation [31–33]. However, much of this forest loss has occurred in the past, and on some islands, the proportion of forest cover has increased in recent years as people abandon agriculture and migrate to towns and cities [34–37]. Although forest cover has increased with abandonment of agriculture on these islands, development and urbanization have increased mostly in the lowlands where habitats continue to face increased development pressures [37,38]. On abandoned lands, some of the regenerating second-growth forests have a mix of native and alien tree species and given this mixture, have been designated as novel forests [39,40]. Although the suitability of novel forests to provide habitat for wintering migrants requires more study, evidence suggests that suitability is related to the specific dominant alien tree species in the forest. For example, a study by Beltrán and Wunderle [41] found that some alien tree species with an abundance of insects were favored foraging sites for some insectivorous birds, including migrants, whereas other alien tree species, some with few insects, were avoided by insectivores. Even cultivated lands with a shade overstory, such as shade coffee plantations, can provide winter habitat for some migrants [26,42], which can be equivalent to natural forest in terms of overwinter site fidelity and annual return [43]. The trend in the Caribbean, however, is to eliminate the shade overstory to permit open grown sun coffee, which is less suitable for many wintering migrants. Thus, the availability of forest or forest-like habitats for migrants wintering in the Caribbean is in flux, suggesting that on some islands availability of winter habitats for migrants may be increasing relative to 50 years ago.

2. Institute Migrant Studies

2.1. Rainfall Effects in the Long-Term Studies of the Winter-Resident Bird Community in the Guánica Dry Forest in Puerto Rico

A long-term bird monitoring program, based on a mist-netting session for three consecutive days each January, was established in the Guánica Commonwealth Forest, a United Nations Biosphere Reserve, in southwestern Puerto Rico. Initiated in 1972, the ongoing, annual netting session has been conducted every January (except in 1977 and 1979) for 45 years in one of the last remaining tracts of nearly pristine subtropical dry forest in the Caribbean (site and methods described in [6,7]). The Guánica migrant bird community is comprised of two diverse sets of seasonal residents [44]: (1) winter residents that are fully integrated into the Guánica forest bird community and consisting of mostly territorial species (e.g., Black-and-white Warbler *Mniotilta varia*, American Redstart and Ovenbird *Seiurus aurocapilla*); and (2) opportunistic species whose numbers vary greatly from year to year and show little site fidelity (Cape May Warbler, *Setophaga tigrina*; Northern Parula, *S. americana*; and Prairie Warbler, *S. discolor*, among others). The Guánica constant-effort mist-netting program has documented dramatic declines in several species of year-round resident birds as well as the dominant set of winter resident migrants (henceforth, "winter residents"), as evidenced by capture rates that are now ~33% of the capture rates recorded 20 years ago [3]. Population estimates for the three most commonly captured winter residents (American Redstart, Black-and-white Warbler, Ovenbird) have declined markedly, and other formerly common migrant species are rarely captured. Despite these dramatic declines in captures of winter residents, annual survival rates of the three most common species have remained constant [3,17], suggesting that declines in migrant captures are driven by declining recruitment into the Guánica forest.

Rainfall, either on the breeding or wintering grounds, affects migrant abundance in the Guánica forest. Total winter resident abundance in Guánica appears to be influenced by rainfall on the breeding grounds. Declines in total winter resident captures following breeding ground droughts were followed by quick recovery in captures when breeding ground rainfall returned to normal [3,17]. Nevertheless, individual species differed in their abundance responses to various rainfall measures, as uncovered in a modeling analysis combined with knowledge of each species' breeding regions (mostly eastern U.S.) as determined with stable hydrogen-isotopes [17]. For example, variation in rainfall in Guánica, measured as total deviation (absolute value) from normal total rainfall, had a positive influence on the abundance of Black-and-white Warblers. Although this finding was unexpected, Duggers et al. [17] speculated that perhaps it resulted from density-dependent effects of permanent resident bird populations (via diffuse competition) on winter resident abundance in Guánica. Ovenbirds, in contrast, were affected by breeding ground rainfall, measured as total annual rainfall in the continental United States, which had a negative effect on their abundance in Guánica. The decline in Ovenbird captures with increased rainfall on the breeding grounds was attributed to reduced reproductive success, perhaps due to flooding of nests or chilling of chicks in this ground-nesting species [17]. The authors also suggested that increased rainfall during fledging and prior to fall migration might contribute disproportionately to juvenile mortality further diminishing the number of Ovenbirds migrating south to Guánica.

Rainfall in Guánica also affects migrant survival, although rainfall effects on survival were only weakly supported for Black-and-white Warbler or for American Redstart [17]. The strongest support for a Puerto Rican rainfall effect on apparent survival for each of these two species was the measure of Guánica rainfall in the first six months of the prior year. Apparent survival for both species declined in response to increased rainfall in the first six months of the prior year, possibly due to diffuse interspecific competition between winter and permanent residents. Previous studies demonstrated that the size of permanent resident populations was positively related to rainfall in

January–June because of dependence on early rains to end the dry season, thereby stimulating insect outbreaks and facilitating resident breeding [7,45,46]. Therefore, more rainfall in the first six months of the year results in more permanent resident insectivores, which in turn, increases competition and reduces habitat quality for the winter insectivorous residents. Further support for the role of competition from permanent resident insectivores comes from the studies of IITF cooperator, Judith Toms, who found that American Redstart abundance was reduced in areas of high density of the insectivorous, permanent resident Adelaide's Warblers (*Setophaga adelaidae*) in Guánica [47]. Given reduced habitat quality (i.e., fewer insects), migrant survival is reduced or migrants abandon Guánica for other wintering sites. In contrast, Ovenbird apparent survival was not influenced by rainfall in Guánica, but rather by summer rainfall in the southeastern U.S., to which adult survival responded positively [17].

In summary, the long-term monitoring study in Guánica has uncovered some diverse responses of migrants to various rainfall measures from the breeding and wintering grounds, depending on the species. Although the responses of the individual species to rainfall were strongest for abundance and weakest for survival, the general patterns for abundance were consistent with the best survival models for the two species with adequate data to model both response variables [17]. For instance, abundance and survival in Ovenbirds and Black-and-white Warblers appeared to be related to breeding ground and wintering ground rainfall, respectively. Although Dugger et al. [17] expected a direct relationship between rainfall, habitat quality, and migrant demography, their modeling efforts with Black-and-white Warblers suggest that rainfall positively affected abundance of permanent residents, thereby causing diffuse competition for food, resulting in a density-dependent effect on winter residents. Therefore, rainfall can affect migrant demography directly (e.g., mortality due to exposure) or indirectly (e.g., increased competition from permanent residents [47]) on either the breeding or wintering grounds, depending on the species. As Dugger et al. [17] note, testing their density-dependent hypothesis will require continued monitoring of wintering and permanent-resident populations to ascertain how those populations fluctuate together over time in the Guánica forest.

2.2. Winter Rainfall Effects on Kirtland's Warblers in the Bahamas Carry Over to the Michigan Breeding Grounds

Evidence for carryover effects of wintering ground rainfall has been found in the Kirtland's Warbler (henceforth, KIWA), a species for which the breeding and wintering grounds are well known. The KIWA winters in the Bahamas archipelago and breeds primarily in Michigan [48], although small breeding colonies have been recently established in Wisconsin and Ontario [49,50]. Evidence for carryover effects from winter rainfall was first recognized by Ryel [51], who found a positive relationship between winter rainfall in the Bahamas and the number of singing males in Michigan. More recently, Rockwell et al. [24] found that after March droughts in the Bahamas, male KIWAs arrived later on the Michigan breeding grounds, where late arriving males had reduced reproductive success. Sensitivity to March rainfall declines was greater in second year males than in older males; later arrivals corresponded with lower March rainfall in the Bahamas. Although the effects of rainfall on the KIWA's food supply in the Bahamas were unknown at the time of the study by Rockwell et al. [24], previous studies elsewhere indicated drought sensitivity in arthropods [52,53] and fruits [54–56].

Studies on the Kirtland's Warbler were initiated by Wunderle and collaborators because the warbler was poorly known on its wintering grounds, despite recognition that wintering ground events could compromise breeding ground conservation efforts for this federally listed endangered species, and thus the need for wintering-ground studies was recognized [57]. Studies were initiated on the island of Eleuthera in the central Bahamas to characterize the KIWA's winter habitat [58] and to determine whether the warblers were susceptible to declines in food resources [59]. Wunderle and collaborators predicted that the KIWAs would be susceptible to food resource declines because of their use of drought-prone habitats on shallow soils on limestone substrates [58] in the dry season of October–April and especially during the driest period in March and April [60], just prior to vernal migration. Moreover, these late-winter droughts are not uncommon in the Caribbean and Middle America, where they have been found to affect migrant body condition [61], as also documented in Guánica [47]. Thus, it was predicted that rainfall or moisture conditions would influence food availability, which in turn would affect KIWA body condition, thereby setting the stage for carryover effects on reproductive success on the breeding grounds.

The study, conducted over four winters on Eleuthera [59], indicated that the warbler's food resources (fruit and arthropods) typically declined during the winter, but varied between winters and study sites. Rainfall was found to be an important driver of variation in fruit abundance, which was not surprising, given the high-water content (60%–70%) of fruits consumed by KIWAs. Despite variation in food availability, the proportion of fruits and arthropods in the KIWA diet (88% of 90 fecal samples contained both) varied little within or among winters, as expected for birds tracking food resources by moving from food-poor to food-rich sites. Supporting this resource-tracking hypothesis was the finding that when KIWAs shifted between study sites within a winter, they moved to sites with higher biomass of ripe fruit and ground arthropods, so that by late-winter, densities of the warbler were positively correlated with biomass of fruits and ground arthropods. Given changes in food abundance in space and time, the researchers expected that the KIWAs would not reside at one site for the entire winter, and indeed the warblers' overwinter site fidelity was low (an average of 43% of the KIWAs stayed on the same site from October to April), and this pattern varied with intensity of the late winter drought.

Evidence for intraspecific resource competition in the KIWAs mediated by dominance hierarchies was found in overwinter site fidelity, which differed by sex (males > females) and age class (adults > juveniles). Sex and age differences in corrected body mass (i.e., body mass scaled to body size) and fat scores were evident from midwinter through late winter, and consistent with outcomes from dominance and experience. Late-winter rains had a positive effect on corrected body mass, suggesting that in drought years the KIWAs might have inadequate body condition for early spring migration and thus arrive late on the breeding grounds, as expected from Rockwell et al.'s carryover effect studies [24].

The data on overwinter site fidelity and annual return of banded KIWAs (2003–2010) combined with the breeding ground site fidelity and annual return data (2006–2011) were used to estimate apparent annual, oversummer, overwinter, and migratory survival for the warbler [62]. These analyses were restricted to males, as sample sizes for females were inadequate for modeling the two mark-recapture data sets. The mean annual survival probability for male KIWAs was 0.58 ± 0.12 SE, a value consistent with annual survival estimates for other migrant warblers (reviewed in [14]). Monthly survival rates for the male KIWAs were relatively high during the stationary periods of the annual cycle (summer = 0.963 ± 0.01 SE; winter = 0.97 ± 0.01 SE) in contrast to the markedly lower monthly survival rates during the migratory period (0.886 ± 0.05 SE), which accounted for ~41% of all annual mortality of adult birds. Using a model selection framework, Rockwell et al. [62] also evaluated the influence of multiple climate variables on annual survival of the KIWA. Their analysis indicated that March rainfall in the Bahamas was the best-supported predictor of annual survival probability. Moreover, March rainfall was positively correlated with KIWA apparent annual survival in the subsequent year, indicating that the late winter rainfall effect carried over to influence an individual's survival probability in later stages of the annual cycle. Thus, the demonstration that March rainfall predicts annual survival of KIWAs corroborates theoretical and empirical evidence that migratory bird populations can be limited by weather on the wintering grounds.

Given the importance of wintering-ground rainfall to the KIWA's food resources, body condition, reproductive success, and annual survival, Wunderle et al. [59] recommended that conservation efforts for the KIWA in the Bahamas archipelago should focus on protecting the least drought-prone habitats. Habitats situated on sites with a shallow freshwater table could be especially important for providing "refugia" for the warblers during late winter droughts. These habitats have been characterized as anthropogenically-disturbed early successional sites (3–28 years post-disturbance) with an abundance of fruit [58,59].

3. Discussion and Conclusions

As studies on the Kirtland's Warblers illustrate, an increase in the length or severity of winter droughts in the Bahamas has the potential to lower the KIWA's annual survival and reduce its reproductive success, causing a two-fold negative impact on population dynamics. The sensitivity of the KIWA's population to winter droughts was evident from projection modeling, which indicated that a decrease in Bahamas March rainfall >12.4% from current mean levels could cause the size of the KIWA population to decrease due to winter droughts alone [62]. Further contributing to concerns about rainfall effects on KIWAs and other migrant birds in the Caribbean are predictions of increased occurrence of droughts under multiple climate change scenarios in the Caribbean [63]. Already, rainfall declines have been documented in Jamaica and recent studies there have found that the timing of rainfall within the dry season, not just the total amount, may be critical for setting migrant vernal departure schedules [64]. Declines in average annual rainfall have also been documented in the Bahamas in the period 1959–1990 [65]. More recently, other studies indicate that the frequency of drought and inter-annual variation in rainfall may also be increasing in the region [66,67]. As the climate changes in the Caribbean, wetter life zones may be replaced by drier life zones, as predicted for Puerto Rico based on analyses using model averaging of statistically downscaled general circulation models by IITF colleagues [68].

Increased variability in rainfall is consistent with expectations that global warming could be increasing the severity of El Niño Southern Oscillation (ENSO) events [69]. If this occurs, Sillett et al. [70] predict that the variance in demographic rates of migratory bird populations could be amplified, which could place small populations at risk of extinction. They demonstrated that demographic rates in Black-throated Blue Warblers (*Setophaga caerulescens*) on the Jamaica wintering grounds and north temperate breeding grounds varied with ENSO fluctuations. During El Niño years (dry in Jamaica), adult survival and fecundity were both lower, and during La Niña years (wet in Jamaica), both were higher. As fecundity increased, the recruitment of new individuals into winter and breeding populations also increased. The take-home message from Sillett et al. [70] is that migratory bird populations can be affected by changes in global climate patterns and that it is important for effective conservation interventions to understand how events occurring throughout a migrant's annual cycle interact to affect population size.

Not only are global climate cycles expected to change in the future, but the frequency of the most powerful tropical cyclones, or hurricanes, in the North Atlantic are expected to increase with global warming [71,72]. In the Caribbean, hurricanes occur with sufficient frequency to be an integral part of the natural disturbance regime [73], and evidence to date suggests that migrant populations that have already completed migration and are residing on their nonbreeding grounds are not strongly influenced by these storms. For example, Hurricane Georges struck the Guánica forest in September 1998 and caused extensive tree limb loss and reduced canopy cover, but the effect on winter resident apparent survival estimates was weak [17]. Of the three migrant species in Guánica with adequate sample size, estimated survival of the Black-and-white Warbler showed the strongest effect of the storm, as survival estimates for the year of the storm were lower than other years. This immediate post-hurricane decrease in Black-and-white Warblers is consistent with loss of their preferred foraging substrate, as demonstrated by Wunderle et al. [74] in Jamaican coffee plantations. It is likely that these local declines were not due to mortality (e.g., hurricanes struck in September before most migrant arrival), but more likely, these warblers shifted to other less-affected sites, a common post-hurricane response in many species [74–76]. Despite Hurricane George's effect on the Guánica forest, captures of winter residents peaked three years afterwards [3], a finding consistent with rapid post-hurricane recovery of wintering migrant populations elsewhere [77–79]. Not only do the effects of wintering-ground hurricanes appear to be relatively mild for most migrant populations, but hurricanes might be beneficial for producing disturbed habitats used by species such as the Kirtland's Warbler [80] or, more generally, be associated with new vegetation growth and thus a pulse of arthropod food resources.

We now recognize that the abundance and distribution of migrant birds are limited by events occurring throughout the annual cycle [16,81] and that events happening in one stage may carry over to influence other stages [20,24,62,70]. Without knowledge of the relative importance of various limiting factors during breeding, migratory, and wintering periods on population dynamics for each species of conservation concern, it will be impossible to implement effective management actions for each of these migrant species [16]. As suggested by Sillett and Holmes [82], it is premature to argue that events on the breeding or wintering grounds or during migration are limiting migrant populations until we understand when and where in the annual cycle various events or factors might limit populations and how these factors may cause carryover effects for multiple migrant species. It is this FAC approach, coupled with continued technological breakthroughs, that will facilitate future identification and mitigation of limiting factors contributing to migrant population declines, which may be exacerbated by global climate change.

As climate change continues we expect that there will be certain migrant species, perhaps some abundant, geographically widespread (on both breeding and nonbreeding grounds), and short-distance migrants, that can adapt to climate change. Adapting to environmental change, however, depends on several conditions including adequate genetic variation (de novo and standing variation), strength and spatial patterning of selection, and requires that the pace of environmental change does not exceed the maximum rate of evolutionary change [83,84]. Nevertheless, even with these stringent conditions for the evolution of adaptation, there are examples of adaptation by migrants, such as found in the geographically widespread European Blackcaps (*Sylvia atricapilla*). Approximately fifty years ago, blackcaps began to overwinter in Ireland and the United Kingdom, where plentiful food in bird feeders and milder winters enticed blackcaps to overwinter rather than migrating to their traditional Iberian Peninsula wintering grounds [85]. Despite now wintering in two distinct wintering quarters, blackcaps return annually to nest in Germany where they pair assortatively based on their wintering areas. The birds wintering further north have a selective advantage, as they produce larger clutches and fledge more young than those wintering south in the Iberian Peninsula. Bearhop et al. [85] noted that most of the behavioral changes in the blackcaps had a genetic basis, which when combined with assortative mating could lead to sympatric speciation. Whether the declines in recruitment of winter residents in Guánica represent adaptive responses to climate change, such as a potential shift to other wintering sites, is unknown and remains to be studied. However, this hypothesis seems unlikely, at least for the declines in American Redstarts wintering in Guánica, which correspond with declines on the breeding grounds [86]. In summary, we recognize that certain species have the potential to adapt to climate or anthropogenic habitat changes, but these are species that are least likely to need conservation intervention. Therefore, the focus of our research is on those migrant species most likely to require conservation intervention. Thus, as emphasized in our essay, these conservation efforts are most likely to be effective if based on research that uses a FAC approach for identification of the factor or factors that limit population growth.

Acknowledgments: We thank David N. Ewert, Ariel E. Lugo, Judith D. Toms, Frank H. Wadsworth, and two anonymous reviewers for constructive comments on the manuscript. Our research has been conducted in cooperation with the University of Puerto Rico.

Author Contributions: Wunderle wrote and revised the manuscript and Arendt added and removed text, provided editorial suggestions and formatted the references.

Conflicts of Interest: The authors declare no conflict of interest.

References

1. Greenberg, R.; Marra, P.P. (Eds.) *Birds of Two Worlds*; John Hopkins University Press: Baltimore, MD, USA, 2005.

2. Sauer, J.R.; Link, W.A. Analysis of the North American breeding bird survey using hierarchical models. *Auk* **2011**, *128*, 87–98. [CrossRef]

3. Faaborg, J.; Arendt, W.J.; Toms, J.D.; Dugger, K.M.; Cox, W.A.; Canals Mora, M. Long-term decline of a winter-resident bird community in Puerto Rico. *Biodivers. Conserv.* **2013**, *22*, 63–75. [CrossRef]

4. Askins, R.A.; Lynch, J.F.; Greenberg, R. Population declines in migratory birds in eastern North America. In *Current Ornithology*; Power, D.M., Ed.; Plenum Press: New York, NY, USA, 1990; pp. 1–57.

5. Wilcove, D.S.; Robinson, S.K. 1990. The impact of forest fragmentation on bird communities in eastern North America. In *Biogeography and Ecology of Forest Bird Communities in Eastern North America*; SPB Academic Publishing: Hague, The Netherlands, 1990; pp. 319–331.

6. Faaborg, J.; Arendt, W.J. Long-term declines in resident warblers in a Puerto Rican dry forest. *Am. Birds* **1989**, *43*, 1226–1230.

7. Faaborg, J.; Arendt, W.J. Long-term declines in winter resident warblers in a Puerto Rican dry forest. In *Ecology and Conservation of Neotropical Migrant Landbirds*; Hagan, J.M., III, Johnston, D.W., Eds.; Smithsonian Institution Press: Washington, DC, USA, 1992; pp. 57–63.

8. Arendt, W.J. Status of North American migrant landbirds in the Caribbean region: A summary. In *Ecology and Conservation of Neotropical Migrant Landbirds*; Hagan, J.M., III, Johnston, D.W., Eds.; Smithsonian Institution Press: Washington, DC, USA, 1992; pp. 143–171.

9. Gauthreaux, S.A., Jr. The use of weather radar to monitor long-term patterns of trans-Gulf migration in spring. In *Ecology and Conservation of Neotropical Migrant Landbirds*; Hagan, J.M., III, Johnston, D.W., Eds.; Smithsonian Institution Press: Washington, DC, USA, 1992; pp. 96–100.

10. Hussell, D.J.T.; Mather, M.H.; Sinclair, P.H. Trends in numbers of tropical- and temperate-winter migrant landbirds in migration at Long Point, Ontario, 1961–1988. In *Ecology and Conservation of Neotropical Migrant Landbirds*; Hagan, J.M., III, Johnston, D.W., Eds.; Smithsonian Institution Press: Washington, DC, USA, 1992; pp. 101–114.

11. Morse, D.H. *Population limitation: Breeding or wintering grounds? In Migrant Birds in the Neotropics*; Keast, A., Morton, E.S., Eds.; Smithsonian Institution Press: Washington, DC, USA, 1980; pp. 505–516.

12. Terborgh, J. *Where Have All the Migrants Gone?* Princeton University Press: Princeton, NJ, USA, 1989.

13. Martin, T.E.; Finch, D.M. (Eds.) *Ecology and Management of Neotropical Migratory Birds*; Oxford University Press: New York, NY, USA, 1995.

14. Faaborg, J.; Holmmes, R.T.; Anders, A.D.; Bildstein, K.L.; Dugger, K.M.; Gauthreaux, S.A., Jr.; Heglund, P.; Hobson, K.A.; Jahn, A.E.; Johnson, D.H.; et al. Conserving migratory land birds in the New World: Do we know enough? *Ecol. Appl.* **2010**, *20*, 398–418. [CrossRef] [PubMed]

15. Robbins, C.S.; Sauer, J.R.; Greenberg, R.S.; Droege, S. Population declines in North American birds that migrate to the neotropics. *Proc. Natl. Acad. Sci. USA* **1989**, *86*, 7658–7662. [CrossRef] [PubMed]

16. Sherry, T.W.; Holmes, R.T. Summer versus winter limitation of populations: What are the issues and what is the evidence? In *Ecology and Management of Neotropical Migratory Birds*; Martin, T.E., Finch, D.M., Eds.; Oxford University Press: New York, NY, USA, 1995; pp. 85–120.

17. Dugger, K.M.; Faaborg, J.; Arendt, W.J.; Hobson, K.A. Understanding survival and abundance of overwintering warblers: Does rainfall matter? *Condor* **2004**, *106*, 744–760. [CrossRef]

18. Smith, J.A.M.; Reitsma, L.R.; Marra, P.P. Moisture as a determinant of habitat quality for a non-breeding Neotropical migratory songbird. *Ecology* **2010**, *91*, 2874–2882. [CrossRef] [PubMed]

19. Stutchbury, B.J.M.; Tarof, S.A.; Done, T.; Grow, E.; Kramer, P.M.; Tautin, J.; Fox, J.W.; Afanasyev, V. Tracking long-distance songbird migration by using geolocators. *Science* **2009**, *323*, 896. [CrossRef] [PubMed]

20. Marra, P.P.; Hobson, K.A.; Holmes, R.T. Linking winter and summer events in a migratory bird by using stable-carbon isotopes. *Science* **1998**, *282*, 1884–1886. [CrossRef] [PubMed]

21. Gill, J.A.; Sutherland, J.W.; Norris, K. Depletion models can predict shorebird distribution at different spatial scales. *Proc. R. Soc. Lond. B* **2001**, *268*, 369–376. [CrossRef] [PubMed]

22. Bearhop, S.; Hilton, G.M.; Votier, S.C.; Waldron, S. Stable isotope ratios indicate that body condition in migrating passerines is influenced by winter habitat. *Proc. R. Soc. Lond. B* **2004**, *271*, S215–S218. [CrossRef] [PubMed]

23. Reudink, M.W.; Studds, C.E.; Marra, P.P.; Kurt Kyser, T.; Ratcliffe, L.M. Plumage brightness predicts non-breeding season territory quality in a long-distance migratory songbird, the American Redstart, *Setophaga ruticilla*. *J. Avian Biol.* **2009**, *40*, 34–41. [CrossRef]

24. Rockwell, S.M.; Bocetti, C.I.; Marra, P.P. Winter rainfall, spring arrival dates and reproductive success in the endangered Kirtland's Warbler (*Setophaga kirtlandii*). *Auk* **2012**, *129*, 744–752. [CrossRef]

25. Latta, S.C.; Cabezas, S.; Mejia, D.A.; Paulino, M.M.; Almonte, H.; Butterworth, C.M.; Bortolotti, G.R. Carry-over effects provide linkages across the annual cycle of a Neotropical bird, the Louisiana Waterthrush *Parkesia motacilla*. *Ibis* **2016**, *158*, 395–406. [CrossRef]

26. Wunderle, J.M., Jr.; Waide, R.B. Distribution of overwintering Nearctic Migrants in The Bahamas and Greater Antilles. *Condor* **1993**, *95*, 904–933. [CrossRef]

27. Terborgh, J.W.; Faaborg, J.R. Factors affecting the distribution and abundance of North American migrants in the Eastern Caribbean region. In *Migrant Birds in the Neotropics*; Keast, A., Morton, E.S., Eds.; Smithsonian Institution Press: Washington, DC, USA, 1980; pp. 145–156.

28. Faaborg, J.; Terborgh, J.W. Patterns of migration in the West Indies. In *Migrant Birds in the Neotropics*; Keast, A., Morton, E.S., Eds.; Smithsonian Institution Press: Washington, DC, USA, 1980; pp. 157–163.

29. Hamel, P.B. *Bachman's Warbler: A Species in Peril*; Smithsonian Institution Press: Washington, DC, USA, 1986.

30. Remsen, J.V. Was Bachman's Warbler a bamboo specialist? *Auk* **1986**, *103*, 216–219.

31. Wunderle, J.M., Jr.; Waide, R.B. Future prospects for Nearctic migrants wintering in Caribbean forests. *Bird Conserv. Int.* **1994**, *4*, 191–207. [CrossRef]

32. Myers, N.; Mittermeier, R.A.; Mittermeier, C.G.; da Fonesca, G.A.B.; Kent, J. Biodiversity hotspots for conservation priorities. *Nature* **2000**, *403*, 853–858. [CrossRef] [PubMed]

33. Mittermeier, R.A.; Turner, W.R.; Larsen, F.W.; Brooks, T.M.; Gascome, C. Global biodiversity conservation: The critical role of hotspots. In *Biodiversity Hotspots*; Zachos, F.E., Habel, J.C., Eds.; Springer: Berlin, Germany, 2011; pp. 3–22.

34. Birdsey, R.A.; Weaver, P.L. *Forest Area Trends in Puerto Rico*; USDA Forest Service, Southern Forest Experiment Station: Ashville, NC, USA, 1987.

35. Helmer, E.H.; Ramos, O.; Del, M.; López, T.; Díaz, W. Mapping the forest type and land cover of Puerto Rico, a component of the Caribbean biodiversity hotspot. *Caribb. J. Sci.* **2002**, *38*, 165–183.

36. Brandeis, T.J.; Helmer, E.H.; Oswalt, S.N. *Puerto Rico's Forests*; Department of Agriculture, Forest Service, Southern Research Station: Asheville, NC, USA, 2007; p. 72.

37. Helmer, E.H.; Kennaway, T.A.; Pedreros, D.H.; Clark, M.L.; Marcano-Vega, H.; Tieszen, L.; Ruzycki, T.R.; Schill, S.R.; Carrington, C.M. Land cover and forest formation distributions for St. Kitts, Nevis, St. Eustatius, Grenada, and Barbados from decision tree classification of cloud-cleared satellite imagery. *Caribb. J. Sci.* **2008**, *44*, 175–198. [CrossRef]

38. Helmer, E.H. Forest conservation and land development in Puerto Rico. *Landsc. Ecol.* **2004**, *19*, 29–40. [CrossRef]

39. Lugo, A.E.; Helmer, E. Emerging forests on abandoned lands: Puerto Rico's new forests. *For. Ecol. Manag.* **2004**, *190*, 145–161. [CrossRef]

40. Lugo, A.E. The outcome of alien tree invasions in Puerto Rico. *Front. Ecol. Evolut.* **2004**, *2*, 265–273. [CrossRef]

41. Beltrán, W.; Wunderle, J.M., Jr. Determinants of tree species preference for foraging by insectivorous birds in novel *Prosopis-Leucaena* woodland in Puerto Rico: The role of foliage palatability. *Biodivers. Conserv.* **2013**, *22*, 2071–2089. [CrossRef]

42. Wunderle, J.M., Jr.; Latta, S.C. Avian abundance in sun and shade coffee plantations and remnant pine forest in the Cordillera Central, Dominican Republic. *Ornitol. Neotrop.* **1996**, *7*, 19–34.

43. Wunderle, J.M., Jr.; Latta, S.C. Winter site fidelity of Nearctic migrants in shade coffee plantations of different sizes in the Dominican Republic. *Auk* **2000**, *117*, 596–614. [CrossRef]

44. Faaborg, J.; Arendt, W.J. Population sizes and philopatry of winter resident warblers in Puerto Rico. *J. Field Ornithol.* **1984**, *55*, 376–378.

45. Faaborg, J.; Arendt, W.J.; Kaiser, M.S. Rainfall correlates of bird population fluctuations in a Puerto Rican dry forest: a nine year study. *Wilson Bull.* **1984**, *96*, 575–593.

46. Dugger, K.M.; Faaborg, J.; Arendt, W.J. Rainfall correlates of bird populations and survival rates in a Puerto Rican dry forest. *Bird Popul.* **2000**, *5*, 11–27.

47. Toms, J.D. Non-Breeding Competition between Migrant American Redstarts (*Setophaga ruticilla*) and Resident Adelaide's Warblers (*Dendroica adelaidae*) in the Guánica Biosphere Reserve, Southwest Puerto Rico. Ph.D. Thesis, University of Missouri, Columbia, MO, USA, 2011.

48. Bocetti, C.I.; Donner, D.M.; Mayfield, H.F. Kirtland's Warbler (*Setophaga kirtlandii*). In *The Birds of North America*; Rodewald, P.G., Ed.; Cornell Lab of Ornithology: Ithaca, NY, USA, 2014.

49. Richard, T. Confirmed occurrence and nesting of the Kirtland's Warbler at CFB Petawawa, Ontario: A first for Canada. *Ont. Birds* **2008**, *26*, 2–15.

50. Trick, J.A.; Greveles, K.; DiTommaso, D.; Robaidek, J. The first Wisconsin nesting record of Kirtland's Warbler (*Dendroica kirtlandii*). *Passeng. Pigeon.* **2008**, *70*, 93–102.

51. Ryel, L.A. Population change in the Kirtland's Warbler. *Jack-Pine Warbler* **1981**, *59*, 76–91.

52. Strong, A.M.; Sherry, T.W. Habitat-specific effects of food abundance on the condition of Ovenbirds wintering in Jamaica. *J. Anim. Ecol.* **2000**, *69*, 883–895. [CrossRef]

53. Johnson, M.D.; Sherry, T.W. Effects of food availability on the distribution of migratory warblers among habitats in Jamaica. *J. Anim. Ecol.* **2001**, *70*, 546–560. [CrossRef]

54. Griz, L.M.S.; Machado, I.C.S. Fruiting phenology and seed dispersal syndromes in caatinga, a tropical dry forest in the northeast of Brazil. *J. Trop. Ecol.* **2001**, *17*, 303–321. [CrossRef]

55. Ramírez, N. Reproductive phenology, life-forms, and habitats of the Venezuelan Central Plain. *Am. J. Bot.* **2002**, *89*, 836–842. [CrossRef] [PubMed]

56. Redwine, J.R.; Sawicki, R.J.; Lorenz, J.J.; Hoffman, W. Ripe fruit availability in the fragmented hardwood forests of the northern Florida Keys. *Nat. Areas J.* **2007**, *27*, 8–15. [CrossRef]

57. United States Fish and Wildlife Service. *Kirtland's Warbler Recovery Plan*, revised ed.US Fish and Wildlife Service: Washington, DC, USA, 1985.

58. Wunderle, J.M., Jr.; Currie, D.; Helmer, E.H.; Ewert, D.N.; White, J.D.; Ruzycki, T.S.; Parresol, B.; Kwit, C. Kirtland's Warblers in anthropogenically disturbed early-successional habitats on Eleuthera, The Bahamas. *Condor* **2010**, *112*, 123–137. [CrossRef]

59. Wunderle, J.M., Jr.; Lebow, P.K.; White, J.D.; Currie, D.; Ewert, D.N. Sex and age difference in site fidelity, food resource tracking, and body condition of wintering Kirtland's Warblers (*Setophaga kirtlandii*) in The Bahamas. *Ornithol. Monogr.* **2014**, *80*, 1–62.

60. Sealey, N.E. *Bahamian Landscapes*, 3rd ed.; Macmillan Education: Oxford, UK, 2006.

61. Sherry, T.W.; Johnson, M.D.; Strong, A.M. *Does winter food limit populations of migratory birds? Birds of Two Worlds*; John Hopkins University Press: Baltimore, MD, USA, 2005; pp. 414–425.

62. Rockwell, S.M.; Wunderle, J.M., Jr.; Sillett, T.S.; Bocetti, C.I.; Ewert, D.N.; Currie, D.; White, J.D.; Marra, P.P. Seasonal survival estimation for a long-distance migratory bird and the influence of winter precipitation. *Oecologia* 2016. [CrossRef] [PubMed]

63. Neelin, J.D.; Münnich, M.; Su, H.; Meyerson, J.E.; Holloway, C.E. Tropical drying trends in global warming models. *Proc. Natl. Acad. Sci. USA* **2006**, *103*, 6110–6115. [CrossRef] [PubMed]

64. Studds, C.E.; Marra, P.P. Linking fluctuations in rainfall to nonbreeding season performance in a long-distance migratory bird, *Setophaga ruticilla*. *Clim. Res.* **2007**, *35*, 115–122. [CrossRef]

65. Martin, H.C.; Weech, P.S. Climate change in the Bahamas? Evidence in the meteorological records. *Bahamas J. Sci.* **2001**, *5*, 22–32.

66. Van der Molen, M.K.; Vugts, H.F.; Bruijnzeel, L.A.; Scatena, F.N.; Pielke Sr, R.A.; Kroon, L.J.M. Meso-scale climate change due to lowland deforestation in the maritime tropics. In *Tropical Montane Cloud Forests: Science for Conservation and Management*; Bruijnzeel, L.A., Scatena, F.N., Hamilton, L.S., Eds.; Cambridge University Press: Cambridge, UK, 2010; pp. 527–537.

67. Comarazamy, D.E.; González, J.E. Regional long-term climate change (1950–2000) in the midtropical Atlantic and its impacts on the hydrological cycle of Puerto Rico. *J. Geophys. Res.* **2011**, *116*, D00Q05. [CrossRef]

68. Khalyani, A.H.; Gould, W.A.; Harmsen, E.; Terando, A.; Quinones, M.; Collazo, J.A. Climate change implications for tropical islands: Interpolating and interpreting statistically downscaled GCM projects for management and planning. *J. Appl. Meteorol. Climatol.* **2016**, *55*, 265–282. [CrossRef]

69. Kerr, R.A. Big El Niños ride the back of slower climate change. *Science* **1999**, *283*, 1108–1109. [CrossRef]

70. Sillett, T.S.; Holmes, R.T.; Sherry, T.W. The El Niño Southern Oscillation impacts population dynamics of a migratory songbird throughout its annual cycle. *Science* **2000**, *288*, 2040–2042. [CrossRef] [PubMed]

71. Emanuel, K.A. Increasing destructiveness of tropical cyclones over the past 30 years. *Nature* **2005**, *436*, 686–688. [CrossRef] [PubMed]

72. Knutson, T.R.; McBride, J.L.; Chan, J.; Emanuel, K.; Holland, G.; Landsea, C.; Held, I.; Kossin, J.P.; Srivastava, A.K.; Sugi, M. Tropical cyclones and climate change. *Nat. Geosci.* **2010**, *3*, 157–163. [CrossRef]

73. Walker, L.R.; Brokaw, N.V.L.; Lodge, D.J.; Waide, R.B. (Eds.) Ecosystem, plant and animal response to hurricanes in the Caribbean. *Biotropica* **1991**, *23*, 13–521.

74. Wunderle, J.M., Jr.; Lodge, D.J.; Waide, R.B. Short-term effects of Hurricane Gilbert on terrestrial bird populations on Jamaica. *Auk* **1992**, *109*, 148–166. [CrossRef]

75. Arendt, W.J. *Adaptations of an Avian Supertramp: Distribution, Ecology, and Life History of the Pearly-Eyed Thrasher (Margarops fuscatus)*; General Technical Report, IITF-27; United States Forest Service: Washington, DC, USA, 2006.

76. Wiley, J.W.; Wunderle, J.M., Jr. The effects of hurricanes on birds, with special reference to Caribbean islands. *Bird Conserv. Int.* **1993**, *3*, 319–349. [CrossRef]

77. Holmes, R.T.; Sherry, T.W.; Reitsma, L. Population structure, territoriality and overwinter survival of two migrant warbler species in Jamaica. *Condor* **1989**, *91*, 545–561. [CrossRef]

78. Wunderle, J.M., Jr. Response of bird populations in a Puerto Rican forest to Hurricane Hugo: The first 18 months. *Condor* **1995**, *97*, 879–896. [CrossRef]

79. Wunderle, J.M., Jr. Population characteristics of Black-throated Blue Warblers wintering in three sites on Puerto Rico. *Auk* **1995**, *112*, 931–946. [CrossRef]

80. Wunderle, J.M., Jr.; Currie, D.; Ewert, D.N. The potential role of hurricanes in the creation and maintenance of Kirtland's Warbler habitat in the Bahamas Archipelago. In Proceedings of the 11th Symposium on the Natural History of the Bahamas. Gerace Research Center, San Salvador, Bahamas, 23–27 June 2005; Rathcke, B.J., Hayes, W.K., Eds.; Gerace Research Center, Ltd.: San Salvador, Bahamas, 2007; pp. 121–129.

81. Latta, S.C.; Baltz, M.E. Population limitation in Neotropical migratory birds: comments. *Auk* **1997**, *114*, 754–762. [CrossRef]

82. Sillett, T.S.; Holmes, R.T. Variation in survivorship of a migratory songbird throughout its annual cycle. *J. Anim. Ecol.* **2002**, *71*, 296–308. [CrossRef]

83. Hendry, A.P.; Farrugia, T.J.; Kinnison, M. Human influences on rates of phenotypic change in wild animal populations. *Mol. Ecol.* **2008**, *17*, 20–29. [CrossRef] [PubMed]

84. Reid, N.H.; Proestou, D.A.; Clark, B.W.; Warren, W.C.; Colbourne, J.K.; Shaw, J.R.; Karchner, S.I.; Hahn, M.E.; Nacci, D.; Oleksiak, M.F.; et al. The genomic landscape of rapid repeated evolutionary adaptation to toxic pollution in wild fish. *Science* **2016**, *354*, 1305–1308. [CrossRef] [PubMed]

85. Bearhop, S.; Fiedler, W.; Furness, R.W.; Votier, S.C.; Waldron, S.; Newton, J.; Bowen, G.J.; Berthold, P.; Farnsworth, K. Assortative mating as a mechanism for rapid evolution of a migratory divide. *Science* **2005**, *310*, 502–504. [CrossRef] [PubMed]

86. Wilson, S.; Ladeau, S.L.; Tøttrup, A.P.; Marra, P.P. Range-wide effects of breeding- and nonbreeding-season climate on the abundance of a neotropical migrant songbird. *Ecology* **2011**, *92*, 1789–1798. [CrossRef] [PubMed]

forests

MDPI

Review

On the Shoulders of Giants: Continuing the Legacy of Large-Scale Ecosystem Manipulation Experiments in Puerto Rico

Tana E. Wood [1,*], Grizelle González [1], Whendee L. Silver [2], Sasha C. Reed [3] and Molly A. Cavaleri [4]

[1] United States Department of Agriculture, Forest Service, International Institute of Tropical Forestry, Jardín Botánico Sur, 1201 Ceiba St.-Río Piedras, San Juan, PR 00926-1119, USA; ggonzalez@fs.fed.us
[2] Department of Environmental Science, Policy, and Management, 130 Mulford Hall #3114, University of California, Berkeley, CA 94720, USA; wsilver@berkeley.edu
[3] U.S. Geological Survey, Southwest Biological Science Center, Moab, UT 84532, USA; screed@usgs.gov
[4] School of Forest Resources and Environmental Science, Michigan Technological University, Houghton, MI 49931, USA; macavale@mtu.edu
* Correspondence: tanawood@fs.fed.us; Tel.: +1-787-764-7935

Received: 3 December 2018; Accepted: 17 February 2019; Published: 27 February 2019

Abstract: There is a long history of experimental research in the Luquillo Experimental Forest in Puerto Rico. These experiments have addressed questions about biotic thresholds, assessed why communities vary along natural gradients, and have explored forest responses to a range of both anthropogenic and non-anthropogenic disturbances. Combined, these studies cover many of the major disturbances that affect tropical forests around the world and span a wide range of topics, including the effects of forest thinning, ionizing radiation, hurricane disturbance, nitrogen deposition, drought, and global warming. These invaluable studies have greatly enhanced our understanding of tropical forest function under different disturbance regimes and informed the development of management strategies. Here we summarize the major field experiments that have occurred within the Luquillo Experimental Forest. Taken together, results from the major experiments conducted in the Luquillo Experimental Forest demonstrate a high resilience of Puerto Rico's tropical forests to a variety of stressors.

Keywords: Luquillo Experimental Forest; tropical; experiments; manipulations; large-scale; Puerto Rico; Caribbean

1. Introduction

The Luquillo Experimental Forest, located in the northeastern corner of Puerto Rico (18° N, 66° W), is one of the oldest reserves in the Western Hemisphere [1] (Figure 1). In 1876, King Alphonso XII proclaimed 10,000 hectares within the Luquillo Mountains a reserve of the Spanish Crown. Just 22 years later, Spain ceded control of the forest to the United States as part of the Treaty of Paris in 1898, and in 1903, President Theodore Roosevelt designated the area a forest reserve under the jurisdiction of what is now the U.S. Department of Agriculture (USDA) Forest Service [1]. The broad diversity of climate, geology, and flora and fauna found within the Luquillo Experimental Forest has attracted a wide range of ecological research throughout its history [1–4]. Within an area of what is now just over 11,000 ha, the elevation spans 100 to 1075 m asl. Across this change in elevation, mean annual rainfall spans 2450 to 4000 mm and average monthly air temperature ranges between 23.5 and 27 °C at lower elevations and 17 and 20 °C at the higher elevations [5]. There are two major bedrocks (marine volcaniclastic and intrusive igneous rocks from the Cretaceous and Tertiary periods), and three major soil orders (Ultisol, Inceptisol, Oxisol) within the Luquillo Experimental Forest [4]. The forest is also highly diverse with 164 animal species (24 endemic), more than 1000 plant species, and 224 tree species (60 endemic) [4]. In light of this range of conditions, it is no surprise that there are five Holdrige life zones found within the Luquillo Experimental Forest: wet forest, rain forest, lower montane wet forest, lower montane rain forest, and a small area in the southwest region that is moist forest life zone [2].

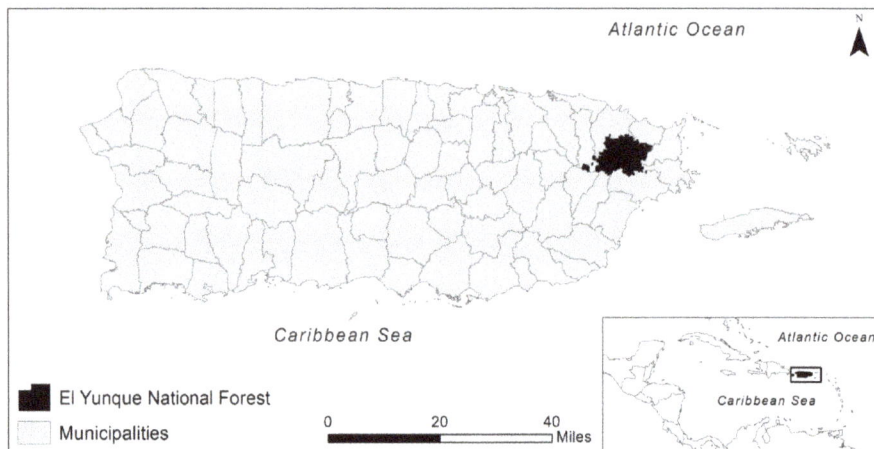

Figure 1. Map of Puerto Rico and the location of the Luquillo Experimental Forest.

Long-term monitoring within the Luquillo Experimental Forest includes assessment of factors such as climate, forest composition, and biomass, and these observations have been a part of the Luquillo Experimental Forest since the early 1900's [1,2]. This long-term monitoring has been critical to establishing fundamental knowledge of the forest system, allowing us to understand basic seasonal patterns, capture the ebbs and flow of forest composition, and to explore the potential for slower, directional changes that might otherwise be missed. However, society often demands answers on timescales much faster than long-term research can provide. While manipulative experiments have numerous challenges [6,7], they nevertheless enable researchers to isolate the effects of changes to individual variables so that we can test mechanistic hypotheses on a more feasible timescale and thus better anticipate the needs, follies, wants, and dreams of man, that are likely to affect the environments of the planet [8]. The manipulative experimental approach can additionally be used to reveal important insights about system responses to extreme, infrequent, or abrupt events [9], thus providing essential insights for societies that must develop coping strategies and management plans for such events.

Throughout its history, the use of and experiments conducted within the Luquillo Experimental Forest have been a reflection of the times. From the early foresters that explored forest management techniques in the 1900's [1] to the more recent large-scale field manipulation experiments [10], these studies provide a window into the greatest interests and concerns of society. The objective of this article is to synthesize the history of field manipulation experiments within the Luquillo Experimental Forest, beginning with the early days of U.S. jurisdiction of the forest. We aim to place these experiments into historical context and to evaluate their contribution to our greater understanding of tropical forest ecology. To understand where we are and where we are going, it is important to take time to reflect on where we have been. As Isaac Newton so aptly said, "If I have seen further, it is by standing on the shoulders of giants [11]".

In the early days of the Luquillo Experimental Forest, the U.S. was keenly interested in maximizing the ability to reforest degraded landscapes and to cultivate and manage forests. In response to this need, the USDA Forest Service began silvicultural experiments on the island of Puerto Rico. Tree plantations were established throughout the Luquillo Experimental Forest from the early 1930's to the late 1950's, and studies of secondary forest succession and forest management practices were implemented [12–14].

Thus began the methodical assessment of the commercial value of timber and the experimentation of forest management practices, such as forest thinning, to promote forest growth. In the 1960's, the development of nuclear technology was well under way and the need for a greater understanding of how radiation exposure might affect our nation's biological resources was recognized [15]. At the same time, scientists were interested in our ability to quantify whole-forest metabolism. With these motivating factors in mind, scientists developed studies of whole-forest energetics [16], evaluated the effects of herbicides [17–19], and implemented one of the three gamma radiation studies conducted in forests in the United States [20]. In 1989 Hurricane Hugo passed over the island of Puerto Rico [21], followed by Hurricane Georges in 1998 [22], which spurred a wealth of research surrounding the effects of hurricanes on forest recovery [22–25]. This included experiments that simulated key aspects of hurricane disturbance, such as debris deposition, at smaller scales [26,27]. In the 21st Century, anthropogenic change and changes to climate emerged as some of the most important research needs of our time. This growing interest in understanding the effects of both long-term directional change and repeated, cyclical disturbance on forest recovery led to the establishment of a wealth of experimental studies, including a clear cutting experiment [28,29], a nitrogen (N) deposition experiment [30], rainfall manipulation studies [31,32], and a forest warming experiment [10]. While there remains much to learn, the Luquillo Experimental Forest represents one of the longest and best-studied tropical forests in the world. Together, the history and wealth of experimental research in the Luquillo Experimental Forest has formulated a greater understanding of tropical forests and their resilience to environmental change, the depth of which has literally filled volumes of books [2,20,33–37]. Below, we highlight the varied and impressive history of manipulation experiments that have taken place within the Luquillo Experimental Forest and provide a general summary of the major conclusions and contributions of each of these studies and their contribution to current understanding.

2. Field Manipulation Experiments in the Luquillo Experimental Forest

2.1. Silviculture (1930s–Today)

By the late 1920's Puerto Rico was largely deforested [38], with descriptions of a landscape that supported "eroded soils, denuded forest areas, sedimented rivers and reservoirs, reduced soil fertility, low crop yields, and an inadequately fed people" [39]. However, following two major hurricanes in 1928 and 1932 combined with a series of economic setbacks, Puerto Rico saw a massive abandonment of agricultural lands as people moved to the cities for work [38]. As a result, much of the island in Puerto Rico entered into early stages of forest succession by the late 1930's [38]. During this period (1933 and 1949) the USDA Forest Service added large amounts of land in the areas surrounding the Luquillo Experimental Forest, in what amounts to approximately 50% of the Luquillo Experimental Forest today [1]. As such, one of the early objectives of the USDA Forest Service was to develop reforestation programs, promote tree growth and cultivate plantation forestry in the tropics. From the early 1930's to the late 1940's more than 1500 hectares of tree plantations were established across the island [12]. These initial tree plantings included mahogany (Swietenia macrophylla) plantations that still grace the Luquillo Experimental Forest today [1]. In 1942, Frank H. Wadsworth, who was to become the Director of the now USDA Forest Service International Institute of Tropical Forestry from 1953 to 1979, arrived in Puerto Rico. In the coming years, Wadsworth would pioneer tropical forestry management and conservation on the island. In 1943 he began establishing a series of 420 0.1 ha plots across the Luquillo Experimental Forest, which ranged from 200 m to 640 m elevation. Plots along this elevation gradient were created in preparation for the development of a land management plan and to satisfy his interest in understanding how trees grow in complex forests [40]. All trees with diameter at breast height of greater than 9.1 cm were tagged, measured, and identified to species. These plots have been re-measured at various points in time from 1947 to present, providing a baseline for understanding the growth and productivity of thousands of trees within the Luquillo Experimental

Forest over a 75-year period [41–47], and a critical foundation against which experimental results have be compared [15,40,48,49].

Around this time, the United States was entering a post-war energy crisis. In response, Wadsworth established new plots in 1945 to evaluate and develop timber management practices. In a portion of these plots, 50% of basal area was removed, which showed that thinning does indeed increase the growth of the remaining trees [3]. Wadsworth later (1957) established approximately 100 0.08 hectare plots across two lower elevation forests, which included a secondary forest and an old growth forest site. In these plots, the commercially valuable trees were marked and measured while the poorest growing trees were removed as part of a pilot management project designed to increase productivity by providing greater canopy freedom [42]. In 1975, 40 of these plots were re-measured. Surprisingly, very little difference between the mean growth rates in these plots versus the old growth forest stands was observed [42]. However, closer analyses of growth data from the various plots established by Wadsworth have revealed the importance of canopy crown diameter and position for determining growth potential [42,45]. The long-term experiment yielded a wide range of plant species that varied within and across plots, and thus the sites were subsequently used to explore the role of biodiversity in ecosystem processes [42,45,50,51]. The known age and composition of the plots, together with the fact that many were established on abandoned pasture land, facilitated the study of long-term carbon dynamics. The sites were resurveyed in 1992, and a subset re-measured in the early 2000's to determine biomass change and soil carbon dynamics with reforestation. Stable carbon isotopes were used to track the soil carbon change over time, giving one of the first estimates of soil carbon gain and loss with tropical reforestation [41]. The initial studies were revolutionary. They represent the first silviculture studies conducted anywhere in Latin America, and are an invaluable resource for long-term research. Many of the tree plantations and experimental plots established by Wadsworth and the International Institute of Tropical Forestry are still present in Puerto Rico today. They serve as the oldest established research plots for evaluating tree growth and production in the forest, providing a baseline for understanding contemporary responses of the Luquillo Experimental Forest to long-term directional changes in climate, and in cyclical disturbance events, such as hurricanes [2,40,43,44,46,48].

2.2. Forest Radiation Experiment (1963–1968)

In the 1960's, nuclear technology was expanding beyond its use for military purposes to include domestic applications such as nuclear energy and excavation of large areas. In particular, the U.S. was considering using a nuclear device for excavating a second Panama Canal between the Pacific and Atlantic Oceans [15]. Earlier radiation experiments conducted in temperate forests in Brookhaven, New York and Dawsonville, Georgia [52] revealed high mortality of pine trees, alleviating the perception that plants were likely resistant to radiation [20]. It thus became apparent that assessments of the consequences of atomic evacuation, nuclear war, and atomic accidents should include effects on natural ecosystems as well as the effects on humans. Scientists recognized that responses to radiation might vary by ecosystem, particularly in lower latitudes, which support highly diverse forests with complex structure [15]. Thus, the Atomic Energy Commission (now the U.S. Department of Energy) funded a 5-year study in 1963 to investigate the effects of radiation on forest processes in the Luquillo Experimental Forest. The prevailing theory was that more complex organisms with larger nuclei would be more sensitive to radiation than less complex organisms with smaller nuclei due to the greater volume of deoxyribonucleic acid (DNA) in large nuclei [53]. It was further hypothesized that the rapid cycling in tropical forests would contribute to a faster response to radiation than what had been observed in temperate systems, and that species diversity and forest complexity were likely to be important factors [20]. To test this theory, a small area within the Luquillo Experimental Forest was exposed to almost continuous gamma radiation (10,000-curie Cesium-137 source; Figure 2) for 24 h a day over a 3-month period (19 January to 27 April, 1965) and was studied for a full year following exposure. Scientists measured the effects of radiation (stress) on plants, seedlings, and seed germination rates, animals, soils, microbes, mineral cycling, forest metabolism and energy flows.

Figure 2. Scientists installing the base and casing for the Cesium radiation source in the Luquillo Experimental Forest. Photo reproduced from [53].

The effects of radiation on the vegetation were not clearly discernable in the initial months following radiation exposure, and in fact, some scientists thought that the canopy leaves in the immediate vicinity of the radiation center might be greener than those nearby. Howard T. Odum, the lead scientist on the project, noted that radiation effects in these initial days were difficult to observe from aerial views, color photographs, and from a nearby mountain slope, despite everyone on the project making daily trips to see the change [53]. However, within a few months, areas within the line-of-site of the radiation source were obviously affected. The ground surrounding the source was bare of living green, moss on the nearby rocks turned a strange blue-black color, many of the surrounding trees began dropping green leaves, and there were observations of albino, chlorotic, and abnormally shaped leaves [53–55]. Scientists additionally noted that the leaves of a *Manilkara* tree close to the radiation center turned bright red before dropping. Radiation exposure also increased susceptibility to insect grazing in the initial months following radiation exposure. The number of leaf holes in the surrounding vegetation increased and bark beetles invaded the most heavily affected trees [53]. Within 7 months of radiation exposure, mortality of the most heavily affected individuals stabilized at less than 10% [55]. However, as the year progressed the canopy continued to open, and by the following year a small gap had opened up in the canopy. Ultimately the majority of trees within the 30-m radius of the radiation source died [28,54]. One of the most dominant plants within the Luquillo Experimental Forest, the sierra palm (*Prestoea montana*), was particularly sensitive to radiation with a 94% decline in population density in the areas surrounding the radiation center [29]. However, some trees were also resistant to radiation. A giant red-trunked *Cyrilla* tree 5-m north of the radiation source was exposed to 100,000 Roentgen of radiation [53]. It was permanently scarred on one side and lost more than half of its leaves; however, the tree survived another 33 years before it died during a major rain storm in 1998 [56].

A year after exposure to radiation, the forest reached the maximum extent of defoliation. It then took over a year for the spread of green cover to return [53]. Recovery of vegetation was found to be more rapid in areas that had been shielded from radiation by rocks and tree trunks [53]. Seedling growth eventually spread throughout the recovery center [57]. Secondary forest species were more resilient than old growth species, demonstrating overall greater resistance to radiation exposure [55]. When considering the multitude of observations following radiation exposure, it became clear that the initial hypothesis that radiation sensitivity would increase with nuclear volume did not hold. For example, the sierra palm (*Prestoea montana*) was much more sensitive to radiation exposure (smaller nuclei) than secondary forest species, which as a group had larger nuclei [53,55]. Forest structure was also considered important when evaluating radiation sensitivity, with more complex forests hypothesized to have greater sensitivity [52,55]. However, the radiation sensitivity of the Luquillo Experimental Forest, which is considered a more complex forest, relative to the response of temperate forests also exposed to radiation, found no particular difference in the response of the two forest types. Rather, they hypothesized that resistance to radiation of some species over others may be due to the simplicity of the physiology rather than the size of the nuclei, which could explain the resistance of the secondary species in the Luquillo Experimental Forest as well as the resistance of the herbaceous vegetation observed in the temperate forests [55].

Twenty-three years later, scientists revisited the radiation site and found slow recovery of the radiated forest relative to natural gaps, as well as an overall loss of seeds in the soil [28]. Scientists have additionally found that the rate of biomass recovery following radiation was much slower than that of biomass recovery following hurricane disturbance and that the depressed recovery has continued to persist in the radiated site as long as 44 years following irradiation [40]. Whether this stunted recovery following radiation exposure is microbially controlled, or due to substantial loss of the seed bank is not clear. While the implementation of the radiation study in the Luquillo Experimental Forest is controversial to this day, at the time it was the most comprehensive study of a tropical forest ever conducted, making the Luquillo Experimental Forest one of the earliest and best-studied tropical forests in the world. Odum used the radiation experiment as an opportunity to study everything he could about the forest [20]. He pioneered studies on ecosystem function and the forest's connections to global cycles of energy, water, nutrients, and carbon, and the resulting book volume compiled by Odum [20], serves as a central reference for scientists studying the Luquillo Experimental Forest to this day.

2.3. Rain Forest Herbicide Experiment (1964–1965)

In the mid-1960's U.S. involvement in the Vietnam War was escalating. Deployment of troops increased from 760 in 1959 to 23,300 in 1964, and reached over 500,000 troops by 1968 (U.S. Department of Defense Manpower Data Center). As a military tactic, the US began using the "Rainbow" herbicides (herbicide mixtures named by the colored identification band painted on their storage barrels), such as Agent Orange, in 1961 to defoliate forests and mangroves, to clear perimeters of military installations and to destroy 'unfriendly' crops as a tactic for decreasing cover and food supplies for the enemy [58]. By 1965 the U.S. had applied approximately two million liters of herbicides in the Republic of Vietnam [58]. In the midst of these events, the USDA Crops Research led a series of herbicide experiments in the Luquillo Experimental Forest from 1964–1965 that were funded by the U.S. Department of Defense Advanced Research Projects Agency. The Department of Army personnel determined that the forests in the Luquillo Experimental Forest were similar to the evergreen forests of Southeast Asia. Thus, tactical herbicides capable of defoliating the tropical vegetation of the Luquillo Experimental Forest could be applied to Southeast Asian forests to reduce the amount of obscuring vegetation, which would reduce the possibility of ambush, and increase the ability to observe the movement of enemy equipment [59].

In January 1964, scientists applied six herbicides (picloram, prometone, bromacil, dicamba, fenac, diuron) to forest soils with a cyclone seeder at rates of 3.7, 10 and 30 kg per ha. They used a randomized complete block design with three replicates for a total of sixty-three 18 m × 24 m plots with 6 m buffers [19]. The effect of herbicide applied to the soils was evident within one month of application and maximum defoliation occurred nine-months post herbicide application [18]. Picloram, a major component of the rainbow herbicide Agent White, was the most effective herbicide, killing between 22% and 71% of vegetation 21 months after treatment [18]. Fenac was the most persistent herbicide in the soil [18], reaching soil depths of 91 to 122 cm within three months of application, suggesting substantial downward movement of herbicides into the soil profile [19]. Forest plant succession occurred within 18 months of herbicide application [19], and there was no effect on the composition of the species that regrew when compared with other secondary forest in the area [18].

Following the initial soil herbicide experiments, scientists explored the effects of foliar application, focusing on the most effective defoliant, picloram, which was applied in various combinations with two other herbicides, paraquat and pyriclor [18]. Foliar application occurred in October, 1965 in separate 0.4 ha plots (53 m × 76 m with a 15 m buffer) established in a randomized block design with two replicates. Herbicides were applied in liquid form at rates of between 6.7 and 20 kg per ha with a Hiller 12-E helicopter that flew in five 11 m swaths at treetop level over each plot. The effects of foliar application on defoliation were much more immediate than either the soil herbicide or the radiation treatments [18], with measurable effects within one week of application and maximum defoliation occurring three-months post-herbicide application (Figure 3). The resulting defoliation substantially affected the light environment and microclimate up to a year following treatment; however, as with soil herbicide application, the forest recovered relatively quickly, and the species composition of succession was not affected [18]. Several months after herbicide application, scientists planted mahogany and teak trees in numerous plots. Both tree species grew well and showed no signs of being affected by herbicides [18]. Thus, while the initial responses to herbicide application were immediate and dramatic, once the herbicides were flushed out of the soil rooting zone, the forest appeared to recover normally.

Overall, picloram proved to be an effective defoliant of trees in the Luquillo Experimental Forest, whether applied directly to the soil or to the vegetation. The relatively rapid rate of regeneration (within 1.5 years) following very high levels of herbicide application to the soil (up to 30 kg per ha) suggests that the effects of the herbicides in this forest are transient and unlikely to 'sterilize' the soil for long periods of time [18]. In addition, the composition of species that comprised initial forest succession following herbicide application did not appear to be affected, with no new 'invasive' species introduced during recovery [19]. Defoliation of the forest occurred more rapidly in response to herbicide application than what was observed in the radiation experiment [18]. At the same time, the forest also recovered from herbicide application at a much faster rate than from radiation exposure [28]. From a military perspective, Agent White, of which picloram is a major component, was found to be less advantageous than Agent Orange because defoliation took several weeks to begin [58]. However, due to changes in chemical market forces in the mid-1960's, Agent Orange production became limited and it was not available in sufficient supplies for military application. Thus, the U.S. transitioned to the use of Agent White in 1966, applying an estimated total of 20,556,525 liters to the forests of the Republic of Vietnam from 1966–1971 [58].

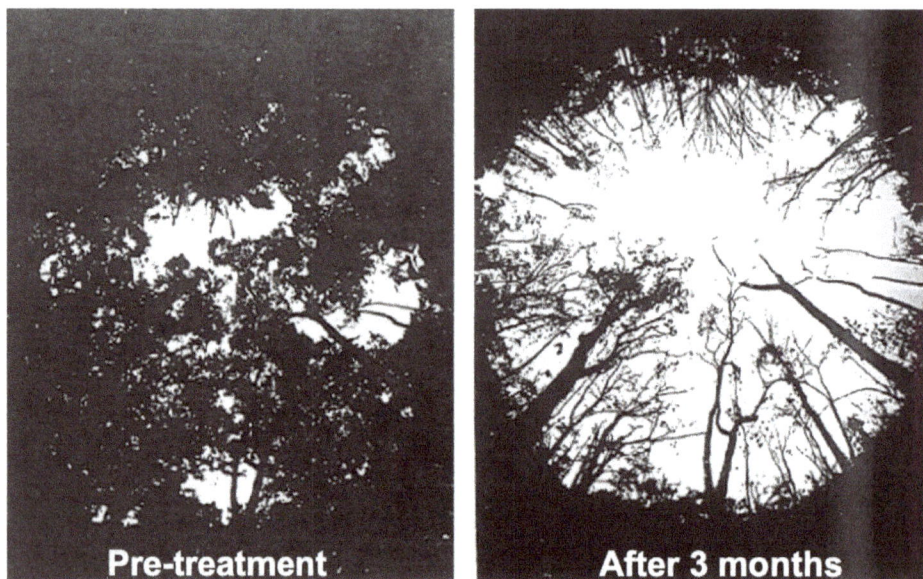

Figure 3. Changes in canopy opening prior to and three months following herbicide application in the Luquillo Experimental Forest. Photograph reproduced from [18].

2.4. Giant Plastic Cylinder Study (1966–1967)

Also supported by the Atomic Energy Commission, Howard T. Odum established the "Giant Plastic Cylinder" study in 1966 in order to better understand whole-forest metabolism and fluxes of water, carbon, and energy [16]. A 17-m tall x 18 m wide plastic chamber (Figure 4) was erected in the Luquillo Experimental Forest with a 22-m canopy access tower in the center, enabling Odum and collaborators to study vertical gradients and partition fluxes into their various components. Six 17-m tall crank up aluminum towers were hauled up the mountain and installed in a hexagonal array on concrete pads with steel wire creating the frame. Plastic material that could be bonded with adhesive was pulled up to the wire frame to form a cylinder. This material lasted a year and a half under forest exposure, which limited the duration of the study [16]. Although the experiment ran for just over one year, the study was immensely valuable for understanding tropical forest function. This study was the first attempt at assessing whole-forest metabolism and provided initial estimates of tropical forest evapotranspiration, forest floor respiration, vertical gradients of photosynthesis, gross photosynthetic rates, leaf area index, and chlorophyll content, as well as an overall assessment of carbon dioxide fluxes. As a whole, the Giant Plastic Cylinder was a visionary prototype, and served as a precursor for future open top chamber and eddy covariance studies.

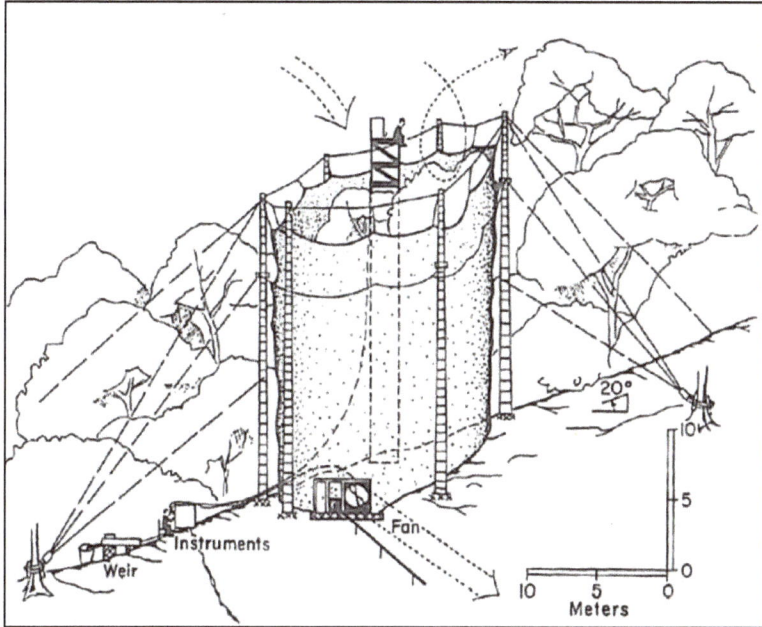

Figure 4. Schematic diagram of the Big Plastic Cylinder Experiment. Reproduced from [16].

2.5. The GAPS Experiment (1988-Today)

By the late 1970's scientists recognized that the area of secondary forests was increasing rapidly [60], and that with few exceptions most tropical countries had a larger land surface cover of secondary vegetation than old growth [61]. It was suggested that the tropics were entering the "era of secondary vegetation" [61]. At the time, studies of carbon dynamics in secondary forests were sparse [60], and to date, most studies have focused on successional changes in vegetation and the drivers, patterns, and consequences of deforestation at large spatial scales [60,62,63]. Fewer studies have followed the *in-situ* biogeochemical effects of deforestation and forest regrowth over time [64]. In 1988, scientists established an experiment to determine the carbon and nutrients effects of tropical deforestation, and to follow soil biogeochemical dynamics and patterns in forest regrowth over time. Three 32 m × 32 m plots (two treatment plots and a split control plot) were established in the Bisley Research Watersheds of the Luquillo Experimental Forest, and surveyed on a 4 m × 4 m grid. Every grid node was sampled for soil (0–10, 10–35, 35–60 cm depths) in the intact forest at the start of the experiment. All trees >10 cm diameter at breast height were measured for basal area, height, and identified to species. In June of 1989, the two treatment plots were clear cut; all aboveground vegetation was weighed and hand-carried off the plots. Trees were measured for allometric equations and subsampled for carbon and nutrient analyses [65]. Hurricane Hugo swept through the forest in September 1989, decimating the control plots. The effects of the hurricane on the treatment plots was minimal: no trees fell into the plots and the added litter was quantified and collected immediately after the storm [65]. The soils and vegetation were then intensively sampled over the next two years, and again at 10-years post-disturbance.

Clear-cutting removed 300 tons of biomass per ha and the large amount of nutrients contained therein [65]. Deforestation led to short-term increases in exchangeable cation concentrations in soils prior to the hurricane, and most soil and forest floor nutrient pools had returned to pre-disturbance levels within 9 weeks. Exchangeable potassium was the only element that declined significantly in soils over this time period. The hurricane approximately doubled the size of the forest floor pool and substantially increased forest floor concentrations of nitrogen, phosphorus, potassium, calcium and magnesium [66]. In the soil, only the concentrations of exchangeable potassium and nitrate (NO_3^-) increased significantly, but these, and the forest floor mass returned to pre-disturbance values within 9 months. Surprisingly there was no effect of the disturbances on soil organic matter content [66]. However, there was a 40% decline in live fine root biomass within 2-months of the clear-cutting, and fine root mortality increased to 70%–77% following the hurricane. Slow fine root decay led to up to 65% mass remaining after one year. Root mortality was associated with high soil NO_3^- concentrations [67]. Forest regrowth was slower in the clear-cut plots than in the surrounding forest 10 years following the disturbances. Fine root biomass recovered within 8-10 years, albeit stocks were more variable in the treatment plots than in the surrounding forest. Soil carbon stocks did not change significantly as a result of the disturbances. Soil phosphorus pools declined during periods of rapid plant regrowth, and then fluctuated over time. Overall, soil carbon and nutrient pools were relatively resilient to disturbance in this forest on a decadal time scale [68]. Deforestation is a continuing problem in tropical regions, resulting in the conversion of 97 million ha of forested land globally from 2001–2012 and the emissions of 47 Gt CO_2 [69]. Research that explores carbon and nutrient cycling in secondary forests continues to be a research need, the results of which would provide invaluable insight into the role that these forests will play in mediating future climate.

2.6. Post-Hurricane Fertilization and Debris-Removal Experiment (1989)

Hurricanes are an important force structuring Puerto Rico's forests, and have been responsible for billions of dollars in damages to U.S. coastal regions and interests [70]. Although much knowledge has been gained about forest responses to hurricanes from observational studies, a mechanistic understanding requires experimental manipulations [71]. In 1989, scientists had just finished their initial set up for a complete fertilization experiment in the Luquillo Experimental Forest when Hurricane Hugo hit Puerto Rico. They took note of the large deposits of green foliage on the forest floor, which was equivalent to over a year's worth of phosphorus being deposited in a 24-h period. In response, scientists added a debris-removal treatment to the experiment resulting in 4 blocks with 3 treatments: fertilization, control, and debris removal (1 month after Hugo). Plots were each 20 m $\times$ 20 m and scientists investigated a wide range of responses including nutrient immobilization [72], litterfall rates, quality and decay [73], earthworm responses [74], effects on understory plants [75]), as well as effects on forest growth and species composition [76]. The fertilization experiment was established in lower- (350–500 m asl) and upper-elevation forests (1050 m asl), where plots had been fertilized with macro-and micronutrients every 3 months since the passage of Hurricane Hugo in 1989 [72,76].

Fertilization stimulated leaf litter production in both forests, but the rate of recovery to pre-hurricane levels was greater at 350–500 m above sea level (20 months after treatment) than those at 1050 m asl (38 months, [64]). Litterfall increased after fertilization and by 2–3 years appeared to reach its maximum [76], with some decreases in magnitude seen a decade later [73]. Experimental removal of litter and woody debris generated by the hurricane (plus any standing stocks present before the hurricane) increased soil nitrogen availability and above-ground productivity by as much as 40% compared to un-manipulated control plots [72]. These increases were similar to those created by quarterly fertilization with inorganic nutrients. Approximately 85% of hurricane-generated debris was woody debris greater than 5 cm diameter. Thus, it appeared that woody debris stimulated nutrient immobilization, resulting in depression of soil nitrogen availability and productivity in control plots [72]. These results together with simulations of an ecosystem model (CENTURY) calibrated for Luquillo Experimental Forest [77] indicated the large wood component of hurricane-generated debris was of sufficiently low quality and of great enough mass to cause the observed effects on productivity. Scientists found no effect of fertilization on the abundance and biomass of earthworms in the upper elevation plots [74]. In the lower-elevation plots, however, the density and biomass of earthworms were significantly greater in the control than in the fertilization treatments. Surprisingly, the removal of hurricane-generated debris significantly increased the density of earthworms. Possible reasons for this positive increase maybe due to increased nitrogen availability in the litter removal soils [72], less leaching of organic compounds from coarse wood [78], and consequently, increased soil pH, and reduction of litter fauna (e.g., frogs, lizards and ants) that function as both competitors for resources and predators [74].

2.7. Canopy Trimming Experiment (2002-Today)

One of the major effects of hurricanes on forests is the defoliation and stem loss of the vegetation resulting in large openings in the canopy and significant deposition of green plant material and coarse woody debris to the forest floor. In an effort to separate the effects of canopy opening and increased debris deposition due to hurricanes, the canopy trimming experiment (CTE) was established in 2002. Canopy branches of 576 trees were trimmed and 32,448 kg of dry mass was distributed over six 30 m × 30 m plots, which is similar to what was observed after Hurricane Hugo. The experiment comprised 3 blocks of 4 full factorial treatments: no trimming and no debris added, trimming performed and debris added, trimming performed and no debris added, no trimming and debris added. Measurements for this experiment are ongoing, and a second canopy trim was conducted in 2014.

Studies within the CTE are diverse, exploring the mechanistic response patterns of tropical forest biota (microbes, plants, animals) and processes (decomposition, herbivory, nutrient cycling, primary production) to canopy and understory disturbance [27,71,79]. As a whole, results from the CTE suggest that cascading effects from canopy openness account for most of the shifts in the forest biota and biotic processes, such as increased plant recruitment and richness, as well as the decreased abundance and diversity of several animal groups [71]. Canopy opening decreased litterfall and litter moisture [80], thereby inhibiting lignin-degrading fungi, which slowed decomposition [79,81]. Debris addition temporarily increased tree basal area [49]. Data also suggest that hurricane disturbance can accelerate the cycling of soil labile organic carbon on a short temporal scale of less than two years [82]. In addition, scientists found that both surface- (0–10 cm) and subsoils (50–80 cm) have the potential to significantly increase carbon and nutrient storage a decade after the sudden deposition of disturbance-related organic debris, suggesting Luquillo Experimental Forest soils can serve as sinks of carbon and nutrients derived from disturbance-induced pulses of organic matter [83].

2.8. Long-Term Nitrogen Addition in Two Forest Types (2002–Today)

Deposition of nitrogen in tropical forests is projected to increase [84] and we know increased anthropogenic nitrogen inputs can have dramatic effects on the structure and function of plant and animal communities [85]. To understand the consequences of increased nitrogen inputs, scientists have been continually applying 50 kg per ha per year of nitrogen fertilizer to twelve 20 m × 20 m plots in two forest types within the Luquillo Experimental Forest (3 fertilized, 3 control per site) since 2002 [30]. Nitrogen additions suppressed nitrogen fixation at both high and low elevation sites [30,86] Thus far, there have been no significant effects of nitrogen fertilization on plant growth, suggesting that plants in this forest are not nitrogen limited. In contrast, significant belowground responses to nitrogen addition included increased soil carbon dioxide fluxes, decline in live root biomass, and an increase in total soil carbon. However, labile soil carbon decreased while mineral-associated carbon increased [30,86]. This work suggests that soil carbon storage may be sensitive to nitrogen deposition even in forests that are not nitrogen limited [30]. Because these plots are some of only a handful of fertilization plots in the tropics, they offer an important opportunity for understanding how changes to nutrient inputs can affect how tropical forests work.

2.9. Throughfall Exclusion Experiment (2008-Today)

In an effort to understand the consequences of drought on the biogeochemistry of tropical soils, small shelters (1.24 × 1.24 m) were installed in the forest understory of the Luquillo Experimental Forest in 2008 to divert water away from soils, effectively reducing soil moisture (Figure 5). Effects on soil gas fluxes, nutrient cycling, and microbial dynamics were studied. Soil carbon dioxide emissions declined and net methane consumption, as well as net nitrous oxide sink behavior, increased in response to reduced soil moisture. Taken together these data suggested drought may decrease greenhouse gas emissions from tropical soils [31,87]. However, microbes showed the capacity to adapt to repeat cycles of drought [32]. Following up on these initial experiments, a new throughfall exclusion experiment was established in 2017 with larger throughfall exclusion shelters (2.4 m × 4.8 m) to further explore the consequences of repeated drought on a range of biogeochemical processes. Shelters were in place for a total of 6 months when Hurricanes Irma and Maria passed over the island of Puerto Rico in September 2017. The shelters were removed during the storms, but following the hurricanes, shelters were re-established and the study was modified to include interactions between soil drought and the recovery of soil biogeochemical responses following hurricane disturbance. Results from this experiment will provide a better understanding of the interactions between droughts and hurricanes, two major drivers of environmental change at the Luquillo Experimental Forest that are projected to increase in frequency under future climate regimes.

Figure 5. Throughfall exclusion shelter installed in the Luquillo Experimental Forest. Photograph by Tana E. Wood.

2.10. TRACE: Tropical Responses to Altered Climate Experiment (2013-Today)

Temperatures are expected to increase significantly in tropical regions over the next two decades [88,89]. How already-warm tropical forests will respond to increasing temperatures remains a critical unknown in our global understanding of climate change effects. For example, whether tropical forests will continue to serve as net sinks for carbon in a warmer world remains highly uncertain [90,91]. In an effort to quantify the effects of increased temperature on tropical forest carbon cycling, scientists established the first field warming experiment in a tropical forested ecosystem in the Luquillo Experimental Forest (Tropical Responses to Altered Climate Experiment (TRACE)) [10]. In 2016, infra-red heaters were deployed in 4-m diameter hexagonal plots to warm understory vegetation and soils by +4 °C above ambient (Figure 6). Pre-treatment measurements were collected for 1-year before warming began in the fall of 2017 [10]. As with Odum's radiation experiment, scientists have taken the opportunity to explore many facets of tropical forest responses to warming, including soil carbon and nutrient fluxes and pools, plant physiology, plant demography, soil microbial communities, and responses of soil microarthropods and native frogs. In 2017, one year following the initiation of warming Hurricanes Irma and Maria passed over the island of Puerto Rico. Warming efforts were paused and scientists capitalized on the opportunity to evaluate whether the prior stress of warming influenced the recovery of forest biomass and processes following hurricane disturbance. New baseline measurements were collected for a full year following the hurricanes and warming was restarted in the fall of 2018. This unique infrastructure offers a potentially once-in-a-lifetime opportunity to assess how warmer tropical plants and soils recover from hurricanes. Experimental research at the TRACE site is ongoing, and publications on the results from the first year of warming are in progress.

Figure 6. Arial photograph of one of the TRACE plots in 2018, 14 months after Hurricane Maria. Photograph by Maxwell Farrington.

3. Conclusions

The Luquillo Experimental Forest has served as a platform for a wealth of innovative and revolutionary experiments, from the Radiation and Giant Plastic Cylinder Experiments initiated in the 1960's to the Canopy Trimming and Tropical Responses to Altered Climate Experiments established in the 2000's. To our knowledge, no other tropical forest has experienced such a range of unique experimentation, or has so clearly marked the history of the fears and interests of our society. While the responses and rates of recovery to these disturbances have been varied, the Luquillo Experimental Forest has demonstrated incredible resistance and recovery in the face of great change. Whether these forests will continue to prevail or if we will see a significant shift in the size and composition of the

Luquillo Experimental Forest as the world's climate and disturbance regimes continue to change is one of the greatest concerns facing scientists today. Current and future experiments conducted in the Luquillo Experimental Forest will continue to address these questions and will provide critical information for the development and refinement of forest management strategies.

Author Contributions: Conceptualization, M.A.C., T.E.W., G.G., and S.C.R.; Writing-Original Draft Preparation, T.E.W., M.A.C., S.C.R., W.L.S., G.G.; Writing-Review & Editing, T.E.W., M.A.C., S.C.R., W.L.S., G.G.

Funding: T.E.W.: M.A.C. and S.C.R. were supported by the U.S. Department of Energy Terrestrial Ecosystem Sciences Program under Award Numbers DE-SC-0011806 and 89243018S-SC-000014. S.C.R. was supported by the U.S. Geological Survey. W.L.S. and G.G. were supported by the NSF Luquillo Critical Zone Observatory (EAR-1331841) and the LTER program (DEB1239764). W.L.S. had additional support from NSF DEB-1457805, DOE TES-DE-FOA-0000749, and the USDA National Institute of Food and Agriculture, McIntire Stennis project CA-B-ECO-7673-MS. The USDA Forest Service's International Institute of Tropical Forestry (IITF) and University of Puerto Rico-Río Piedras provided additional support. All research at IITF is done in collaboration with the University of Puerto Rico.

Acknowledgments: We would like to thank Olga Ramos for creating the map of Puerto Rico used in this manuscript as well as Ariel E. Lugo and Jess Zimmerman for valuable suggestions on an earlier version of the manuscript. We are also grateful to all of the scientists who have worked in Puerto Rico to build an understanding of how these important ecosystems work. Any use of trade, firm, or product names is for descriptive purposes only and does not imply endorsement by the U.S. Government.

Conflicts of Interest: The authors declare no conflict of interest.

References

1. Weaver, P.L. *The Luquillo Mountains: Forest Resources and Their History*; United States Department of Agriculture, Forest Service International Institute of Tropical Forestry: San Juan, PR, USA, 2012.
2. Harris, N.L.; Lugo, A.E.; Brown, S.; Heartsill Scalley, T. *Luquillo Experimental Forest: Research History and Opportunities*; U.S. Department of Agriculture: Washington, DC, USA, 2012; p. 152.
3. Brown, S.; Lugo, A.E.; Silander, S.; Liegel, L. *Research History and Opportunities in the Luquillo Experimental Forest*; General Technical Report (GTR)-SRS-044; U.S. Dept of Agriculture, Forest Service, Southern Forest Experiment Station: New Orleans, LA, USA, 1983; p. 132.
4. Quiñones, M.; Parés-Ramos, I.K.; Gould, W.A.; Gonzalez, G.; McGinley, K.; Ríos, P. *El Yunque National Forest Atlas*; General Technical Report; International Institute of Tropical Forestry: San Juan, PR, USA, 2018.
5. Garcia-Martino, A.R.; Warner, G.S.; Scatena, F.N.; Civco, D.L. Rainfall, runoff and elevation relationships in the Luquillo Mountains of Puerto Rico. *Caribb. J. Sci.* **1996**, *32*, 413–424.
6. Aronson, E.L.; McNulty, S.G. Appropriate experimental ecosystem warming methods by ecosystem, objective, and practicality. *Agric. For. Meteorol.* **2009**, *149*, 1791–1799. [CrossRef]
7. Leuzinger, S.; Luo, Y.; Beier, C.; Dieleman, W.; Vicca, S.; Körner, C. Do global change experiments overestimate impacts on terrestrial ecosystems? *Trends Ecol. Evol.* **2011**, *26*, 236–241. [CrossRef] [PubMed]
8. Odum, H.T. Rain forest structure and mineral cycling homeostasis. In *A Tropical Rain Forest: A Study of Irradiation and Ecology at El Verde*; Odum, H.T., Pigeon, R.F., Eds.; U.S. Atomic Energy Commission: Oak Ridge, TN, USA, 1970.
9. Jentsch, A.; Kreyling, J.; Beierkuhnlein, C. A new generation of climate-change experiments: Events, not trends. *Front. Ecol. Environ.* **2007**, *5*, 365–374. [CrossRef]
10. Kimball, B.A.; Alonso-Rodríguez, A.M.; Cavaleri, M.A.; Reed, S.C.; González, G.; Wood, T.E. Infrared heater system for warming tropical forest understory plants and soils. *Ecol. Evol.* **2018**, *8*, 1932–1944. [CrossRef] [PubMed]
11. Newton, I. Isaac Newton Letter to Robert Hooke 1675. Available online: https://digitallibrary.hsp.org/index.php/Detail/objects/9792 (accessed on 27 November 2018).
12. Marrero, J. A Survey of the Forest Plantations in the Caribbean National Forest. Master's Thesis, School of Forestry and Conservation, University of Michigan, Ann Arbor, MI, USA, 1947.
13. Wadsworth, F. Growth in the lower montane rain forest of Puerto Rico. *Caribb. For.* **1947**, *8*, 27–35.
14. Wadsworth, F.H. Forest management in the Luquillo mountains, III. Selection of products and silvicultural policies. *Caribb. For.* **1952**, *13*, 93–142.

15. Odum, H.T. The Rainforest and Man: An Introduction. In *A Tropical Rain Forest: A Study of Irradiation and Ecology at El Verde*; Atomic Energy Commission, Division of Technical Information: Oak Ridge, TN, USA, 1970; pp. A-5–A-11.

16. Odum, H.T.; Jordan, C.F. Metabolism and Evapotranspiration of the Lower Forest in a Giant Plastic Cylinder. In *A Tropical Rain Forest: A Study of Irradiation and Ecology at El Verde*; Atomic Energy Commission, Division of Technical Information: Oak Ridge, TN, USA, 1970; p. I-165.

17. Dowler, C.; Tschirley, F. Evaluation of herbicides applied to foliage of four tropical woody species. *J. Agric. Univ. P. R.* **1970**, *54*, 676–682.

18. Dowler, C.C.; Tschirley, F.H. Effects of Herbicide on a Puerto Rican Rain Forest. In *A Tropical Rain Forest: A Study of Irradiation and Ecology at El Verde*; Atomic Energy Commission, Division of Technical Information: Oak Ridge, TN, USA, 1970; p. B-315.

19. Dowler, C.C.; Forestier, W.; Tschirley, F. Effect and persistence of herbicides applied to soil in Puerto Rican forests. *Weed Sci.* **1968**, *16*, 45–50.

20. Odum, H.T.; Pigeon, R.F. *A Tropical Rain Forest: A Study of Irradiation and Ecology at El Verde, Puerto Rico*; Atomic Energy Commission, Division of Technical Information: Oak Ridge, TN, USA, 1970.

21. Scatena, F.; Larsen, M. Physical aspects of hurricane Hugo in Puerto Rico. *Biotropica* **1991**, *23*, 317–323. [CrossRef]

22. Boose, E.R.; Serrano, M.I.; Foster, D.R. Landscape and regional impacts of hurricanes in Puerto Rico. *Ecol. Monogr.* **2004**, *74*, 335–352. [CrossRef]

23. Beard, K.H.; Vogt, K.A.; Vogt, D.J.; Scatena, F.N.; Covich, A.P.; Sigurdardottir, R.; Siccama, T.G.; Crowl, T.A. Structural and functional responses of a subtropical forest to 10 years of hurricanes and droughts. *Ecol. Monogr.* **2005**, *75*, 345–361. [CrossRef]

24. Scatena, F.; Moya, S.; Estrada, C.; Chinea, J. The first five years in the reorganization of aboveground biomass and nutrient use following Hurricane Hugo in the Bisley Experimental Watersheds, Luquillo Experimental Forest, Puerto Rico. *Biotropica* **1996**, *28*, 424–440. [CrossRef]

25. Comita, L.S.; Uriarte, M.; Thompson, J.; Jonckheere, I.; Canham, C.D.; Zimmerman, J.K. Abiotic and biotic drivers of seedling survival in a hurricane-impacted tropical forest. *J. Ecol.* **2009**, *97*, 1346–1359. [CrossRef]

26. Shiels, A.B.; Zimmerman, J.K.; García-Montiel, D.C.; Jonckheere, I.; Holm, J.; Horton, D.; Brokaw, N. Plant responses to simulated hurricane impacts in a subtropical wet forest, Puerto Rico. *J. Ecol.* **2010**, *98*, 659–673. [CrossRef]

27. Shiels, A.B.; González, G. Understanding the key mechanisms of tropical forest responses to canopy loss and biomass deposition from experimental hurricane effects. *For. Ecol. Manag.* **2014**, *332*, 1–10. [CrossRef]

28. Taylor, C.M.; Silander, S.; Waide, R.B.; Pfeiffer, W.J. Recovery of a tropical forest after gamma irradiation: A 23-year chronicle. In *Tropical Forests: Management and Ecology*; Springer: New York, NY, USA, 1995; pp. 258–285, ISBN 1-4612-7563-6.

29. McCormick, J.F. Growth and Survival of the Sierra Palm Under Radiation Stress in Natural and Simulated Environments. In *A Tropical Rain Forest: A Study of Irradiation and Ecology at El Verde*; Atomic Energy Commission, Division of Technical Information: Oak Ridge, TN, USA, 1970; p. D-193.

30. Cusack, D.F.; Torn, M.S.; McDowell, W.H.; Silver, W.L. The response of heterotrophic activity and carbon cycling to nitrogen additions and warming in two tropical soils. *Glob. Change Biol.* **2010**, *16*, 2555–2572. [CrossRef]

31. Wood, T.E.; Silver, W.L. Strong spatial variability in trace gasdynamics following experimental drought in a humid tropical forest. *Glob. Biogeochem. Cycles* **2012**, *26*. [CrossRef]

32. Bouskill, N.J.; Chien Lim, H.; Borglin, S.; Salve, R.; Wood, T.E.; Silver, W.L.; Brodie, E.L. Pre-exposure to drought increases the resistance of tropical forest soil bacterial communities to extended drought. *Int. Soc. Microb. Ecol.* **2012**, *7*, 384–394. [CrossRef] [PubMed]

33. Brokaw, N. *A Caribbean Forest Tapestry: The Multidimensional Nature of Disturbance and Response*; Oxford University Press: New York, NY, USA, 2012; ISBN 0-19-533469-8.

34. González, G.; Willig, M.R.; Waide, R.B. *Ecological Gradient Analyses in a Tropical Landscape: Multiple Perspectives and Emerging Themes*; International Institute of Tropical Forestry: San Juan, PR, USA, 2013; pp. 13–20.

35. Buss, H.L.; Gould, W.A.; Larsen, M.C.; Liu, Z.; Martinuzzi, S.; Murphy, S.F.; Stallard, R.F.; Pares-Ramos, I.K.; White, A.F.; Zou, X. *Water Quality and Landscape Processes of Four Watersheds in Eastern Puerto Rico*; US Geological Survey: Reston, VA, USA, 2012.

36. Reagan, D.P.; Waide, R.B. *The Food Web of a Tropical Rain Forest*; University of Chicago Press: Chicago, IL, USA, 1996; ISBN 0-226-70599-4.

37. Walker, L.R. Summary of the effects of Caribbean hurricanes on vegetation. *Biotropica* **1991**, *23*, 442–447.

38. Rudel, T.K.; Perez-Lugo, M.; Zichal, H. When fields revert to forest: development and spontaneous reforestation in post-war Puerto Rico. *Prof. Geogr.* **2000**, *52*, 386–397. [CrossRef]

39. Koenig, N. *A Comprehensive Agricultural Program for Puerto Rico*; United States Department of Agriculture: Washington, DC, USA, 1953.

40. Lugo, A.E. Heartsill-Scalley Research in the Luquillo Experimental Forest has advanced understanding of tropical forests and resolved management issues. In *USDA Forest Service Experimental Forests and Ranges*; Springer: New York, NY, USA, 2014; pp. 435–461.

41. Crow, T.R. A rainforest chronicle: a 30-year record of change in structure and composition at El Verde, Puerto Rico. *Biotropica* **1980**, *12*, 42–55. [CrossRef]

42. Crow, T.R.; Weaver, P.L. *Tree Growth in a Moist Tropical Forest of Puerto Rico*; Institute of Tropical Forestry: Río Piedras, PR, USA, 1977.

43. Drew, A.P.; Boley, J.D.; Zhao, Y.; Johnston, M.H.; Wadsworth, F.H. Sixty-two years of change in subtropical wet forest structure and composition at El Verde, Puerto Rico. *Interciencia* **2009**, *34*, 34.

44. Heartsill Scalley, T. Insights on Forest Structure and Composition from Long-Term Research in the Luquillo Mountains. *Forests* **2017**, *8*, 204. [CrossRef]

45. Parresol, B.R. Basal area growth for 15 tropical tree species in Puerto Rico. *For. Ecol. Manag.* **1995**, *73*, 211–219. [CrossRef]

46. Shugart, H.H. *A Theory of Forest Dynamics. The Ecological Implications of Forest Succession Models*; Springer: New York, NY, USA, 1984; ISBN 0-387-96000-7.

47. Wadsworth, F.H.; Englerth, G.H. Effects of the 1956 hurricane on forests in Puerto Rico. *Caribb. For.* **1959**, *20*, 38–51.

48. Shiels, A.B.; González, G.; Willig, M.R. Responses to canopy loss and debris deposition in a tropical forest ecosystem: Synthesis from an experimental manipulation simulating effects of hurricane disturbance. *For. Ecol. Manag.* **2014**, *332*, 124–133. [CrossRef]

49. Zimmerman, J.K.; Hogan, J.A.; Shiels, A.B.; Bithorn, J.E.; Carmona, S.M.; Brokaw, N. Seven-year responses of trees to experimental hurricane effects in a tropical rainforest, Puerto Rico. *For. Ecol. Manag.* **2014**, *332*, 64–74. [CrossRef]

50. Silver, W.L.; Brown, S.; Lugo, A.E. Effects of changes in biodiversity on ecosystem function in tropical forests. *Conserv. Biol.* **1996**, *10*, 17–24. [CrossRef]

51. Silver, W.L.; Kueppers, L.M.; Lugo, A.E.; Ostertag, R.; Matzek, V. Carbon sequestration and plant community dynamics following reforestation of tropical pasture. *Ecol. Appl.* **2004**, *14*, 1115–1127. [CrossRef]

52. Woodwell, G.M. Effects of ionizing radiation on terrestrial ecosystems. *Science* **1962**, *138*, 572–577. [CrossRef] [PubMed]

53. Odum, H.T.; Murphy, P.; Drewry, G.; McCormick, J.F.; Schinhan, C.; Morales, E.; McIntyre, J.A. Effects of Gamma Radiation on the Forest at El Verde. In *A Tropical Rain Forest: A Study of Irradiation and Ecology at El Verde*; Atomic Energy Commission, Division of Technical Information: Oak Ridge, TN, USA, 1970; p. D-3.

54. Murphy, P.G. Tree Growth at El Verde and the Effects of Ionizing Radiation. In *A Tropical Rain Forest: A Study of Irradiation and Ecology at El Verde*; Atomic Energy Commission, Division of Technical Information: Oak Ridge, TN, USA, 1970; p. D-141.

55. Smith, R.F. The Vegetation Structure of a Puerto Rican Forest Before and After Short-Term Gamma Radiation. In *A Tropical Rain Forest: A Study of Irradiation and Ecology at El Verde*; Atomic Energy Commission, Division of Technical Information: Oak Ridge, TN, USA, 1970; p. D-103.

56. Scatena, F.N. *The Death of a Luquillo Giant*; Annual Letter; U.S.D.A. Forest Service International Institute of Tropical Forestry: San Juan, PR, USA, 1998.

57. McCormick, J.F. Direct and Indirect Effects of Gamma Radiation on Seedling Diversity and Abundance in a Tropical Forest. In *A Tropical Rain Forest: A Study of Irradiation and Ecology at El Verde*; Atomic Energy Commission, Division of Technical Information: Oak Ridge, TN, USA, 1970; p. D-201.

58. Stellman, J.M.; Stellman, S.D.; Christian, R.; Weber, T.; Tomasallo, C. The extent and patterns of usage of Agent Orange and other herbicides in Vietnam. *Nature* **2003**, *422*, 681. [CrossRef] [PubMed]

59. Young, A.L. *The History of the US Department of Defense Programs for the Testing, Evaluation, and Storage of Tactical Herbicides*; Office of the Under Secretary of Defense: Arlington, VA, USA, 2006; pp. 1–81.

60. Brown, S.; Lugo, A.E. Tropical secondary forests. *J. Trop. Ecol.* **1990**, *6*, 1–32. [CrossRef]

61. Gómez-Pompa, A.; Vazquez-Yanes, C. Studies on the secondary succession of tropical lowlands: The life cycle of secondary species. In Proceedings of the First International Congress of Ecology: Structure, Functioning and Management of Ecosystems, The Hague, The Netherlands, 8–14 September 1974.

62. Baccini, A.; Goetz, S.; Walker, W.; Laporte, N.; Sun, M.; Sulla-Menashe, D.; Hackler, J.; Beck, P.; Dubayah, R.; Friedl, M. Estimated carbon dioxide emissions from tropical deforestation improved by carbon-density maps. *Nat. Clim. Chang.* **2012**, *2*, 182. [CrossRef]

63. Giam, X. Global biodiversity loss from tropical deforestation. *Proc. Natl. Acad. Sci.* **2017**, *114*, 5775–5777. [CrossRef] [PubMed]

64. Silver, W.; Ostertag, R.; Lugo, A. The potential for carbon sequestration through reforestation of abandoned tropical agricultural and pasture lands. *Restor. Ecol.* **2000**, *8*, 394–407. [CrossRef]

65. Scatena, F.N.; Silver, W.; Siccama, T.; Johnson, A.; Sanchez, M.J. Biomass and Nutrient Content of the Bisley Experimental Watersheds, Luquillo Experimental Forest, Puerto Rico, Before and After Hurricane Hugo, 1989. *Biotropica* **1993**, *25*, 15–27. [CrossRef]

66. Silver, W.L.; Scatena, F.N.; Johnson, A.H.; Siccama, T.G.; Watt, F. At What Temporal Scales Does Disturbance Affect Belowground Nutrient Pools? *Biotropica* **1996**, *28*, 441–457. [CrossRef]

67. Silver, W.L.; Vogt, K.A. Fine-root dynamics following single and multiple disturbances in a subtropical wet forest ecosystem. *J. Ecol.* **1993**, *81*, 729–738. [CrossRef]

68. Teh, Y.A.; Silver, W.L.; Scatena, F.N. A decade of belowground reorganization following multiple disturbances in a subtropical wet forest. *Plant Soil* **2009**, *323*, 197–212. [CrossRef]

69. Busch, J.; Engelmann, J. Cost-effectiveness of reducing emissions from tropical deforestation, 2016–2050. *Environ. Res. Lett.* **2017**, *13*, 015001. [CrossRef]

70. Strobl, E. The economic growth impact of hurricanes: evidence from US coastal counties. *Rev. Econ. Stat.* **2011**, *93*, 575–589. [CrossRef]

71. Shiels, A.B.; Gonzalez, G.; Lodge, D.J.; Willig, M.R.; Zimmerman, J.K. Cascading effects of canopy opening and debris deposition from a large-scale hurricane experiment in a tropical rain forest. *Bioscience* **2015**, *65*, 871–881. [CrossRef]

72. Zimmerman, J.; Pulliam, W.; Lodge, D.; Quinones-Orfila, V.; Fetcher, N.; Guzman-Grajales, S.; Parrotta, J.; Asbury, C.E.; Walker, L.; Waide, R. Nitrogen immobilization by decomposing woody debris and the recovery of tropical wet forest from hurricane damage. *Oikos* **1995**, *72*, 314–322. [CrossRef]

73. Yang, X.; Warren, M.; Zou, X. Fertilization responses of soil litter fauna and litter quantity, quality, and turnover in low and high elevation forests of Puerto Rico. *Appl. Soil Ecol.* **2007**, *37*, 63–71. [CrossRef]

74. Gonzalez, G.; Li, Y.; Zou, X. Effects of post-hurricane fertilization and debris removal on earthworm abundance and biomass in subtropical forests in Puerto Rico. In *Minhocas na America Latina: Biodiversidade e Ecologia*; Brown, G.G., Fragoso, C., Eds.; International Institute of Tropical Forestry: San Juan, PR, USA, 2007; pp. 99–108.

75. Halleck, L.F.; Sharpe, J.M.; Zou, X.Z. Understorey fern responses to post-hurricane fertilization and debris removal in a Puerto Rican rain forest. *J. Trop. Ecol.* **2004**, *20*, 173–181. [CrossRef]

76. Walker, L.R.; Zimmerman, J.K.; Lodge, D.J.; Guzman-Grajales, S. An altitudinal comparison of growth and species composition in hurricane-damaged forests in Puerto Rico. *J. Ecol.* **1996**, 877–889. [CrossRef]

77. Sanford, R.L., Jr.; Parton, W.J.; Ojima, D.S.; Lodge, D.J. Hurricane effects on soil organic matter dynamics and forest production in the Luquillo Experimental Forest, Puerto Rico: Results of simulation modeling. *Biotropica* **1991**, *23*, 364–372. [CrossRef]

78. Zalamea, M.; González, G.; Ping, C.-L.; Michaelson, G. Soil organic matter dynamics under decaying wood in a subtropical wet forest: effect of tree species and decay stage. *Plant Soil* **2007**, *296*, 173–185. [CrossRef]

79. González, G.; Lodge, D.J.; Richardson, B.A.; Richardson, M.J. A canopy trimming experiment in Puerto Rico: The response of litter decomposition and nutrient release to canopy opening and debris deposition in a subtropical wet forest. *For. Ecol. Manag.* **2014**, *332*, 32–46. [CrossRef]

80. Silver, W.L.; Hall, S.J.; González, G. Differential effects of canopy trimming and litter deposition on litterfall and nutrient dynamics in a wet subtropical forest. *For. Ecol. Manag.* **2014**, *332*, 47–55. [CrossRef]

81. Lodge, D.J.; Cantrell, S.A.; González, G. Effects of canopy opening and debris deposition on fungal connectivity, phosphorus movement between litter cohorts and mass loss. *For. Ecol. Manag.* **2014**, *332*, 11–21. [CrossRef]

82. Liu, X.; Zeng, X.; Zou, X.; Lodge, D.; Stankavich, S.; González, G.; Cantrell, S. Responses of Soil Labile Organic Carbon to a Simulated Hurricane Disturbance in a Tropical Wet Forest. *Forests* **2018**, *9*, 420. [CrossRef]

83. Gutiérrez del Arroyo, O.; Silver, W.L. Disentangling the long-term effects of disturbance on soil biogeochemistry in a wet tropical forest ecosystem. *Glob. Chang. Biol.* **2018**, *24*, 1673–1684. [CrossRef] [PubMed]

84. Matson, P.; Lohse, K.A.; Hall, S.J. The globalization of nitrogen deposition: consequences for terrestrial ecosystems. *AMBIO J. Hum. Environ.* **2002**, *31*, 113–119. [CrossRef]

85. Bobbink, R.; Hicks, K.; Galloway, J.; Spranger, T.; Alkemade, R.; Ashmore, M.; Bustamante, M.; Cinderby, S.; Davidson, E.; Dentener, F. Global assessment of nitrogen deposition effects on terrestrial plant diversity: A synthesis. *Ecol. Appl.* **2010**, *20*, 30–59. [CrossRef] [PubMed]

86. Cusack, D.F.; Silver, W.; McDowell, W.H. Biological nitrogen fixation in two tropical forests: ecosystem-level patterns and effects of nitrogen fertilization. *Ecosystems* **2009**, *12*, 1299–1315. [CrossRef]

87. Wood, T.E.; Detto, M.; Silver, W.L. Sensitivity of soil respiration to variability in soil moisture and temperature in a humid tropical forest. *PLoS ONE* **2013**, *8*, e80965. [CrossRef] [PubMed]

88. Diffenbaugh, N.; Scherer, M. Observational and model evidence of global emergence of permanent, unprecedented heat in the 20th and 21st centuries. *Clim. Chang.* **2011**, *107*, 615–624. [CrossRef] [PubMed]

89. Mora, C.; Frazier, A.G.; Longman, R.J.; Dacks, R.S.; Walton, M.M.; Tong, E.J.; Sanchez, J.J.; Kaiser, L.R.; Stender, Y.O.; Anderson, J.M.; et al. The projected timing of climate departure from recent variability. *Nature* **2013**, *502*, 183–187. [CrossRef] [PubMed]

90. Wood, T.E.; Cavaleri, M.A.; Reed, S.C. Tropical forest carbon balance in a warmer world: a critical review spanning microbial- to ecosystem-scale processes. *Biol. Rev.* **2012**, *87*, 912–927. [CrossRef] [PubMed]

91. Cavaleri, M.A.; Reed, S.C.; Smith, W.K.; Wood, T.E. Urgent need for warming experiments in tropical forests. *Glob. Chang. Biol.* **2015**, *21*, 2111–2121. [CrossRef] [PubMed]

forests MDPI

Review

Novelty and Its Ecological Implications to Dry Forest Functioning and Conservation

Ariel E. Lugo [1,*] and Heather E. Erickson [1,2]

[1]   International Institute of Tropical Forestry, USDA Forest Service, 1201 Ceiba, Jardín Botánico sur, Río Piedras 00926-1115, Puerto Rico; ericksonheather@yahoo.com
[2]   Consulting Research Ecology, 2947 NE 31st Ave., Portland, OR 97212, USA
*   Correspondence: alugo@fs.fed.us; Tel.: +1-787-764-7743

Academic Editor: Timothy A. Martin
Received: 24 February 2017; Accepted: 6 May 2017; Published: 10 May 2017

Abstract: Tropical and subtropical dry forest life zones support forests with lower stature and species richness than do tropical and subtropical life zones with greater water availability. The number of naturalized species that can thrive and mix with native species to form novel forests in dry forest conditions in Puerto Rico and the US Virgin Islands is lower than in other insular life zones. These novel dry forests are young (<60 years) with low structural development, high species dominance, and variable species density. Species density is low during initial establishment and increases with age. At the 1-ha scale, novel forests can have greater species density than mature native forests. Species groups, such as nitrogen-fixing species, and other naturalized species that dominate novel dry forests, have a disproportional influence on forest element stoichiometry. Novel dry forests, compared to the mean of all forest species assemblages island-wide, tend to have fallen leaf litter with lower than average manganese and sodium concentrations and lower than average C/N and C/P ratios. After accounting for significant differences in stand age, geology, and or precipitation, novel dry forests compared to native dry forests have higher C anomalies, lower Ca and Na anomalies, and lower C/N ratio anomalies. Taken together, these characteristics may influence litter decomposition rates and the species composition, diversity, and food web dynamics in litter and soil. Novel dry forests also contribute to the conservation of native plant species on highly degraded lands.

Keywords: novel forests; stoichiometry of leaf litter; nitrogen fixing trees; naturalized species; C/N; C/P; and N/P ratios; Puerto Rico; Caribbean; element concentration in leaf litter; succession; species dominance

1. Introduction

Humans are attracted to tropical and subtropical dry forest life zones (sensu lato [1]; dry life zone(s) from now on) because the climate is favorable to their health, agricultural activity, and fuelwood production [2,3]. The consequences of human activity to dry forests are well documented, as these forests are converted to non-forest land covers such as pastures or agriculture, or their aboveground stem wood biomass is unsustainably removed to satisfy fuelwood demand [2,4]. Soil degradation in the form of compacted, nutrient-depleted, or eroded soils, is a common outcome of intensive human activity in dry life zones [4]. This activity tends to fragment the landscapes of dry life zones [5].

When land use pressure is reduced on deforested dry forest landscapes, successional processes allow for the re-establishment of dry forests, as has been observed in Central America, the Caribbean, and other tropical countries [6,7]. In Puerto Rico, recurring island-wide forest inventories uncovered the phenomena that forest succession after abandonment of agricultural use resulted in forest stands dominated by introduced species [8]. Hobbs et al. [9] identified these forests as novel forests, because they are a consequence of human activity, result in new species combinations, and are expanding in land cover throughout the world [10]. Novel forests are "the new wild" of the Anthropocene Epoch [11], and represent a "new world order" [10]; in Puerto Rico, 75 percent of the forest cover is now novel [12].

One of the conservation challenges of the Anthropocene is to characterize novel forests and identify their structure and functioning. Our objective is to assess the ecological implications of novelty in Puerto Rico's dry forests through a synthesis of published information coupled with a new analysis of island-wide fallen leaf chemistry, first reported in Erickson et al. [13]. To accomplish this goal, we need to place dry forests in environmental and historical contexts so that the adaptive role of novelty can be revealed. Therefore, we first consider the effects of climate and land use history on Puerto Rican dry forests and then summarize novel dry forest structure and leaf litter chemistry in relation to the species composition of stands. We end with a discussion on the implications of novelty to dry forest functioning and conservation.

2. Methods

We review the dry forest literature for Puerto Rico and the Caribbean with particular attention to forest structure and functioning. To display the climatic conditions of dry forest assemblages, we used a moisture availability index applied to forests throughout Puerto Rico and the US Virgin Islands by Brandeis et al. in their study of forest species assemblages in those islands [14]. The index (C/mm) is the quotient of air temperature in degrees centigrade (C) and rainfall in millimeters (mm). It was based on 30-year average annual rainfall and air temperature for each of the 22 species assemblages (used here interchangeably with "forest") in their analysis. Brandeis et al. [14] analyzed the forest communities of Puerto Rico and the US Virgin Islands using the results of island-wide forest inventories. From their Tables 8 and 9 [14], we selected nine novel and eight native dry forest species assemblages for comparisons of forest structure. These assemblages were all successional and of similar age (<60 years), the main difference being their species composition.

Species Importance Value (IV) curves (sensu Whittaker [15,16]) are used to establish species dominance in forests and infer levels of stress. Ranking species according to their IV, which is an index that includes the species basal area and stem density relative to those of the stands where they occur, assesses species dominance. Whittaker [15,16] showed that the steep IV curves approach geometric series, while the flatter IV curves approach lognormal distributions and suggested that steeper curves reflected communities under stress. Another structural index that we used to assess dry forest stature was the Holdridge Complexity Index. Holdridge [1] used this Index (based on forest structural measures and number of species) to show that forest complexity increased with moisture availability.

For the analysis of element chemistry and stoichiometry of novel and native dry forests, we build on the study of Erickson et al. [13] who analyzed fallen leaf litter mass and chemistry (11 elements) in 140 plots located across Puerto Rico within 14 of the species assemblages described by Brandeis et al. [14]. Using the same data set as in Erickson et al. [13], we selected the five driest forest communities for a total of 41 plots. Although these communities are commonly found in the Dry Forest Life Zone (sensu Holdridge), each community contains plots located in wetter life zones. We note that there are novel forest communities with greater mean annual precipitation than these five that are considered in the Erickson et al. [13] paper but not here, where the emphasis is on drier novel and native communities. Modeled mean annual precipitation (cf. [13]) for individual plots in this study ranges from 787 mm to 2322 mm. Although the species assemblages used in this analysis are identified by the dominant tree species, the leaf litter samples that were chemically analyzed represent the litter of all the species in the stand, not necessarily a monospecific leaf litter.

To calculate concentration anomalies for the five forest assemblages (three novel and two native), for each element and ratio we subtracted a mean value based on all plots within the 14 island-wide assemblages (roughly 139, depending on element) from the mean of each assemblage. Thus, the concentration anomalies of fallen leaves establish the stoichiometry of dry forests (novel and native) in relation to the corresponding mean for all plots in all species assemblages island-wide. These concentration anomaly comparisons are conservative given that the island-wide averages include the dry forest means. Concentration anomalies for all novel plots together and all native plots together were compared to island-wide means (zero on the anomaly graphs) for each element using *T*-Tests (SAS version 9.4, SAS Institute, Cary, NC, USA).

We tested for individual differences in concentration anomalies among the five forest assemblages, which would suggest idiosyncratic community-scale responses, using ANOVA followed by a post-hoc Tukey–Kramer analysis (SAS Institute, Cary, NC, USA). We also tested whether anomalies differed between novel and native dry forest assemblages using contrast statements. Variables, except for carbon, nitrogen, sulfur, calcium, magnesium and the N/P ratio, which were normally distributed, were log-transformed to meet assumptions of tests. Occasional extreme outliers were removed to further improve normality. The presence of karst has been shown to influence fallen leaf C, Ca, Mn, Al, and Fe chemistry [13], and only about a third of novel plots were located on karst compared to about 75 percent of the native plots. Similarly, forest assemblages differed in mean annual precipitation and mean midpoint age (cf. [13]), also shown to influence fallen leaf chemistry for some elements [13]. Accordingly, we retained mean annual precipitation and stand age as co-variates and accounted for presence/absence of karst in models if significant at the 0.05 level.

3. Results

3.1. Dry Forests in General

Dry forest environmental conditions affect forest structural development and species composition. Dry forests have the lowest complexity, species richness, and stature among tropical forests. Brown and Lugo [17] showed that carbon accumulation in vegetation and soil and litterfall were lower in tropical and subtropical dry forests compared to tropical and subtropical moist, wet, and rain forests. Martínez Yrizar [18] found a positive relationship between rainfall and aboveground biomass and litterfall for dry forests from different tropical locations. For Puerto Rico, biomass was low in forests with the lower moisture availability conditions (Figure 1). Moreover, Gentry [19] found that the dry forest flora of the Neotropics was less diverse than that of moist and wet forests, and that Caribbean dry forests were at the lower end of the species richness gradient of dry forests. He thought that the climate of dry forests coupled to limestone substrates in the Caribbean limited the diversification of its dry forest flora.

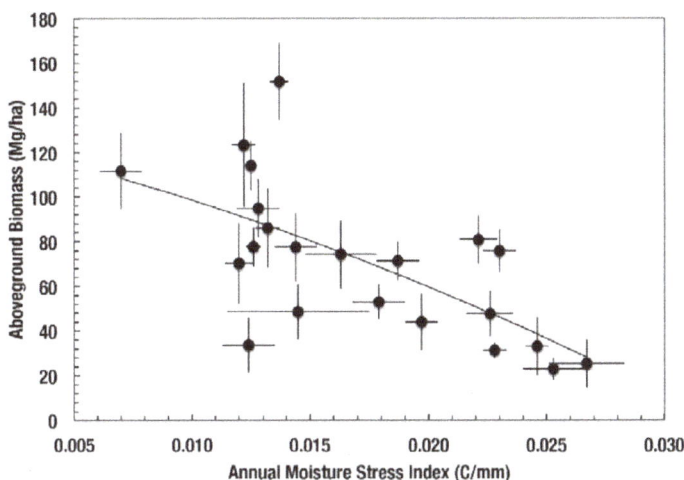

Figure 1. Moisture Availability Index and aboveground biomass of Puerto Rican forests [14]. Vertical and horizontal bars are standard error of the mean reported by Brandeis et al. [14].

3.2. Naturalized Species and the Structure of Novel Dry Forests

Fifteen percent of the plant species of the Caribbean are introduced species [20], most of which naturalize, i.e., establish self-sustaining wild populations. Introduced species can represent between 43 to 110 percent of the native insular flora of individual oceanic islands in the Pacific [21]. In Puerto Rico, the number of naturalized tree species varies with life zone, peaking in the moist and wet forests with lower numbers in rain and dry forest climates. Francis and Liogier [22] listed 118 tree species as naturalized to Puerto Rico (about 18 percent of the tree flora). Of these, approximately 29 species grow in the dry life zone (annual rainfall below 1000 mm), compared to about 43 in the moist life zone (between 1000 and 2000 mm annual rainfall). Thirty-one other tree species grow in an annual rainfall range between 1500 and 3800 mm. This means that the number of tree species available to colonize degraded sites and remix with native species to form novel forests is reduced in the dry life zone compared to moist or wet life zones.

Novel dry forests in Puerto Rico and the US Virgin Islands are characterized by lower tree density, basal area, aboveground biomass, and tree height than native dry forests of similar age (<60 years) (Table 1). The age of dry forest native stands tends to be on the higher end of the age range but they are also secondary forests recovering from agricultural disturbance. The age category for most novel forest stands in Brandeis et al. [14] was 23 to 49 years, which helps explain the low level of structural development. The level of dominance of the top ranked species was similar in both native and novel dry forests, but native stands tended to have more species than novel ones [14,23,24].

Table 1. Average structural parameters of dry forests in Puerto Rico and the US Virgin Islands [14]. Averages are based on nine novel and eight native species assemblages of similar age. The p value denotes the level of significance in a t-test comparison of averages.

Forest Status	Stem Density (stems/ha)	Basal Area (m²/ha)	Aboveground Biomass (Mg/ha)	Tree Height (m)
Novel	3103	9.71	36.43	5.83
Native	4348	16.4	70.33	7.3
p value	0.012	0.011	0.0002	0.002

Early succession novel dry forests (<60 years), such as those studied by Molina Colón et al. [23], support few species with very high dominance (up to 90 percent), reflecting the initial colonization of deforested sites by a few species. Successional dry novel forests exhibit higher dominance than nearby mature historic dry forests (>80 years), and their species density is lower [24,25]. Older novel dry forests support more species than mature native dry forests at the 1-ha scale, but have less species density than mature historic forests at smaller sampling scales [26]. Native species that are unable to colonize degraded sites are able to grow in sites colonized by non-native tree species, thus increasing the diversity of novel forests [24–26].

3.3. Stoichiometry of Leaf Litter

The overarching pattern in the comparisons of element concentration anomalies between novel and native dry forests and with island-wide forests is the absence of a consistent pattern among elements. After accounting for differences in stand age, C concentration anomalies ($p = 0.044$) were greater in novel dry forests than in native dry forests (Figure 2a) although differences among individual assemblages also existed. For example, native stands dominated by *Citharexylum* had significantly lower carbon concentration anomalies ($p = 0.004$, Figure 2a) than other assemblages. Novel dry forests tended to show lower Ca ($p = 0.051$) and Na ($p = 0.059$) anomalies than native dry forests Figure 2e,h). Nitrogen, sulfur, phosphorus, magnesium, potassium, manganese, aluminum, and iron concentration anomalies did not differ among the forest assemblages or between novel and native dry forests (Figure 2b–d,f,g,i–k).

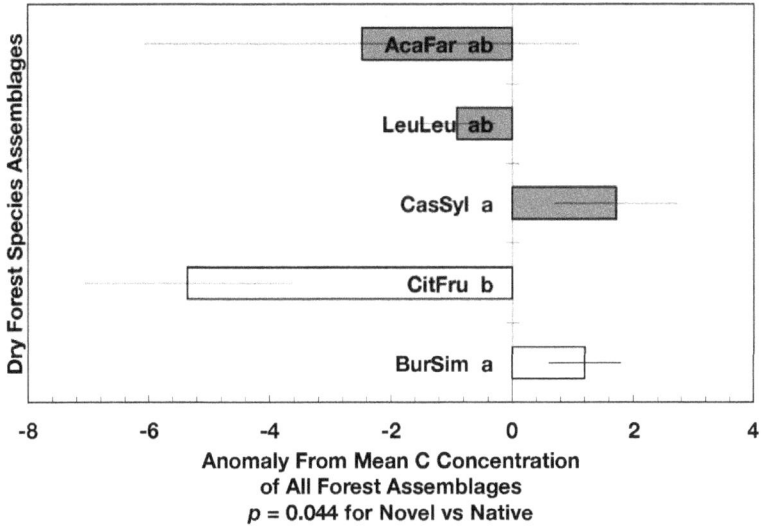

(a)

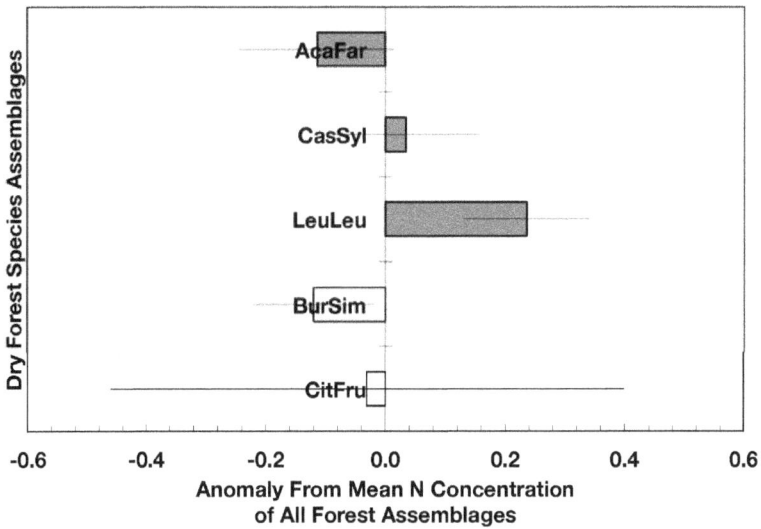

(b)

Figure 2. *Cont.*

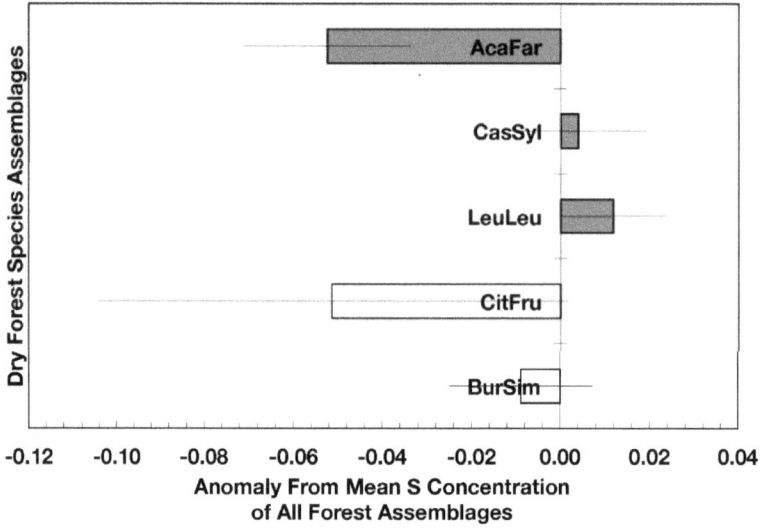

(c)

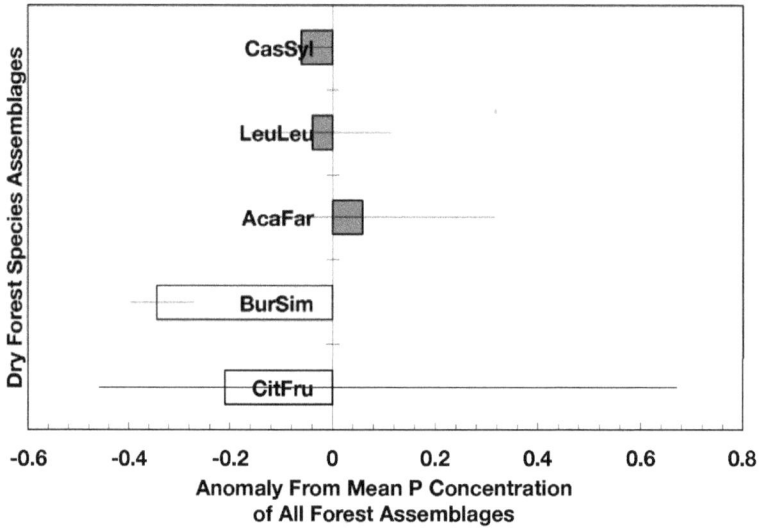

(d)

Figure 2. *Cont.*

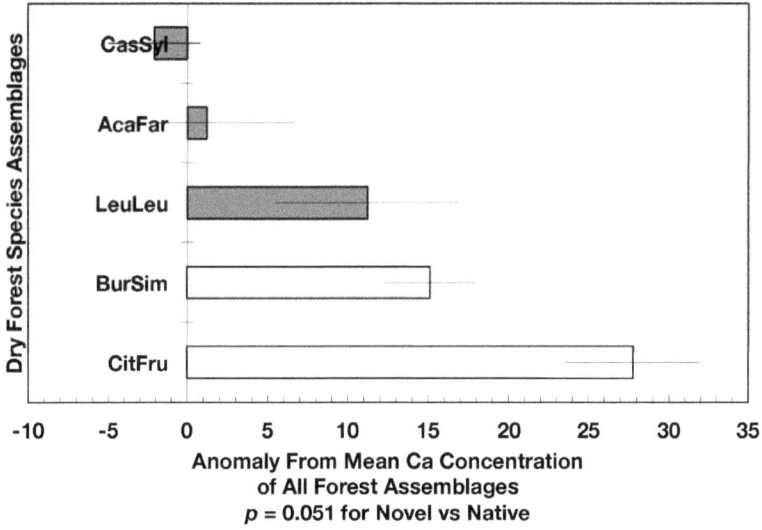

(e)

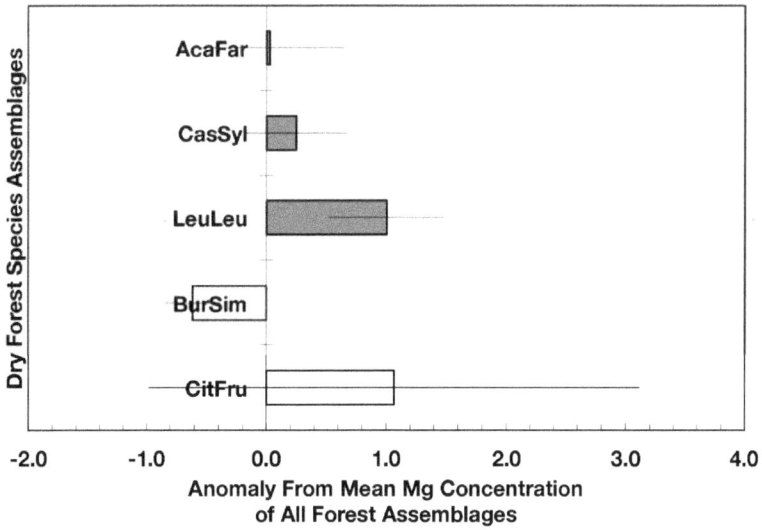

(f)

Figure 2. *Cont.*

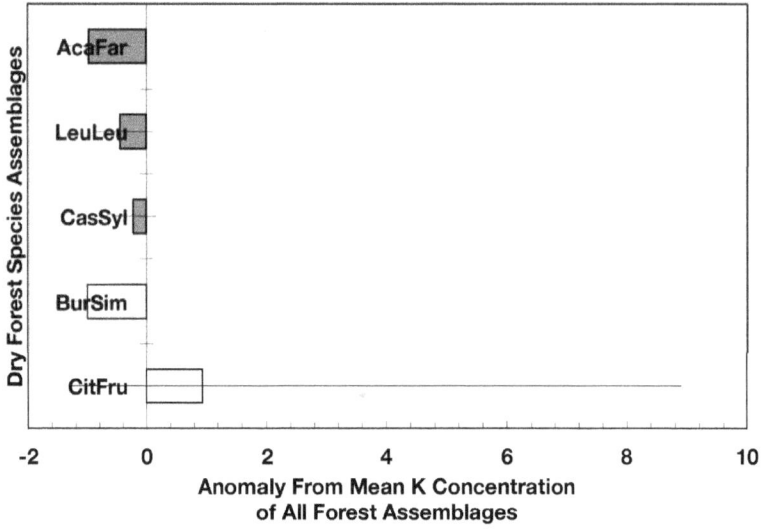

(g)

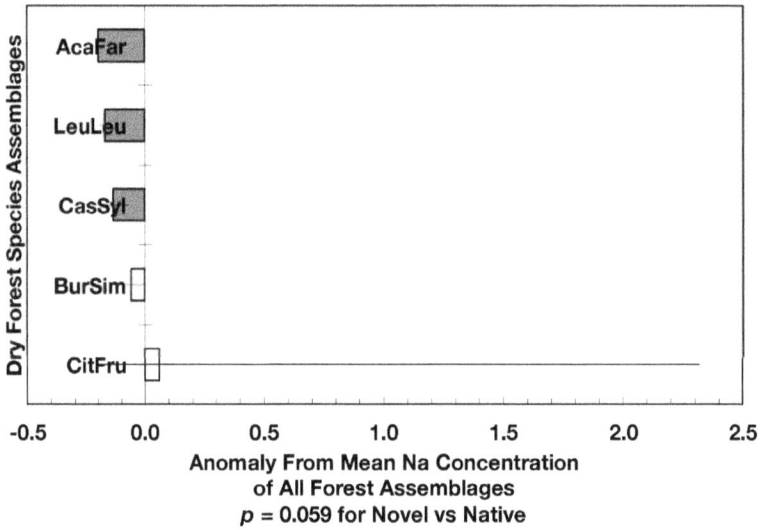

(h)

Figure 2. *Cont.*

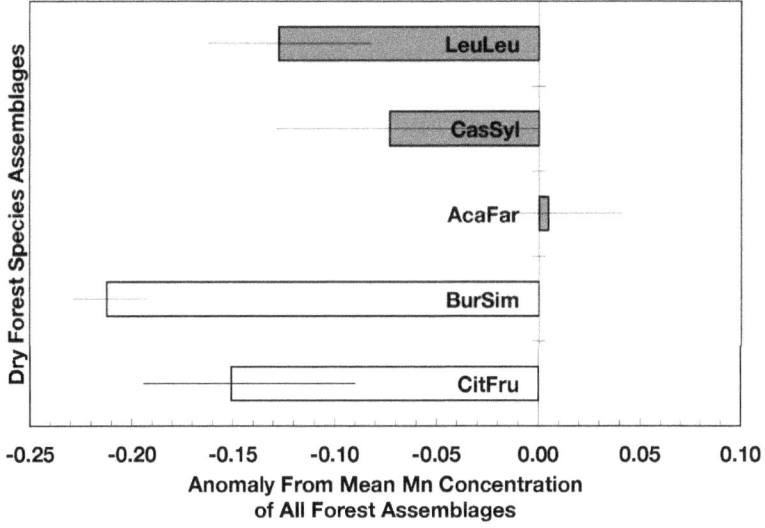

(i)

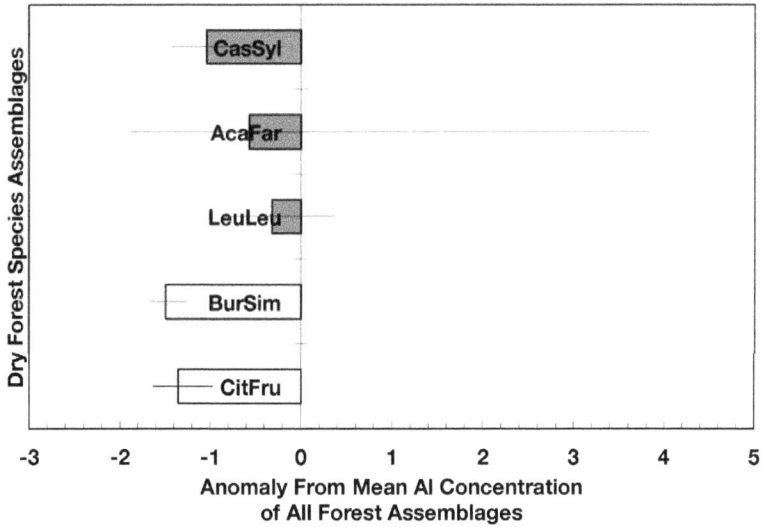

(j)

Figure 2. *Cont.*

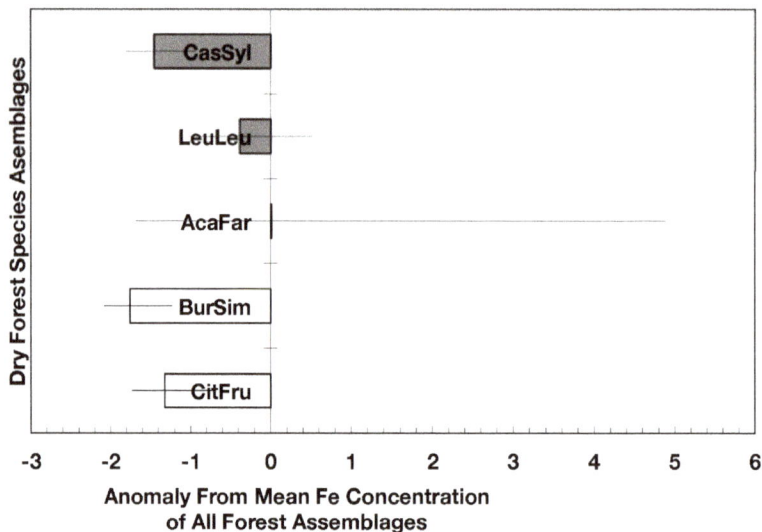

(k)

Figure 2. Mean (±standard error) element concentration anomalies in fallen leaves of dry forest assemblages of Puerto Rico. Units are in percent for carbon, nitrogen, and sulfur and mg/g for all others. Elements include carbon (**a**), nitrogen (**b**), sulfur (**c**), phosphorus (**d**), calcium (**e**), magnesium (**f**), potassium (**g**), sodium (**h**), manganese (**i**), aluminum (**j**) and iron (**k**). Codes for species assemblages and the number of replicates in parenthesis are (novel assemblages in bold and shaded bars): CitFru (3)—*Citharexylum fruticosum*, BurSim (9)—*Bursera simaruba*, **CasSyl (14)**—*Casearia sylvestris*, **LeuLeu (10)**—*Leucaena leucocephala*, and **AcaFar (5)**—*Acacia farnesiana*. Anomalies are based on a mean of all plots within 14 island-wide species assemblages. Anomalies with the same letter in (**a**) indicate forest assemblages that are not significantly different from each other. Means and standard errors in (**d**), (**g**), (**h**), (**i**), and (**k**) were back-transformed from ln-transformed data.

Native dry forest concentration anomalies tended to differ from island-wide averages more than novel dry forest concentration anomalies. Native forests had significantly lower P, Al, and Fe concentration anomalies than island-wide averages (p = 0.015, 0.0007, and 0.026, respectively), while novel forest anomalies for these elements did not differ from island-wide averages (Figure 2d,j,k). Native dry forests had higher Ca concentration anomalies than the island-wide average (p < 0.0001, Figure 2e). Both native and novel dry forests had lower Mn anomalies (p = 0.0001 and 0.053, respectively) than the island-wide average Figure 2i). Novel dry forests had significantly lower Na anomalies compared to island-wide averages (p = 0.0006, Figure 2h). These comparisons to island-wide averages do not take into account differences in geology, precipitation, or mean stand age among the assemblages. For example, around 75 percent of the native plots compared to a third of the novel plots were located on karst substrates, which potentially explains their positive Ca and negative P, Al, and Fe anomalies (cf. [13]). As well, native plots tended (p = 0.073) to be older than novel plots (28 versus 15 years). While older plots island-wide have been shown to have lower Al, Fe, and P concentrations in fallen leaves than younger plots [13]; these older plots are greater than 50 years, and only four of the native and three of the novel plots were in this age class. Nonetheless, these tests show how plot location influences, to some degree, the differences that may exist between novel and native forest stands within this dataset; the ANOVA results on the other hand account for effects of the co-variables.

Novel dry forest assemblages had lower C/N ratio anomalies than native assemblages (p = 0.045, Figure 3a); these anomalies were also significantly lower than the island-wide average (p = 0.0017) while those from native dry forests did not differ from the island-wide average. The C/P ratio anomalies did not differ between novel and native dry forest assemblages but novel dry forest assemblages as a group did show significantly lower anomalies ($p = 0.0013$) than the island-wide average (Figure 3b). The N/P ratio anomalies did not differ between novel and native dry forests (nor between the forest assemblages) though there was a slight indication that native forest N/P anomalies were higher than the island-wide average ($p = 0.061$; Figure 3c).

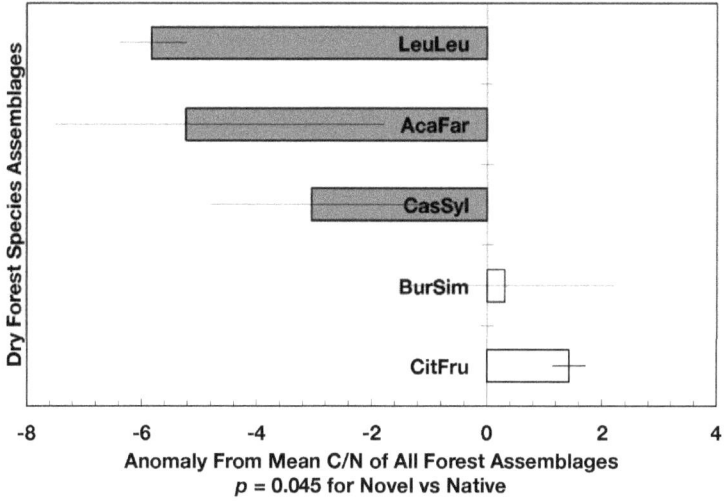

(a)

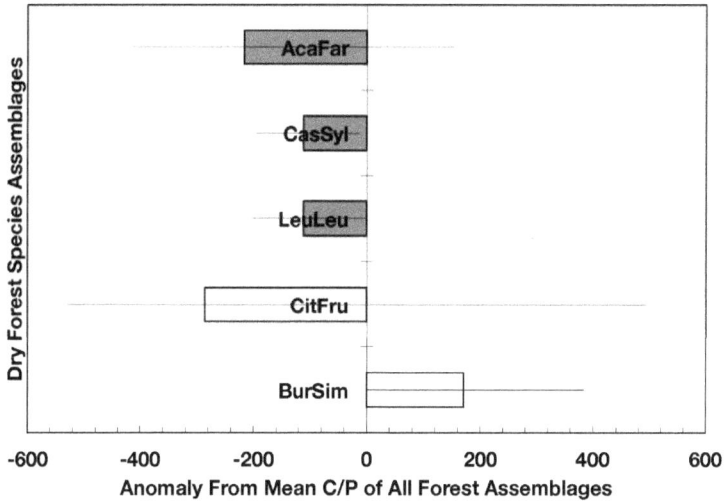

(b)

Figure 3. *Cont.*

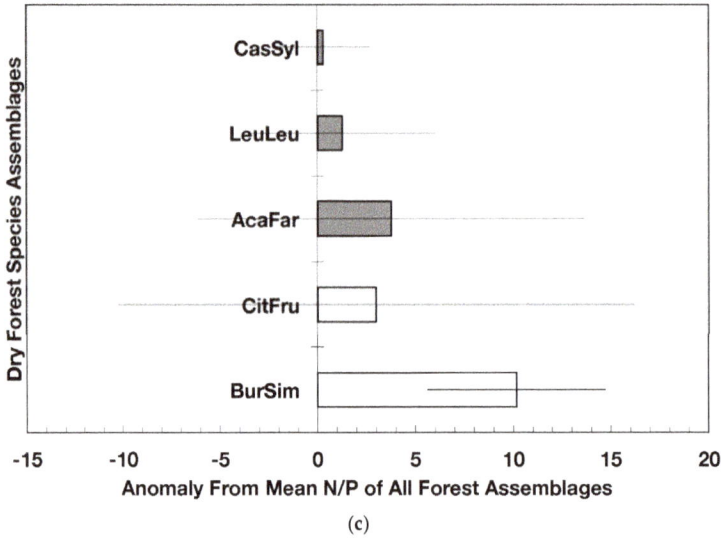

(c)

Figure 3. Anomalies for element ratios in dry forests assemblages of Puerto Rico: carbon to nitrogen (**a**), carbon to phosphorus (**b**), and nitrogen to phosphorus (**c**). Values are the mean (±standard error). Codes for species assemblages and the number of replicates in parenthesis are (novel assemblages in bold and shaded bars): CitFru (3)—*Citharexylum fruticosum*, BurSim (9)—*Bursera simaruba*, **CasSyl (15)**—*Casearia sylvestris*, **LeuLeu (10)**—*Leucaena leucocephala*, and **AcaFar (5)**—*Acacia farnesiana*. Anomalies are based on a mean of all plots within 14 island-wide species assemblages. Means and standard errors in (**a**) and (**b**) were back-transformed from ln-transformed data.

Table 2 shows the variation of element and mass accumulation by dry forest species assemblage reported by Erickson et al. [13]. They found no differences in forest floor or fallen leaf mass among species assemblages, but differences in concentration among assemblages (their Table 2) made a difference in the absolute accumulation of phosphorus, calcium, manganese, and aluminum.

Table 2. Mass (g/m^2) of forest floor litter (Total), fallen leaf mass, and chemical element mass (g/m^2) in total litter for dry forest species assemblages (those in bold are novel assemblages), and subtropical dry forest life zone forests. Other codes for species assemblages are: CitFru—*Citharexylum fruticosum*, BurSim—*Bursera simaruba*, **CasSyl**—*Casearia sylvestris*, **LeuLeu**—*Leucaena leucocephala*, **AcaFar**—*Acacia farnesiana*. Data are from Erickson et al. [13].

Species Assemblage and Replicates	Total Mass	Leaf Mass	C	N	S	P	Ca	Mg	K	Na	Mn	Al	Fe
CitFru (3)	285	94	131	4.2	1.2	0.18	7.9	0.79	0.55	0.07	0.04	0.49	0.65
BurSim (9)	666	434	333	10.1	1.9	0.26	19.2	1.39	0.9	0.23	0.09	0.87	0.63
CasSyl (14)	465	234	224	7.5	1.4	0.29	8.1	1.29	1.06	0.09	0.10	0.98	0.83
LeuLeu (10)	691	225	325	11.7	1.9	0.44	16.7	2.05	1.33	0.08	0.11	0.69	0.46
AcaFar (5)	325	89	154	4.8	1.2	0.15	7.6	0.66	0.44	0.03	0.03	0.68	0.70
Life Zone (12)	543	145	253	6.7	1.4	0.21	14.1	1.22	0.69	0.04	0.05	0.76	0.82

4. Discussion

4.1. Species Dominance, Density, and Novelty

Broadly speaking, the forests of Puerto Rico and the Caribbean are characterized by high species dominance and low species density [27]. Typically, the IV curves for Puerto Rican forests are steep with a short tail (few species) in contrast to those of continental tropical forests, which are less steep and have longer tails. Research in Puerto Rico shows that tree species dominance increases with increasing environmental stress such as low or excessive moisture availability or nutrient limitations [27]. As stands mature, tree species dominance decreases through succession. However, even in undisturbed and mature native forests, tree species dominance is higher in Puerto Rico than measured in Amazonian forests [28] but similar to other Caribbean islands [29].

While biogeographical isolation plays a role in the ability of species to disperse from continental regions to islands [30], also at play in the sharp differences in community structure and species density between the Caribbean and Amazonia is the recurrent hurricane disturbance regime of the Caribbean [31]. Lugo [31] argued that hurricane disturbances are partially responsible for the low and similar level of tree species density across all forest types in Puerto Rico (about 45 to 55 tree species per ha), and for the high dominance of species in insular forests [27,32]. However, irrespective of explanation, the empiric reality is that the IV curves for historic Puerto Rican forests are steep with short tails, reflecting high species dominance by a few species, and low species density. How do these characteristics change with novelty?

Novel forests contain novel mixtures of native and non-native tree species, while the naturalized species component is missing in the native stands. The addition of naturalized species increases the species density in novel systems at the 1-ha scale. The age of forest stands also affects species dominance and density. Tree species dominance decreases and species density increases in novel and native dry forests as they age, reflecting the accumulation of mostly native species through succession [23–25].

Puerto Rican novel forests in dry and wetter life zones are younger (<60 years) than island-wide historic native forests (>80 years) [33] because their establishment follows land abandonment, which island-wide started in the 1960s, accelerated over the succeeding decades, and finally slowed down at the onset of the 21st century [34]. Geographic scale is important because species area curves of forests in Puerto Rico are steep [26,27], which means that species density changes rapidly with area sampled. However, the species area curve for novel forests is initially less steep than that of historic forests, but reaches a higher plateau than historic forests at the 1-ha scale [26]. We believe that more research is needed in the species/area relationships during the colonizing phases of novel forests, as well as the landscape aspects of native species persistence in novel landscapes.

4.2. The Importance of Species

As moisture conditions change from wet to moist to dry, fewer species are available to colonize degraded sites and those that colonize exert more dominance over their communities. Dominant species occupy more space and process and accumulate more resources than non-dominant species. Thus, the characteristics of novel dry forests will be more dependent on the characteristics of naturalized species colonizing abandoned, degraded sites, because a small group of those species will become very abundant and dominant in the emerging forests.

The ways in which individual species can influence the structure and functioning of forests is through their natural history traits, including their physiognomy, growth characteristics, phenological rhythms, and stoichiometry. Hulshof et al. [35] discuss how species functional traits vary and influence dry forest structure and functioning, as did Lopezaraiza-Mikel et al. [36] for the phenology of dry forest tree species. We illustrate the role of species in forest ecosystem functioning with the stoichiometry of fallen leaf litter.

4.3. Stoichiometry of Novel and Native Dry Forests

The results on the influence of species assemblages on forest-level characteristics (fallen leaf chemistry) underscore the point made earlier that the species composition and age of forests can have relevance to mass and nutrient fluxes of forests. The high chemical quality of fallen leaf litter in novel dry forests (low C/N and C/P) compares favorably with the chemical quality of species assemblages known for their high primary productivity and rapid nutrient cycling [37]. Thus, the fallen leaf chemistry of novel dry forests reflects substrates that are favorable for rapid decomposition and recycling (Table 2).

Erickson et al. [13] found that all dry forests (novel or native) in the dry life zone had significant differences in fallen leaf chemistry when compared to forests in other life zones of Puerto Rico. The dry forests fallen leaves had the lowest concentration of carbon on the island (42 percent), the highest concentrations of calcium (29 mg/g) and magnesium (3.6 mg/g), and the lowest concentrations of sodium (0.11 mg/g). When Puerto Rico's dry forest results are compared to ten leaf litter chemistry values reported by Jaramillo and Sanford [38] for dry forests in Mexico, India, Belize, Australia, and Puerto Rico (site not included in Erickson et al. [13]), the following patterns emerge: the potassium and magnesium concentrations in Puerto Rico's dry life zone are lower than those in all but two locations, while the calcium, nitrogen, and phosphorus concentrations are higher than those of all but one location. This comparison suggests more labile leaf litter in Puerto Rican dry forests given the higher nitrogen and phosphorus concentrations. Jaramillo and Sanford [38] also report total mass and nutrient accumulation in litter for the same locations discussed above. Comparisons show lower mass accumulation and similar nutrient content in the leaf litter of Puerto Rican dry forests. A lower leaf litter mass is consistent with more labile leaf litter.

Erickson et al. [13] also found that the youthful age of forests (<60 years) resulted in higher phosphorus, iron, and aluminum concentrations and lower carbon concentrations and C/P and N/P in fallen leaf litter compared to older stands island-wide. They concluded that the availability of nitrogen (discussed below) and phosphorus might be greater in today's island's forests than before deforestation. Do novel dry forests exhibit the same element concentrations as native dry forests? The answer to this question is not categorical because the pattern of element concentration for fallen leaf litter at times varied with forest assemblage, mean stand age, geological substrate, or precipitation. Nevertheless, we did find trends and significant differences in element concentrations and ratios between novel and native dry forests in Puerto Rico (Figures 2 and 3). These differences suggest potential acceleration of nutrient and mass fluxes due to the faster decomposition of leaf litter with lower C/N and Ca, i.e., more labile leaf litter. Similarly, the concentration and ratio anomalies for the dry forest communities compared to the grand mean of all plots across the island also suggest potential acceleration of nutrient and mass fluxes as a result of differences between species assemblages. For example, the lower than average C/N and C/P in novel dry forests (Figure 3a,b) support the notion of a faster leaf decomposition rate, which is reflected in a lower accumulation of leaf litter in these forests (Table 2). As we discuss below, species composition and dominance in relation to age and geographic scale play a role not only in the stoichiometry of forest stands but also in their influence in the restoration of degraded sites and continuing process of community assembly.

4.4. Implications of Novelty to Functioning and Conservation of Dry Forests

The landscape of Puerto Rico, like Anthropocene landscapes everywhere, is dynamic and in constant change as a result of human activity [39]. Nevertheless, long-term processes such as the forest transition from a deforested to a forested landscape exert their influence on the age (young) of forests and the particular successional stage in which they find themselves. Molina Colón et al. [23] found that the early stages of dry forest establishment result in patches of dissimilar novel species assemblages (similarity index of 26 percent), whose species composition is dictated by seed sources and their dispersal vectors. The spatial diversity of forest patches may share common species, such as *Leucaena*, a nitrogen fixing tree, that grows throughout the dry life zone independently of the type of past land use [40]. However, the combination of all patches support more species than found in individual patches, even when they are in close proximity to each other.

Ramjohn et al. [5] found that up to 86 percent of the tree species in a protected area of the dry life zone, were found in novel forest fragments dispersed through the landscape. Fragments of 0.04 ha contained about 45 percent of the protected area tree species while fragments of 33 ha contained 75 percent of the reference species. Thus, the dissimilarity of species combinations among patches counteracts the high dominance and low species density of individual forest patches so that the landscape as a whole conserves the species richness of the dry forest formation. Therefore, novelty of species composition contributes to the colonization of degraded sites and the conservation of native species that otherwise could not colonize degraded sites abandoned after agricultural activities.

Nitrogen-fixing species such as *Leucaena* have the capacity to colonize highly degraded soils and re-start arrested succession [13,40,41]. Erickson et al. [13] found that the basal area of nitrogen fixing species in 14 species assemblages in Puerto Rico was positively related to leaf litter phosphorus concentration and negatively correlated with leaf litter C/N ratio. Compared to the native dry forests, the lower C/N ratios (Figure 3a) found in fallen leaf litter from novel forests suggest faster litter decomposition rates [42], which illustrate how species stoichiometry influences ecosystem processes. Moreover, novel dry forests contain more nitrogen fixing species than native dry forests (Figure 4). Because of the abundance of nitrogen fixers and high dominance of naturalized species in novel forests, an individual or group of species with particular concentrations of nutrients can determine the nutrient and carbon fluxes and stoichiometric ratios in forests (Figure 3).

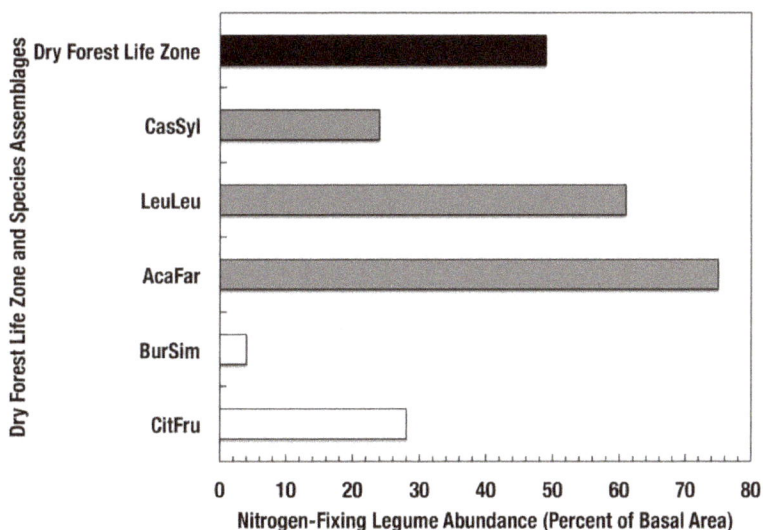

Figure 4. Percent of stand basal area of nitrogen-fixing species in dry forest assemblages of Puerto Rico [13]. Codes for species assemblages are (shaded bars correspond to novel assemblages identified in bold below): CitFru—*Citharexylum fruticosum*, BurSim—*Bursera simaruba*, **CasSyl**—*Casearia sylvestris*, **LeuLeu**—*Leucaena leucocephala*, and **AcaFar**—*Acacia farnesiana*. Clear bars are native dry forests and the solid bar represents forests of the subtropical dry forest life zone.

There is no a priori reason to expect that species from different geographical origins would have particular nutrient concentrations or even unique life history traits. However, naturalized species that dominate novel forests do so under strong environmental selective pressure. In dry forests, these include the low moisture conditions (which limit the number of species that overcome this limitation), a deforested landscape (which selects for pioneer species), and degraded soil conditions, which select for particular nutrient use efficiencies and life history strategies. Thus, the stoichiometry of a species can set in motion a cascade effect through nutrient pathways that can influence rate of carbon fluxes (e.g., decomposition or primary productivity) and associated faunal and microbial organisms. This has already been shown in monospecific dry forest stands under extreme drought conditions by Barberena Arias [43].

Barberena Arias [43] found that the air temperature and humidity condition under the canopy of individual dry forest species was particular to the species. The tree species also regulated the amount and concentration of elements accumulating on the forest floor. These two conditions, in turn, had an effect on the richness, species composition, and trophic condition of soil arthropods decomposing litter. The N/P ratio was inversely related to arthropod density. On a landscape with a patchwork of dissimilar novel and native forest communities, the "priming" chemical effect of the dominant species might create "islands" with diverse nutrient levels and proportions that, in turn, affect soil organisms and ecosystem processes. This is a hypothesis to be tested with future research.

In summary, the implications of novelty to the ecology and conservation of Puerto Rico's dry forests include (1) increases in species dominance and density at the 1-ha scale; (2) restoring forest conditions on degraded sites; (3) facilitating the regeneration and growth of native species on degraded sites; (4) establishing litter patches with diverse concentrations of chemical elements; (5) increasing the presence of nitrogen-fixing species and the availability of nitrogen in forest stands; (6) increasing the concentration of some elements in leaf litter above those observed in native forest stands; (7) decreasing the C/N ratio below those observed in native dry forest stands; and (8) potentially accelerating the flux of mass and nutrients by accelerating litter decomposition or primary productivity rates.

Acknowledgments: This study was conducted in cooperation with the University of Puerto Rico. We benefited from the collaborative research and discussions with numerous students and colleagues. Gisel Reyes collaborated with the literature search. We thank Ernesto Medina for substantial comments on the manuscript.

Author Contributions: Both authors contributed equally to the text and interpretation of results with Lugo focusing on the literature review and Erickson on the statistical analyses.

Conflicts of Interest: Authors declare no conflict of interests.

References

1. Holdridge, L.R. *Life Zone Ecology*; Tropical Science Center: San José, Costa Rica, 1967.
2. Murphy, P.G.; Lugo, A.E. Ecology of tropical dry forest. *Annu. Rev. Ecol. Syst.* **1986**, *17*, 67–88. [CrossRef]
3. Murphy, P.G.; Lugo, A.E. Dry forests of Central America and the Caribbean. In *Seasonally Dry Tropical Forests*; Bullock, S.H., Mooney, H.A., Medina, E., Eds.; Cambridge University Press: Cambridge, UK, 1995; pp. 9–34.
4. Maass, J.M. Conversion of tropical dry forest to pasture and agriculture. In *Seasonally Dry Tropical Forests*; Bullock, S.H., Mooney, H.A., Medina, E., Eds.; Cambridge University Press: Cambridge, UK, 1995; pp. 399–422.
5. Ramjohn, I.A.; Murphy, P.G.; Burton, T.M.; Lugo, A.E. Survival and rebound of Antillean dry forests: Role of forest fragments. *For. Ecol. Manag.* **2012**, *284*, 124–132. [CrossRef]
6. Rudel, T.K.; Lugo, M.P.; Zichal, H. When fields revert to forest: Development and spontaneous reforestation in post-war Puerto Rico. *Prof. Geogr.* **2000**, *52*, 386–397. [CrossRef]
7. Rudel, T.K. *Tropical Forests: Regional Paths of Destruction and Regeneration in the Late Twentieth Century*; Columbia University Press: New York, NY, USA, 2005.
8. Lugo, A.E.; Helmer, E. Emerging forests on abandoned land: Puerto Rico's new forests. *For. Ecol. Manag.* **2004**, *190*, 145–161. [CrossRef]
9. Hobbs, R.J.; Arico, S.; Aronson, J.; Baron, J.S.; Bridgewater, P.; Cramer, V.A.; Epstein, P.R.; Ewel, J.J.; Klink, C.A.; Lugo, A.E.; et al. Novel ecosystems: Theoretical and management aspects of the new ecological world order. *Glob. Ecol. Biogeogr.* **2006**, *15*, 1–7. [CrossRef]
10. Hobbs, R.J.; Higgs, E.S.; Hall, C.M. (Eds.) *Novel Ecosystems: Intervening in the New Ecological World Order*; Willey-Blackwell: West Sussex, UK, 2013.
11. Pearce, F. *The New Wild: Why Invasive Species Will Be Nature's Salvation*; Beacon Press: Boston, MA, USA, 2015.
12. Martinuzzi, S.; Lugo, A.E.; Brandeis, T.J.; Helmer, E.H. Geographic distribution and level of novelty of Puerto Rican forests. In *Novel Ecosystems: Intervening in the New Ecological World Order*; Hobbs, R.J., Higgs, E.S., Hall, C., Eds.; Wiley: Oxford, UK, 2013; pp. 81–87.
13. Erickson, H.E.; Helmer, E.H.; Brandeis, T.J.; Lugo, A.E. Controls of fallen leaf chemistry and forest floor element masses in native and novel forests across a tropical island. *Ecosphere* **2014**, *5*, 48. [CrossRef]
14. Brandeis, T.J.; Helmer, E.; Vega, H.M.; Lugo, A.E. Climate shapes the novel plant communities that form after deforestation in Puerto Rico and the U.S. Virgin Islands. *For. Ecol. Manag.* **2009**, *258*, 1704–1718. [CrossRef]

15. Whittaker, R.H. Dominance and diversity in land plant communities: Numerical relations of species express the importance of competition in community function and evolution. *Science* **1965**, *147*, 250–260. [CrossRef] [PubMed]

16. Whittaker, R.H. *Communities and Ecosystems*; The Macmillan Company: Toronto, ON, Canada, 1970.

17. Brown, S.; Lugo, A.E. The storage and production of organic matter in tropical forests and their role in the global carbon cycle. *Biotropica* **1982**, *14*, 161–187. [CrossRef]

18. Martínez, Y.A. Biomass distribution and primary productivity of tropical dry forests. In *Seasonally Dry Tropical Forests*; Bullock, S.H., Mooney, H.A., Medina, E., Eds.; Cambridge University Press: Cambridge, UK, 1995; pp. 326–345.

19. Gentry, A.H. Diversity and floristic composition of neotropical dry forests. In *Seasonally Dry Tropical Forests*; Bullock, S.H., Mooney, H.A., Medina, E., Eds.; Cambridge University Press: Cambridge, UK, 1995; pp. 146–194.

20. Acevedo-Rodríguez, P.; Strong, M.T. Floristic richness and affinities in the West Indies. *Bot. Rev.* **2008**, *74*, 5–36. [CrossRef]

21. Kiehn, M. Invasive alien species and islands. In *The Biology of Island Floras*; Bramwell, D., Caujapé-Castells, J., Eds.; Cambridge University Press: Cambridge, UK, 2011; pp. 365–384.

22. Francis, J.K.; Liogier, H.A. *Naturalized Exotic Tree Species in Puerto Rico*; General Technical Report SO-82; USDA, Forest Service, Southern Forest Experiment Station: New Orleans, LA, USA, 1991.

23. Molina Colón, S.; Lugo, A.E.; Ramos, O. Novel dry forests in southwestern Puerto Rico. *For. Ecol. Manag.* **2011**, *262*, 170–177. [CrossRef]

24. Molina-Colón, S.; Lugo, A.E. Recovery of a subtropical dry forest after abandonment of different land uses. *Biotropica* **2006**, *38*, 354–364. [CrossRef]

25. Molina Colón, S. Long-Term Recovery of a Caribbean Dry Forest after Abandonment of Different Land Uses in Guánica, Puerto Rico. Dissertation, University of Puerto Rico, Río Piedras, Puerto Rico, 1998.

26. Lugo, A.E.; Brandeis, T.A. New mix of alien and native species coexist in Puerto Rico's landscapes. In *Biotic Interactions in the Tropics. Their Role in the Maintenance of Species Diversity*; Burslem, D., Pinard, M., Hartley, S., Eds.; Cambridge University Press: Cambridge, UK, 2005; pp. 484–509.

27. Lugo, A.E. Los bosques. In *Biodiversidad de Puerto Rico. Vertebrados Terrestres y Ecosistemas*; Joglar, R.L., Ed.; Editorial del Instituto de Cultura Puertorriqueña: San Juan, PR, USA, 2005; pp. 395–548.

28. Jardim, F.C.S.; Hosokawa, R.T. Estrutura da floresta equa- torial úmida da Estação Experimental de Silvicultura Tropical do INPA. *Acta Amazon.* **1986**, *16–17*, 411–508.

29. Beard, J.S. *The Natural Vegetation of the Windward and Leeward Islands. Oxford Forestry Memoirs 21*; The Clarendon Press: Oxford, UK, 1949.

30. Whittaker, R.J. *Island Biogeography: Ecology, Evolution, and Conservation*; Oxford University Press: Oxford, UK, 1998.

31. Lugo, A.E. Visible and invisible effects of hurricanes on forest ecosystems: An international review. *Austral Ecol.* **2008**, *33*, 368–398. [CrossRef]

32. Lugo, A.E. Dominancia y diversidad de plantas en Isla de Mona. *Acta Cient.* **1991**, *5*, 65–71.

33. Kennaway, T.; Helmer, E.H. The forest types and ages cleared for land development in Puerto Rico. *GISci. Remote Sens.* **2007**, *44*, 356–382. [CrossRef]

34. Brandeis, T.J.; Turner, J.A. *Puerto Rico's Forests, 2009*; USDA Forest Service Resource Bulletin SRS-191; Southern Research Station: Ashville, NC, USA, 2013.

35. Hulshof, C.M.; Yrízar, A.M.; Burques, A.; Boyle, B.; Enquist, B.J. Plant functional trait variation in tropical dry forests: A review and synthesis. In *Tropical Dry Forests in the Americas: Ecology, Conservation, and Management*; Sánchez-Azofeifa, A., Powers, J.S., Fernandes, G.W., Quesada, M., Eds.; CRC Press, Taylor & Francis Group: Boca Raton, FL, USA, 2014; pp. 129–140.

36. Lopezaraiza-Mikel, M.; Quesada, M.; Álvarez-Añorve, M.; Ávila-Cabadilla, L.; Martén-Rodriguez, S.; Calvo-Alvarado, J.; do-Santo, M.M.; Fernandes, G.W.; Sánchez-Azofeifa, A.; de J. Aguilar-Aguilar, M.; et al. Phenological patterns of tropical dry forests along latitudinal and successional gradients in the Neotropics. In *Tropical Dry Forests in the Americas*; Sánchez-Azofeifa, A., Powers, J.S., Fernandes, G.W., Quesada, M., Eds.; CRC Press, Taylor & Francis Group: Boca Raton, FL, USA, 2014; pp. 101–128.

37. Abelleira Martínez, O.; Lugo, A.E. Post sugarcane succession in moist alluvial sites in Puerto Rico. In *Post-Agricultural Succession in the Neotropics*; Myster, R.W., Ed.; Springer: New York, NY, USA, 2008; pp. 73–92.

38. Jaramillo, V.J.; Sanford, R.L. Nutrient cycling in tropical deciduous forests. In *Seasonally Dry Tropical Forests*; Bullock, S.H., Mooney, H.A., Medina, E., Eds.; Cambridge University Press: Cambridge, UK, 1995; pp. 346–361.

39. Lugo, A.E. Can we manage tropical landscapes? An answer from the Caribbean perspective. *Landsc. Ecol.* **2002**, *17*, 601–615. [CrossRef]

40. Parrotta, J.A. *Leucaena leucocephala (Lam.) de Wit. Leucaena*; SO-ITF-SM-52; International Institute of Tropical Forestry, USDA Forest Service: Río Piedras, Puerto Rico, 1992.

41. Santiago García, R.J.; Colón, S.M.; Sollins, P.; van Bloem, S.J. The role of nurse trees in mitigating fire effects on tropical dry forest restoration: A case study. *Ambio* **2008**, *37*, 604–608. [CrossRef] [PubMed]

42. Berg, B.; McClaugherty, C. *Plant Litter: Decomposition, Humus Formation, Carbon Sequestration*; Springer: Berlin, Germany, 2003.

43. Barberena-Arias, M.F. Single Tree Species Effects on Temperature, Nutrients and Arthropod Diversity in Litter and Humus in the Guánica Dry Forest. Ph.D. Thesis, University of Puerto Rico, Río Piedras, Puerto Rico, 2008.

![forests logo] *forests*

MDPI

Article

Traits and Resource Use of Co-Occurring Introduced and Native Trees in a Tropical Novel Forest

Jéssica Fonseca da Silva [1,2,3,*], Ernesto Medina [1,3,4] and Ariel E. Lugo [1]

1   International Institute of Tropical Forestry, USDA Forest Service, Río Piedras 00926-1115, Puerto Rico; medinage@gmail.com (E.M.); alugo@fs.fed.us (A.E.L.)
2   Center for Applied Tropical Ecology and Conservation, University of Puerto Rico, San Juan 00931, Puerto Rico
3   Department of Biology, University of Puerto Rico-Río Piedras, San Juan 00936-8377, Puerto Rico
4   Centro de Ecología, Instituto Venezolano de Investigaciones Científicas, Caracas 1020-A, Venezuela
*   Correspondence: jefonsecasilva@gmail.com

Received: 1 July 2017; Accepted: 6 September 2017; Published: 12 September 2017

Abstract: Novel forests are naturally regenerating forests that have established on degraded lands and have a species composition strongly influenced by introduced species. We studied ecophysiological traits of an introduced species (*Castilla elastica* Sessé) and several native species growing side by side in novel forests dominated by *C. elastica* in Puerto Rico. We hypothesized that *C. elastica* has higher photosynthetic capacity and makes more efficient use of resources than co-occurring native species. Using light response curves, we found that the photosynthetic capacity of *C. elastica* is similar to that of native species, and that different parameters of the curves reflected mostly sun light variation across the forest strata. However, photosynthetic nitrogen use-efficiency as well as leaf area/mass ratios were higher for *C. elastica*, and both the amount of C and N per unit area were lower, highlighting the different ecological strategies of the introduced and native plants. Presumably, those traits support *C. elastica*'s dominance over native plants in the study area. We provide empirical data on the ecophysiology of co-occurring plants in a novel forest, and show evidence that different resource-investment strategies co-occur in this type of ecosystem.

Keywords: introduced species; leaf C and N densities; novel forests; photosynthetic nitrogen use-efficiency; leaf mass per area

1. Introduction

The Anthropocene Epoch is associated with rapidly changing environmental conditions and high rates of species introductions, leading to the formation of novel forests [1]. These emerging forests contain species assemblages that include co-occurring introduced and native tree species [2,3]. Novel forests comprise about 35% of global terrestrial ecosystems [4] and are expected to become more common in the future. There has been much debate in recent years about the implications of novel forests for biodiversity. However, little empirical data are yet available to understand how the tree biota might respond to changing environmental conditions and how this might affect the functioning of present and future forests [5,6].

It is well known that introduced species—commonly the dominant tree species in novel forests—are generally considered a risk for biodiversity due to their ability to outperform native species in terms of productivity, reproductive capacity, and recruitment (e.g., [7,8]). In general, studies report faster growth, higher maximum assimilation rate at saturating light intensities, higher dark respiration and transpiration, more efficient use of resources, and faster nutrient cycling for introduced species compared to native species [9–17]. Usually, introduced species act as pioneers [18] during succession, giving them an advantage in the colonization of disturbed and degraded sites. The ecophysiology of native species can be influenced by the effects that introduced species have on novel ecosystems [19], which may be large when introduced species are abundant and dominate the plant community [19,20].

Less is known about species traits and mechanisms that allow native plants to survive and thrive when a novel forest is formed. In a community containing co-occurring native and introduced species, those able to compete efficiently for the same resources, and/or use them in different ways are more likely to persist [10,15]. Presumably, native and introduced species occupy different positions in the leaf economic spectrum [10,21–24], which describes the nutrient and organic matter investment of plants on leaf structure and functioning. However, we are unaware of empirical studies demonstrating this pattern in tropical mature novel forests.

We chose a novel forest in Puerto Rico dominated by the introduced tree *Castilla elastica* Sessé, to study the ecophysiology of co-occurring native and introduced trees. Based on previous studies in the area [20,25] and on the literature, we anticipate that *C. elastica* has a strong influence on the abiotic conditions in this community, which might have effects on the ecophysiology of native species. We hypothesize that *C. elastica* has higher photosynthetic capacity than native species, probably related to its highly efficient use of resources; and that *C. elastica* and native species occupy different regions of the leaf economic spectrum, particularly regarding to leaf area/mass ratios and concentrations of C and N per unit leaf area. To test our hypotheses, we use photosynthesis light-response curves, and resource use and resource investment indexes to compare species in the community. We also measure the light availability across the forest.

2. Materials and Methods

2.1. Castilla elastica—A Dominant Introduced Tree

Castilla elastica is one example of a dominant introduced naturalized tree in Puerto Rico, originally from Central and South America [26,27]. The introduction of *C. elastica* to Puerto Rico happened at the beginning of 20th century according to the local Agricultural Experiment Station in Mayagüez, Puerto Rico. Originally, the government attempted unsuccessfully to produce latex from *C. elastica*, but also used it as shade tree in coffee plantations. Today, novel forests of *C. elastica* are present throughout Puerto Rico, concentrating in the humid northwest region. These forests covered about 100 hectares in the 1990s, corresponding to less than one percent of the country's land area [26].

2.2. Study Area

We studied plants at the biological reserve El Tallonal located in the municipality of Arecibo (18°24′27″ N 66°43′53″ W), which is classified as a subtropical moist forest [28]. The predominant soil type at the sinkholes of El Tallonal, where *C. elastica* is dominant, is the Oxisol of the series *Almirante* [29]. The annual mean temperature and precipitation are 25.5 °C and 1295 mm, respectively. The dry season is from January to April, and the wettest months are July to September.

Agriculture and cattle grazing were common activities at El Tallonal until the 1950s, and it is likely that *C. elastica* was introduced in the area around 1940s and then abandoned few years later. Forest regeneration occurred naturally after land abandonment, and aerial photographs show that areas covered by novel forests of *C. elastica* had been growing for about 50 years by 2005. Currently, *C. elastica* has a mean species Importance Value Index of 37% (a composite index of relative density, cover, and frequency) in these forests, indicating that the species occupies a dominant position [25]. The forest is referred to hereafter as *Castilla* novel forest. Modifications on species composition and functioning in the study area have been associated with the dominance of this species [20,25].

2.3. Sampling

We measured in situ photosynthetic light responses of leaves of *C. elastica* and of co-occurring native species (Table 1). To that end, we took advantage of two 26 m-tall meteorological towers standing in the study site that allowed data collection at different heights or forest strata, i.e., canopy = 25 m in height, subcanopy = 15 m in height and understory = ground level to two meters in height. Large trees ($\geq$10 cm of diameter at breast height (DBH)), assumed as adult trees, were measured in the canopy and in the subcanopy. Saplings (1.5 $\leq$ DBH < 2.5 cm , between one and two m in height), and juvenile plants (seedlings and young individuals of 10 to 40 cm in height) were measured in the understory. The number of trees and species that could be measured and sampled from the towers was limited, imposing restrictions on the statistical analyses. However, the same trees could be repeatedly measured throughout an entire year, compensating partially for the small number of individuals measured.

Sampling campaigns were performed during periods of contrasting rainfall and temperature. For practical purposes, below we refer to each measuring period as follows: December to January as December 2008; March to April as March 2009; June to July as June 2009; and October to November as November 2009.

In the canopy, we selected two trees of *C. elastica* and two of native species. In the subcanopy, we measured a *C. elastica* tree and one tree of native species (the only one present at this forest stratum). In the understory, we measured four saplings: two of *C. elastica* and two of native species. For juvenile plants, we selected six individuals on each sampling event (three of *C. elastica* and three of native species), and measured one leaf per individual, instead of two. Each pair of leaves and trees (native and introduced) was measured at the same level to ensure they were exposed to a similar light environment. In total, we obtained at least two light response curves, per species, per measuring period (except in the case of a few juvenile plants that were represented by a single curve), for a total of 92 light response curves. Plants in the subcanopy and understory were only measured during December 2008, March 2009, and June 2009.

Table 1. List of species analyzed, their botanical families, and the forest stratum at novel *Castilla* forests. Taxonomic classification follows [26,30].

Species	Family	Forest Stratum
Casearia guianensis (Aubl.) Urban [N]	Flacourtiaceae	Us
Casearia sylvestris Sw. [N]	Flacourtiaceae	SC
Castilla elastica Sessé [In]	Moraceae	C, SC, Us
Chrysophyllum argenteum Jacques [N]	Sapotaceae	Us
Cordia alliodora (Ruiz & Pav.) [N]	Boraginaceae	C
Faramea occidentalis (L.) A. Rich [N]	Rubiaceae	Us
Guarea guidonia (L.) Sleumer [N]	Meliaceae	Us
Ocotea floribunda (Sw.) Mez [N]	Lauraceae	C
Ocotea leucoxylon (Sw.) Mez [N]	Lauraceae	C
Thouinia striata Radlk [Ne]	Sapindaceae	Us
Trichilia pallida Sw [N]	Meliaceae	Us

In = introduced naturalized, N = native, and Ne = native endemic species; C = Canopy, SC = Subcanopy and Us = Understory.

Environmental data were recorded using a HOBO micro-station data logger (H21-002, Onset Computer Corporation, Bourne, MA, USA). Three micro stations were installed in each tower and forest strata: canopy (26 m in height), subcanopy (15 m) and understory (1.5 m). Air temperature ($^\circ$C), photosynthetic photon flux density (PPFD, μmol m^{-2} s^{-1}) and air relative humidity (percent) were recorded from December 2008 to November 2009. Care was taken to ensure that environmental data recorded by loggers were representative of leaf conditions.

2.4. Light Response Curves and Leaf Harvesting Protocol

Light response curves were measured in the field, using a portable infrared gas analyser (LC*pro*+, ADC BioScientific Ltd., Hertfordshire, UK). The LC*pro*+ analyses the difference between ambient CO_2 concentration and the concentration of CO_2 in a leaf chamber (Δc), and calculates CO_2 assimilation rate (A, in μmol m^{-2} s^{-1}) and stomatal conductance to water vapour (gs, in mol m^{-2} s^{-1}).

Gas exchange measurements were performed in the morning, after leaves had received natural illumination for at least two hours. In the canopy and subcanopy, we measured CO_2 assimilation in response to increasing light intensity from dark conditions (zero) to 2000 µmol m^{-2} s^{-1} of photosynthetic photon flux density (PPFD), using the following sequence: 0, 100, 250, 500, 1000, 1500 and 2000 µmol m^{-2} s^{-1}. In the understory, we used PPFD from zero to 600 µmol m^{-2} s^{-1}, using the following sequence: 0, 50, 100, 200, 400 and 600 µmol m^{-2} s^{-1}. We allowed enough time between changes in light intensity to ensure leaf equilibration. Temperature was kept constant at 25 °C and CO_2 concentration was maintained at about 380 volume per million (vpm) to avoid short term variations that would render measurements meaningless. After conducting each light response curve, leaves were harvested to measure the leaf area, leaf mass and leaf carbon and nitrogen concentrations.

Light response curve data were then analysed using a non-linear mixed model as follows: $A = A_{max} * (1 - e^{-\alpha \, (PPFD-LCP)})$ [31]. Sigma Plot (v.11.0, Systat Software Inc., San Jose, CA, USA) was used to fit the curves and calculate the following parameters: maximum photosynthetic rate at saturating light intensities (A_{max}), light compensation point (LCP), light saturation point (LSP), dark respiration and quatum yield. We also calculated the stomatal conductance of water vapour at maximum A rates (gs_{max}, in mol m^{-2} s^{-1}), the intrinsic water use efficiency at maximum rates (WUE$_i$) as the molar ratio of A_{max} and gs_{max} (in µmol mol^{-1}), and the photosynthetic nitrogen use efficiency (PNUE) per unit of leaf nitrogen at maximum assimilation rates (PNUE = A_{max}/nitrogen content, in µmol mol^{-1} s^{-1}).

2.5. Samples Processing

Leaf area of fresh leaves was measured using a leaf area meter (LI-3100, LI-COR Biosciences, Lincoln, NE, USA). After that, leaves were dried in the oven at 65 °C for at least three days, and then weighed to obtain their dry mass. Leaf surface area (m^2), leaf dry mass (kg), and leaf mass per area (LMA in kg m^{-2}), were determined to compare leaf structure among species and groups [32]. Leaf carbon and nitrogen concentrations were measured by macro dry combustion using a LECO CNS-2000 analyser (LECO Corporation, St. Joseph, MI, USA). Molar N concentrations are given on both area (m^2) and mass basis (kg).

2.6. Statistical Analyses

All analyses were performed using JMP 13.0 (SAS Institute Inc., Cary, NC, USA). Differences in environmental conditions between forest strata were tested using an Analysis of Variance (ANOVA) and posterior Tukey test for differences among measuring periods. Normality tests showed that biological parameters were not normally distributed, therefore we used nonparametric tests analogous to one-way analysis of variance (Wilcoxon/Kruskal-Wallis test, at maximum $p = 0.05$), or multiple pair-wise contrasts (Wilcoxon z), for comparing the photosynthetic capacity parameters, LMA, and C and N concentrations.

The set of native species available for measurement around the tower area was treated as a single group as there were no significant statistical differences in their photosynthetic parameters. This group was compared to the set of *C. elastica* measurements at every sampling season and stratum (canopy, subcanopy and understory). Linear regression analysis was used to test the relationship between leaf area and leaf mass, and the concentrations of C and N.

3. Results

3.1. Environmental Conditions

Mean precipitation and temperature range for each period were the following: December (149 mm, and 18 to 27 °C), March (33 mm, and 20 to 28 °C), June (144 mm, and 23 to 31 °C) and November (189 mm, and 22 to 30 °C). Throughout the year, mean temperature varied from 20 to 24 °C with a peak in June. The relative humidity was over 80% for almost the entire study period.

Light intensity (PPFD$_{mean}$) varied from the understory to the canopy by two orders of magnitude: from about 6 to more than 700 μmol m^{-2} s^{-1} (Figure 1). The average PPFD$_{mean}$ per day in the canopy was 686 μmol m^{-2} s^{-1}, and the average daily PPFD$_{sum}$ was 30 mol m^{-2} day^{-1}. The PPFD$_{mean}$ received in the subcanopy and in the understory were only 10% and 1% of that received in the canopy, respectively. The photoperiod (the total hours in a day during which the PAR sensors record incident and diffuse sunlight in each stratum), varied throughout the year between 12 and 13 h in the canopy and subcanopy, and between 8 and 11 in the understory. Both the PPFD$_{mean}$ and PPFD$_{sum}$ did not differ between dates in the canopy, but they were significantly higher during March and June season in both the subcanopy and the understory, coinciding with the pronounced leaf shedding of *C. elastica* trees [20,25].

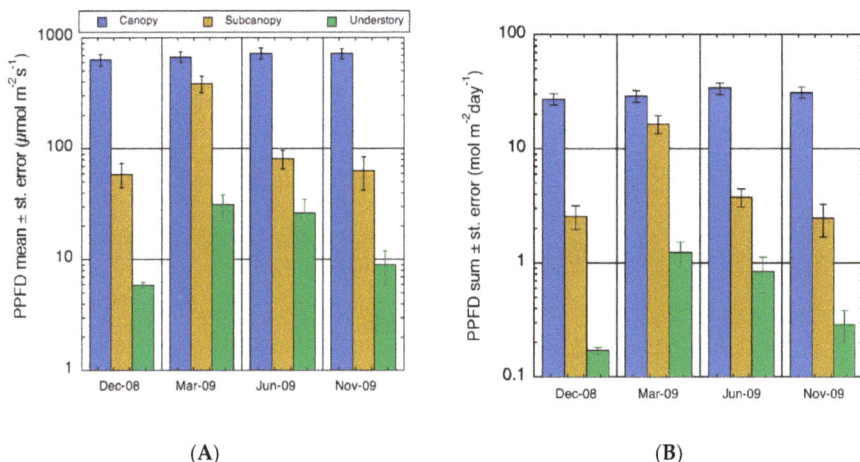

Figure 1. Environmental variables at *Castilla* novel forest. (**A**) Mean photosynthetic photon flux density (PPFD$_{mean}$) and (**B**) total photosynthetic photon flux density (PPFD$_{sum}$) per day for the different forest strata during four measuring periods (*n* = 60–62).

3.2. Leaf Dimensions

Castilla elastica produced larger and heavier leaves compared to the group of native species (Figure 2). In addition, their LMA was lower and decreased 3.3 times from the canopy to the understory. For the group of native species, the same pattern was observed but the decrease in LMA was less pronounced (2.2 times), indicating lower plasticity of this parameter.

3.3. Concentration of C and N, and C:N Ratios

Median C concentration of the native species was 43 mol kg^{-1}, ranging from 35 to 47 mol kg^{-1}, whereas *C. elastica* had a narrower range (34 to 40 mol kg^{-1}) and a lower median (38 mol kg^{-1}) (Groups differed significantly: Wilcoxon/Kruskal-Wallis χ^2 = 35, $0.001 < \chi^2 < p$). In the case of leaf N concentration, grouped native species had a median of 1.7 (range: 1.2 to 2.2 mol kg^{-1}), whereas *C. elastica* had a significantly larger median leaf N concentration (2.05 mol kg^{-1}) and a range between 1.3 and 2.4 mol kg^{-1} (Wilcoxon/Kruskal-Wallis χ^2 = 16, $0.001 < \chi^2 < p$).

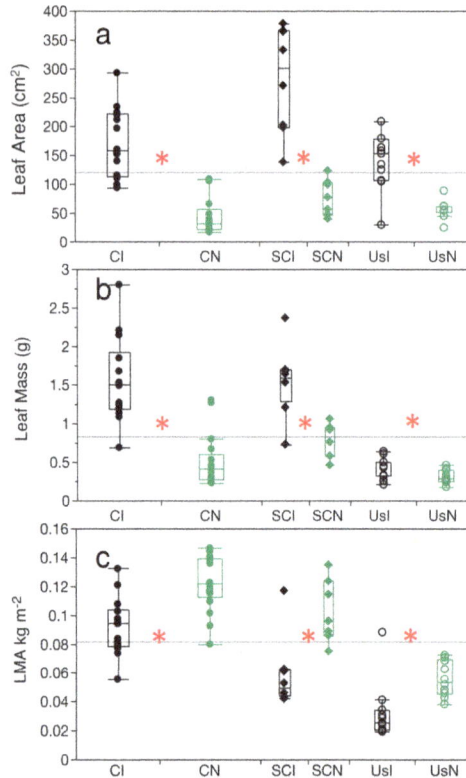

Figure 2. (**a**) Leaf area; (**b**) leaf mass and (**c**) leaf mass per area (LMA) of *C. elastica* (I) (black symbols) and group of native species (N) (green symbols) measured in the canopy (C), subcanopy (SC), and understory (Us) of a *Castilla* novel forest. Number of samples: CI = 14, CN = 16, SCI = 8, SCN = 7, UsI = 11, UsN = 10. The red asterisks indicate significant differences between groups within each stratum (Wilcoxon z, *p* < 0.01).

The LMA values of both groups increased from the understory to the canopy, and as expected, C concentration per unit area was linearly correlated with the LMA in both groups, with a slightly higher slope for grouped native species (Figure 3). The same pattern was observed in the case of N concentration per area (N_{area}), although the models explained lower percentages of data variance in comparison to those for C concentrations. The continuous and rapid increase of C and N per unit area from the understory to the canopy in both groups is expected as a response to the higher light energy available for photosynthesis in the upper forest strata.

Within each stratum, median C:N ratios were significantly lower for *C. elastica* (Figure 4). For this species, the ratio decreased markedly from the canopy to the understory, whereas no uniform pattern was observed for the grouped native species.

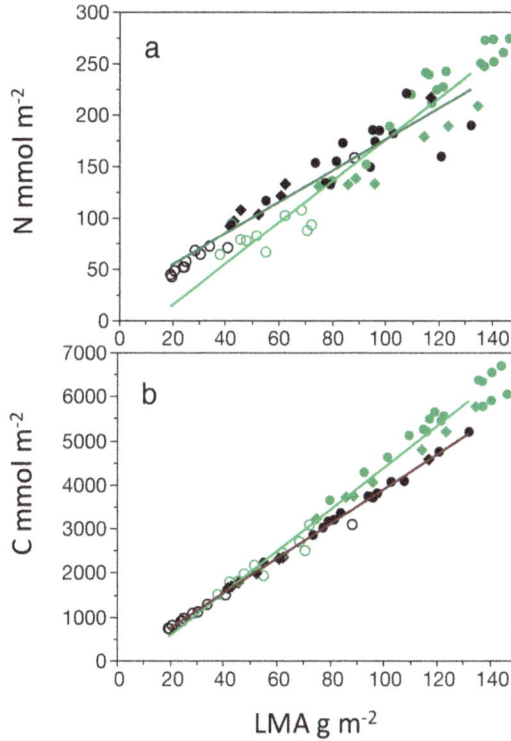

Figure 3. Variations of the relationship between leaf mass per area (LMA) and (**a**) N and (**b**) C concentrations per unit area in the vertical profile of the *Castilla* novel forest. Green symbols: native species; black symbols: *C. elastica*. Open circles (Us), squares (SC), closed circles (C). For, native species: $C = -380 + 47,000$ LMA, $R^2 = 0.97$; and $N = -26 + 2000$ LMA, $R^2 = 0.92$ ($n = 34$). For *C. elastica*: $C = -53 + 39,000$ LMA, $R^2 = 0.99$; and $N = 23.4 + 1500$ LMA, $R^2 = 0.91$ ($n = 33$).

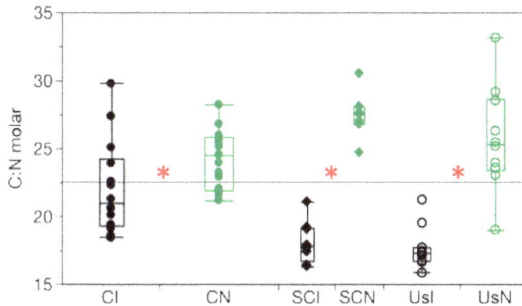

Figure 4. Vertical variation in leaf C:N ratios for *C. elastica* (I) (black symbols) and grouped native species (N) (green symbols) measured in the canopy (C), sub-canopy (SC), and understory (Us) of a *Castilla* novel forest. The stars indicate significant differences between groups within each stratum (Wilcoxon test, $p < 0.05$). Overall comparison between *C. elastica* and the group of native species: Wilcoxon/Kruskal-Wallis test $\chi^2 = 44$, $0.001 < \chi^2 < p$.

3.4. Comparisons Among Photosynthetic Capacity and Other Physiological Traits

We found no differences among species and across measuring periods when comparing photosynthetic parameters from the controlled light response curves. However, these parameters varied between forest strata and were always higher in the canopy than in the subcanopy and understory, respectively (Table 2). The lack of differences among groups within each forest stratum indicates similar adaptability of the photosynthetic apparatus in both introduced and native species. In addition, both gs_{max} (0.24 to 0.40 mol m^{-2} s^{-1} for all species) and WUE_i (89 to 51 and 11 to 63 µmol mol^{-1}, for *C. elastica* and grouped native species respectively) overlapped in both the introduced and native species groups, across all strata and measuring periods.

Photosynthetic rate per area was linearly and positively related to the N concentration per unit area in both groups (Figure 5a), with a slightly greater gradient for *C. elastica*. Unexpectedly, the regression between A_{max} and N concentration per unit mass was significant only for *C. elastica* (Figure 5b).

Table 2. Median $\pm$ Median Absolute Deviation (number of observations) for the main photosynthetic parameters in the Castilla forest at El Tallonal. Medians differed between group of species per stratum, except for A_{max} per unit mass.

Stratum	Amax		LCP	LSP
	µmol kg^{-1} s^{-1}	µmol m^{-2} s^{-1}	µmol m^{-2} s^{-1}	
Canopy	104.4 $\pm$ 18.8 (31)	11.3 $\pm$ 1.6 (34)	22.6 $\pm$ 3.2	927 $\pm$ 57
Subcanopy	113.8 $\pm$ 25.4 (15)	7.6 $\pm$ 1.7 (19)	15.2 $\pm$ 3.5	775 $\pm$ 122
Understory	127.1 $\pm$ 42.6 (21)	4.9 $\pm$ 1.0 (26)	9.6 $\pm$ 1.9	275 $\pm$ 57
χ^2 Wilcoxon/Kruskal-Wallis	$p = 0.462$	$p < 0.0001$	$p < 0.0001$	$p < 0.0001$

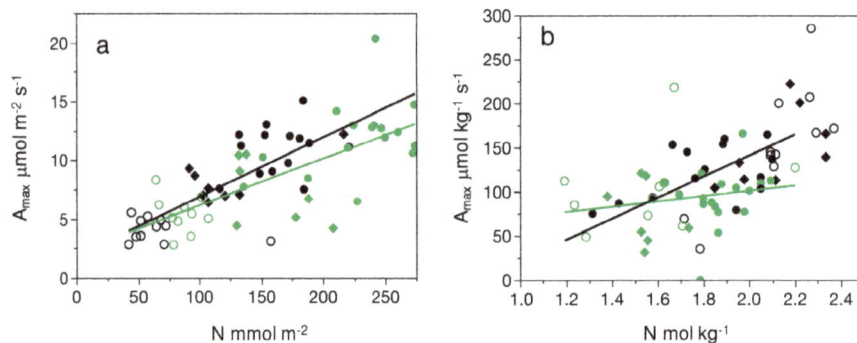

Figure 5. Relationship between A_{max} and N concentration per unit area ($A_{max\ area}$, (**a**)), and unit mass ($A_{max\ mass}$, (**b**)). Symbols as in Figure 3. For *C. elastica*, $A_{max\ area}$ = 1.87 + 0.05 N_{area}, R^2_{adj} = 0.59, F = 47, $p > F < 0.0001$, $n = 33$; $A_{max\ mass}$ = −97.9 + 119.4 N_{mass}, R^2_{adj} = 0.38, F = 21, $p > F < 0.0001$. For the group of native species, $A_{max\ area}$ = 2.18 + 0.04 N_{area}, R^2_{adj} = 0.50, F = 31, $p > F < 0.0001$, $n = 33$; R^2_{adj} = 0.005, not significant.

3.5. Photosynthetic Nitrogen Use-Efficiency

Within each stratum, *C. elastica* had median values above the overall mean, but significant differences between *C. elastica* and the native species group were found only at the canopy level (Figure 6a). *Castilla elastica* showed a higher PNUE compared to all native species when data was pooled together (median of 67.5 and 53.7 µmol CO_2 mol^{-1} N s^{-1}, for *C. elastica* and grouped native species, respectively) (Figure 6b).

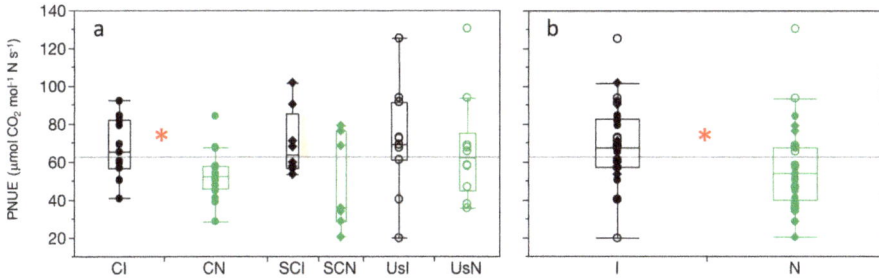

Figure 6. Comparison of photosynthetic nitrogen use-efficiencies for *Castilla elastica* = (I) (black symbols) and group of native species (N) (green symbols). The red asterisk indicates significant differences between groups of species (Wilcoxon and Wilcoxon/Kruskal-Wallis tests). (**a**) Photosynthetic nitrogen use efficiency per stratum and groups of species (CI > CN, Wilcoxon z value = 2.66, *p* = 0.0078) and (**b**) comparison between I and N (Wilcoxon/Kruskal-Wallis χ^2 = 8.8, $p > \chi^2$ = 0.0031).

4. Discussion

4.1. Castilla elastica Influences the Irradiance Below the Canopy

The novel forest under study has been regenerating for at least 60 years and already presents a complex canopy structure [25]. This complex canopy drives the vertical differences in PPFD across strata. The understory only receives a tiny proportion of the total irradiance that reaches the canopy (~1%). Clark et al. [33], in a study of mature tropical forests, also reported that only 1% to 2% of the photosynthetically active radiation reaches the understory.

Below the canopy, the PPFD varied across seasons, probably because of changes in leaf area index. The fact that *C. elastica* is a deciduous tree, and loses its leaves during the dry season [20,26,34], promotes high variation in the irradiance received in the subcanopy (8–50%) and in the understory throughout the year (<1% to 4%). These changes probably influence the ecophysiology of the whole plant community in the area.

4.2. Similar Photosynthetic Capacity and Water Use Among Species

Species groups varied little in their photosynthetic capacities and water use (A_{max}, gs_{max}, and WUE_i) when comparing plants within the same forest stratum. This resulted in similar carbon gain among species. This similarity in photosynthetic capacities was unexpected, because introduced species often have higher photosynthetic capacity and resource gain than native species [9,15,35]. Instead, the differences that were found across forest strata demonstrated leaf adjustment to the large differences in irradiance and reflected the different stages of plant development.

Photosynthetic characteristics of upper canopy leaves of *C. elastica* in moist forests in Panama were studied in detail by Kitajima et al. [36]. Leaves developed in the early wet season were compared to those developed in the pre-dry season. The study showed that *C. elastica* did not exhibit seasonal phenotypes. There were no significant differences from early wet season to pre-dry season for the following indicators: light saturated oxygen evolution, nitrogen content per unit mass, photosynthetic N use efficiency, and LMA. Maximum CO_2 assimilation rates and nitrogen concentrations of *C. elastica* canopy leaves in Panama overlap the values reported here. However, LMA values are much lower in Panama (0.065–0.077 kg m^{-2}) than in our study (0.093 kg m^{-2}), which can probably be explained by the different methods used for measuring the leaves.

4.3. Nitrogen Use-Efficiency and Leaf Area Are Advantages for Castilla elastica

Castilla elastica adult trees showed the most efficient use of nitrogen in photosynthesis, particularly at the canopy and subcanopy strata, where light energy was not limiting photosynthesis. Thus, similarly to introduced species in general, *C. elastica* outperformed native species in this aspect. This feature could contribute to explain *C. elastica*'s dominance in this novel forest. It is remarkable that we found significant linear relationships between N and photosynthesis per unit area but not per unit mass, as is usually reported in other sites [37]. Values of PNUE for the native species group reported here are comparable to those of late successional species elsewhere (29–84 vs. 56 μmol CO_2 mol^{-1} N s^{-1}) [37], while those of *C. elastica* are much lower than those of early successional species (41–92 vs. 216 μmol CO_2 mol^{-1} N s^{-1}).

Leaf mass per area of *C. elastica* was significantly lower than that of native species. This shows that *C. elastica*'s leaves are less expensive because they attain similar photosynthetic rates as that of native species while investing a lot less C and N. Moreover, *C. elastica* showed higher plasticity than native species, by adjusting to the different irradiance across forest strata, as suggested by the different leaf mass/area relationships across strata. Phenotypic plasticity is usually high in introduced species for a number of traits, including LMA [38]. Although high plasticity does not always indicate better performance, this is usually the case and can give an advantage to introduced species in their new habitat [14,39]. In the case of *C. elastica*, plasticity in LMA seems to have contributed to its dominance in the new range.

Leaf mass per area of *C. elastica* adult plants was in the range of variation for those reported by Reich et al. [39] for deciduous and pioneer woody plants in tropical forests (0.03–0.4 kg m^{-2}). We found even lower LMA for saplings and juvenile plants of *C. elastica* (0.015–0.02 kg m^{-2}). Low LMA is often associated with a high relative growth rate and invasiveness [40–47]. Investing resources in less expensive leaves (i.e., low LMA) might result in more efficient light interception for *C. elastica* in comparison to native species.

5. Conclusions

Overall, photosynthetic capacity was unexpectedly similar among species in the *Castilla* novel forest. High PNUE and low LMA support *C. elastica*'s higher competitive capacity over native species and could explain its dominance in this novel ecosystem. These results contradict our hypothesis of higher photosynthetic capacity for *C. elastica*, but support the hypothesis of more efficient use of resources (C and N) by the introduced species. The results also indicate that introduced and native species do not occupy overlapping positions in the leaf economic spectrum.

Acknowledgments: This research project was supported by the International Institute of Tropical Forestry (Institute), USDA-Forest Service, the Center for Applied Tropical Ecology and Conservation, the University of Puerto Rico, and Ciudadanos del Karso. The work was done in collaboration with the University of Puerto Rico. We are grateful to Abel Vale, the owner of El Tallonal, who allowed us to conduct the study in his property. We thank Humberto Robles, Oscar Abelleira, Jasmine Shaw, Seth Rifkin, Dylan Rhea and Omar Gutiérrez for helping in the sampling and field data collection. We are grateful to Humberto Robles, Maricel Beltrán, Ariana Beltrán and Edgardo Valcárcel who helped to process the leaf samples. Thanks to the staff of the Institute's Chemistry Lab for conducting the chemical analyses of leaf material. We are also grateful to Pascal Bugnion for English language editing on this manuscript.

Author Contributions: J.F.d.S., E.M. and A.E.L. conceived the ideas and designed the methodology; J.F.d.S. collected the data; J.F.d.S. and E.M. analyzed the data and led the writing of the manuscript. All authors contributed critically to the drafts and gave final approval for publication.

Conflicts of Interest: The authors declare no conflict of interest.

References

1. Malhi, Y.; Gardner, T.A.; Goldsmith, G.R.; Silman, M.R.; Zelazowski, P. Tropical forests in the Anthropocene. *Annu. Rev. Environ. Resour.* **2014**, *39*, 125–159. [CrossRef]

2. Lugo, A.E.; Helmer, E. Emerging forests on abandoned land: Puerto Rico's new forests. *For. Ecol. Manag.* **2004**, *190*, 145–161. [CrossRef]

3. Hobbs, R.J.; Arico, S.; Aronson, J.; Baron, J.S.; Bridgewater, P.; Cramer, V.A.; Epstein, P.R.; Ewel, J.J.; Klink, C.A.; Lugo, A.E.; et al. Novel ecosystems: Theoretical and management aspects of the new ecological world order. *Glob. Ecol. Biogeogr.* **2006**, *15*, 1–7. [CrossRef]

4. Hobbs, R.J.; Higgs, E.S.; Hall, C. *Novel Ecosystems: Intervening in the New Ecological World Order*, 1st ed.; John Wiley & Sons Ltd.: Chichester, UK, 2013.

5. Higgs, E. Novel and designed ecosystems. *Restor. Ecol.* **2017**, *25*, 8–13. [CrossRef]

6. Corlett, R.T. The Anthropocene concept in ecology and conservation. *Trends Ecol. Evol.* **2015**, *30*, 36–41. [CrossRef] [PubMed]

7. Alpert, P. The advantages and disadvantages of being introduced. *Biol. Invasions* **2006**, *8*, 1523–1534. [CrossRef]

8. Pyšek, P.; Jarošík, V.; Hulme, P.E.; Pergl, J.; Hejda, M.; Schaffner, U.; Vilà, M. A global assessment of invasive plant impacts on resident species, communities and ecosystems: The interaction of impact measures, invading species' traits and environment. *Glob. Chang. Biol.* **2012**, *18*, 1725–1737. [CrossRef]

9. Pattison, R.R.; Goldstein, G.; Ares, A. Growth, biomass allocation and photosynthesis of invasive and native Hawaiian rainforest species. *Oecologia* **1998**, *117*, 449–459. [CrossRef] [PubMed]

10. Baruch, Z.; Goldstein, G. Leaf construction cost, nutrient concentration, and net CO_2 assimilation of native and invasive species in Hawaii. *Oecologia* **1999**, *121*, 183–192. [CrossRef] [PubMed]

11. McDowell, S.C.L. Photosynthetic characteristics of invasive and noninvasive species of *Rubus* (Rosaceae). *Am. J. Bot.* **2002**, *89*, 1431–1438. [CrossRef] [PubMed]

12. Burns, J.H. A comparison of invasive and noninvasive dayflowers (Commelinaceae) across experimental nutrient and water gradients. *Divers. Distrib.* **2004**, *10*, 387–397. [CrossRef]

13. Leicht Young, S.A.; Silander, J.A., Jr.; Latimer, A.M. Comparative performance of invasive and native *Celastrus* species across environmental gradients. *Oecologia* **2007**, *154*, 273–282. [CrossRef] [PubMed]

14. Funk, J.L. Differences in plasticity between invasive and native plants from a low resource environment. *J. Ecol.* **2008**, *96*, 1162–1173. [CrossRef]

15. Peñuelas, J.; Sardans, J.; Llusià, J.; Owen, S.M.; Carnicer, J.; Giambelluca, T.W.; Rezende, E.L.; Waite, M.; Niinemets, Ü. Faster returns on "leaf economics" and different biogeochemical niche in invasive compared with native plant species. *Glob. Chang. Biol.* **2010**, *16*, 2171–2185. [CrossRef]

16. Cavaleri, M.A.; Sack, L. Comparative water use of native and invasive plants at multiple scales: A global meta-analysis. *Ecology* **2010**, *91*, 2705–2715. [CrossRef] [PubMed]

17. Lamarque, L.J.; Delzon, S.; Lortie, C.J. Tree invasions: A comparative test of the dominant hypotheses and functional traits. *Biol. Invasions* **2011**, *13*, 1969–1989. [CrossRef]

18. Whitmore, T.C. Canopy gaps and the two major groups of forest trees. *Ecology* **1989**, *70*, 536–538. [CrossRef]

19. Didham, R.K.; Tylianakis, J.M.; Gemmell, N.J.; Rand, T.A.; Ewers, R.M. Interactive effects of habitat modification and species invasion on native species decline. *Trends Ecol. Evol.* **2007**, *22*, 489–496. [CrossRef] [PubMed]

20. Fonseca da Silva, J. Dynamics of novel forests of *Castilla elastica* in Puerto Rico: From species to ecosystems. *Ecol. Evol.* **2015**, *5*, 3299–3311. [CrossRef] [PubMed]

21. Wright, I.J.; Reich, P.B.; Westoby, M.; Ackerly, D.D.; Baruch, Z.; Bongers, F.; Cavender-Bares, J.; Chapin, T.; Cornelissen, J.H.C.; Diemer, M.; et al. The worldwide leaf economics spectrum. *Nature* **2004**, *428*, 821–827. [CrossRef] [PubMed]

22. Vitousek, P.M. Biological invasions and ecosystem processes: Towards an integration of population biology and ecosystem studies. *Oikos* **1990**, *57*, 7–13. [CrossRef]

23. Funk, J.L.; Vitousek, P.M. Resource-use efficiency and plant invasion in low-resource systems. *Nature* **2007**, *446*, 1079–1081. [CrossRef] [PubMed]

24. Heard, M.J.; Sax, D.F. Coexistence between native and exotic species is facilitated by asymmetries in competitive ability and susceptibility to herbivores. *Ecol. Lett.* **2013**, *16*, 206–213. [CrossRef] [PubMed]

25. Fonseca da Silva, J. Species Composition, Diversity and Structure of Novel Forests of *Castilla elastica* in Puerto Rico. *Trop. Ecol.* **2014**, *55*, 231–244.

26. Francis, J.K.; Liogier, H.A. *Naturalized Exotic Tree Species in Puerto Rico*; General Technical Report SO-82; USDA Forest Service: New Orleans, LA, USA, 1991.

27. Sautu, A.; Baskin, J.M.; Baskin, C.C.; Condit, R. Studies on the seed biology of 100 native species of trees in a seasonal moist tropical forest, Panama, Central America. *For. Ecol. Manag.* **2006**, *234*, 245–263. [CrossRef]

28. Ewel, J.; Whitmore, J.L. *The Ecological life Zones of Puerto Rico and the U.S. Virgin Islands*; Forest Service Research Paper ITF-18; USDA Forest Service, International Institute of Tropical Forestry: Río Piedras, Puerto Rico, 1973.

29. Viera-Martínez, C.A.; Abelleira Martínez, O.J.M.; Lugo, A.E. Estructura y química del suelo en un bosque de *Castilla elastica* en el Carso del norte de Puerto Rico: Resultados de una calicata. *Acta Cient.* **2008**, *22*, 29–35.

30. Axelrod, F.S. *A Systematic Vademecum to the Vascular Plants of Puerto Rico*; Botanical Research Institute of Texas: Fort Worth, TX, USA, 2011.

31. Peek, M.S.; Russek-Cohen, E.; Wait, D.A.; Forseth, I.N. Physiological response curve analysis using nonlinear mixed models. *Oecologia* **2002**, *132*, 175–180. [CrossRef] [PubMed]

32. Evans, G.C.; Hughes, A.P. Plant growth and the aerial environment. I. Effect of artificial shading on *Impatiens parviflora*. *New Phytol.* **1961**, *60*, 150–180. [CrossRef]

33. Clark, D.B.; Clark, D.A.; Rich, P.M.; Weiss, S.; Oberbauer, S.F. Landscape-scale evaluation of understory light and canopy structures: Methods and application in a Neotropical lowland rain forest. *Can. J. For. Res.* **1996**, *26*, 747–757. [CrossRef]

34. Sakai, S. Thrips pollination of androdioecious *Castilla elastica* (Moraceae) in a seasonal tropical forest. *Am. J. Bot.* **2001**, *88*, 1527–1534. [CrossRef] [PubMed]

35. Mozdzer, T.J.; Zieman, J.C. Ecophysiological differences between genetic lineages facilitate the invasion of non-native *Phragmites australis* in North American Atlantic coast wetlands. *J. Ecol.* **2010**, *98*, 451–458. [CrossRef]

36. Kitajima, K.; Mulkey, S.S.; Wright, S.J. Seasonal leaf phenotypes in the canopy of a tropical dry forest: Photosynthetic characteristics and associated traits. *Oecologia* **1997**, *109*, 490–498. [CrossRef] [PubMed]

37. Reich, P.B.; Walters, M.B.; Ellsworth, D.S.; Uhl, C. Photosynthesis-nitrogen relations in Amazonian tree species. *Oecologia* **1994**, *97*, 62–72. [CrossRef] [PubMed]

38. Davidson, A.M.; Jennions, M.; Nicotra, A.B. Do invasive species show higher phenotypic plasticity than native species and, if so, is it adaptive? A meta-analysis. *Ecol. Lett.* **2011**, *14*, 419–431. [CrossRef] [PubMed]

39. Palacio López, K.; Gianoli, E. Invasive plants do not display greater phenotypic plasticity than their native or non-invasive counterparts: A meta-analysis. *Oikos* **2011**, *120*, 1393–1401. [CrossRef]

40. Reich, P.B.; Walters, M.B.; Ellsworth, D.S. From tropics to tundra: Global convergence in plant functioning. *Proc. Natl. Acad. Sci. USA* **1997**, *94*, 13730–13734. [CrossRef] [PubMed]

41. Davis, M.A.; Grime, J.P.; Thompson, K. Fluctuating resources in plant communities: A general theory of invasibility. *J. Ecol.* **2000**, *88*, 528–534. [CrossRef]

42. Grotkopp, E.; Rejmánek, M.; Rost, T.L. Toward a causal explanation of plant invasiveness: Seedling growth and life-history strategies of 29 pine (Pinus) species. *Am. Nat.* **2002**, *159*, 396–419. [CrossRef] [PubMed]

43. Shea, K.; Chesson, P. Community ecology theory as a framework for biological invasions. *Trends Ecol. Evol.* **2002**, *17*, 170–176. [CrossRef]

44. Lake, J.C.; Leishman, M.R. Invasion success of exotic plants in natural ecosystems: The role of disturbance, plant attributes and freedom from herbivores. *Biol. Conserv.* **2004**, *117*, 215–226. [CrossRef]

45. Hamilton, M.A.; Murray, B.R.; Cadotte, M.W.; Hose, G.C.; Baker, A.C.; Harris, C.J.; Licari, D. Life-history correlates of plant invasiveness at regional and continental scales. *Ecol. Lett.* **2005**, *8*, 1066–1074. [CrossRef]

46. Leishman, M.R.; Thomson, V.P. Experimental evidence for the effects of additional water, nutrients and physical disturbance on invasive plants in low fertility Hawkesbury Sandstone soils, Sydney, Australia. *J. Ecol.* **2005**, *93*, 38–49. [CrossRef]

47. Grotkopp, E.; Rejmánek, M. High seedling relative growth rate and specific leaf area are traits of invasive species: Phylogenetically independent contrasts of woody angiosperms. *Ecology* **2007**, *94*, 526–532. [CrossRef] [PubMed]

forests

MDPI

Article

Substrate Chemistry and Rainfall Regime Regulate Elemental Composition of Tree Leaves in Karst Forests

Ernesto Medina [1,2,3,*], Elvira Cuevas [3] and Ariel E. Lugo [1]

[1]   International Institute of Tropical Forestry (IITF), USDA Forest Service, Rio Piedras 00926-1115, Puerto Rico; alugo@fs.fed.us

[2]   Centro de Ecología, Instituto Venezolano de Investigaciones Científicas, Altos de Pipe 1020A, Venezuela

[3]   Departmento de Biología, Universidad de Puerto Rico, San Juan 00926; Puerto Rico; epcuevas@gmail.com

*   Correspondence: medinage@gmail.com; Tel.: +1-787-755-5637

Academic Editor: Timothy A. Martin

Received: 15 April 2017; Accepted: 20 May 2017; Published: 25 May 2017

Abstract: Forests on calcareous substrates constitute a large fraction of the vegetation in Puerto Rico. Plant growth on these substrates may be affected by nutrient deficiencies, mainly P and Fe, resulting from high pH and formation of insoluble compounds of these elements. The occurrence of these forests in humid and dry areas provides an opportunity to compare nutrient relations, water use efficiency, and N dynamics, using biogeochemical parameters. We selected sites under humid climate in the north, and dry climate in the southwest of Puerto Rico. Adult, healthy leaves of species with high importance values were collected at each site and analyzed for their elemental composition and the natural abundance of C and N isotopes. Calcium was the dominant cation in leaf tissues, explaining over 70% of the ash content variation, and Al and Ca concentration were positively correlated, excepting only two Al-accumulating species. Karst vegetation consistently showed high N/P ratios comparable to forests on P-poor soils. Dry karst sites had significantly higher $\delta^{13}C$ and $\delta^{15}N$ ratios. We conclude that forests on karst are mainly limited by P availability, and that mechanisms of nutrient uptake in the rhizosphere lead to linear correlations in the uptake of Ca and Al. Isotope ratios indicate higher water use efficiency, and predominant denitrification in dry karst forest sites.

Keywords: tropical karst; element concentration; N/P ratios; Ca/Al relationship; $\delta^{13}C$; $\delta^{15}N$

1. Introduction

Limestone areas are widespread in the Caribbean under dry to subhumid climatic conditions. Those areas developed mostly during the middle Oligocene to the middle Pliocene and have been documented extensively in Jamaica, Cuba, Santo Domingo, and Puerto Rico [1–6]. Several forest types, from dry to moist, covering large areas of those geologic formations, have been described in detail in many locations [4,7–9].

In Puerto Rico, karst areas constitute about 27% of the territory and are separated as moist karst forming a northern belt crossing the island from east to west, and a dry karst formation in the southwest, under semiarid climate [4]. These formations have a similar geological age (Pliocene-Oligocene) but are under different rainfall regimes [10]. The limestone formations are covered throughout the island by sediments constituted by non-calcareous material that, in the northern karst belt, accumulate in depressions between haystack (*mogotes*) hills [10,11]. On top of the *mogotes* in the karst belt, or on the calcareous hills of the southern karst, this material may accumulate within crevices and fractures of the calcareous strata. Surface soils in karstic areas, when present, are highly heterogeneous, frequently shallow, and, in many cases, difficult to sample. Soils derived from the calcareous substrate itself are infrequent. On ridge tops and slopes, soil is almost inexistent, and substrate is a rocky, partially fragmented surface, covered with organic residues produced by the vegetation. Fragments of the carbonaceous rock and organic matter slide downhill and accumulate in the valleys [4,11,12]. On the valleys, true soils develop constituted by calcareous rocks fragments and the remaining material transported from volcaniclastic areas during the time when the carbonate rocks were under water (see [11] for an explanation of the composition of blanket sands). Assessment of nutrient availability for plants on the ridge top sites using soils as reference is therefore not a practical approach. The alternative is the analysis of plant organs, particularly leaves [13].

Nutrient availability for plant growth in soils derived from carbonate rocks is usually limited, particularly in P, as it tends to be immobilized as Ca compounds of low solubility [14]. High soil pH determines lower mobility of elements such as Mn and Fe, resulting in potential limitation for plant growth (Ca induced iron chlorosis). The nutrient relationships of plants from calcareous substrates have been mainly studied in temperate climates [15] and little is known from tropical vegetation [16].

Typical vegetation of moist karst forests has been described by Alvarez et al. [17] and Acevedo-Rodríguez and Axelrod [18] and that of the dry karst forest was studied by Murphy and Lugo [19], and documented in detail by Monsegur [20].

Our study took advantage of detailed phytosociological analysis of the northern karst vegetation undertaken by Chinea [21] and Aukema et al. [22]. In the latter, the authors quantified the composition of forest units according to species dominance relationships and geomorphological positions on the *mogotes* (top, slopes, valleys). This study was critical to the identification of tree assemblages on the top of *mogotes*. In addition, we used the vegetation analysis of Murphy and Lugo [19] and Molina and Lugo [23] for the Guánica dry forest in southern Puerto Rico. The dominant tree species in typical sites of northern and southern karts areas, as defined by their importance value, reveals the occurrence of common and restricted species. For example, *Coccoloba diversifolia* (Polygonaceae) and *Gymnanthes lucida* (Euphorbiaceae) occur as dominant species in both moist and dry karst, whereas *Pisonia albida* (Nyctaginaceae), *Thouinia portoricensis* (Sapindaceae), and *Pictetia aculeata* (Fabaceae) seem to be restricted to dry karst forests, and *Lonchocarpus lancifolius* (Fabaceae), and *Prunus myrtifolia* (Rosaceae) are restricted to the northern moist karst belt [24].

The objective of the present study was to provide a geochemical characterization of the karst areas in the island of Puerto Rico through the elemental analysis of substrate and plant leaves. Leaf analyses provide information on both the availability of specific elements in the soil solution, and reveal physiological properties such as exclusion or facilitation of element uptake at the root level [13]. We tested hypotheses related to the leaf elemental composition, substrate geochemistry, and environmental humidity: (1) Ca is the main component of leaf ash due to its expected high availability in the calcareous substrate; (2) Lower total P concentrations and high N to P ratios reflect relative limited availability of P, probably associated with the formation of insoluble Ca compounds; (3) Concentrations of heavy metals (Fe and Mn) and Al, are probably low compared to species from forests on non-calcareous substrates; (4) Environmental humidity determined by seasonality and magnitude of rainfall affect plant water use efficiency and rate of organic N mineralization in the substrate.

2. Materials and Methods

For the study of the nutritional relationships of tree species strictly associated with karstic substrates, we selected sites in the humid karst belt and in the semiarid south west karst area in Puerto Rico. The sampling was restricted to plants growing on the top of haystack hills in the northern karst belt, and on ridge sites and coastal carbonate pavement in the southwestern karst area.

We selected six *mogote* tops sites in the moist northern karst belt and two sites in the dry Guánica forest in southwestern Puerto Rico (Table 1). The former sites were located along a west-east line on the northern karst belt, on top of the Aymamon limestone type, whereas those on the dry karst areas are on the Ponce limestone type [10]. The northern karst belt receives, on average, annual rainfall amounts of 1000–1500 mm, whereas in the southern karst areas long-term annual rainfall averages lie below 1000 mm. This contrasting rainfall pattern is clearly exemplified by two stations, Utuado (158 m), located in the middle of the Aymamon karst [10] in the northern karst belt, and Ensenada (46 m), on the southwestern coast, representative for the rainfall regime of the southern karst belt (Figure 1).

Table 1. Karst site locations (Lat, Long) and approximate altitude.

	Karst Site	North Lat (°)	West Long (°)	Altitude (m)
Moist sites	Guajataca	18.42	66.83	230
	El Tallonal	18.41	66.73	160
	Cambalache	18.45	66.59	45
	Río Lajas	18.40	66.26	50
	Nevarez	18.41	66.25	35
	Hato Tejas	18.40	66.19	100
Dry sites	Guánica ridge	17.97	66.87	160
	Guánica Dwarf	17.95	66.83	13

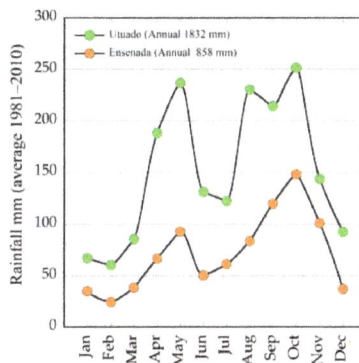

Figure 1. Average rainfall pattern of humid and dry karst sites.

The species sampled at each site were those more common, with high importance values [16,19,22,23] (Table 2). The number of species varied from site to site due to different areal extension, and the number of tree species growing on the *mogote* top. In total, we sampled 30 species from the northern, moist sites, and 26 species from the dry sites, representing 29 families and 18 orders. Only five species were sampled from both humidity sites.

For elemental analyses, we sampled at each site healthy mature leaves, without visible herbivore consumption, of trees from the upper forest canopy. Leaves were dried in the laboratory in a ventilated oven at 65 °C (3–7 days) and ground to pass an 18-mesh screen. Rock fragments on the forest floor were collected from each site. In the laboratory, rock samples were dissolved in 2N HCl for elemental analysis.

Elemental composition of samples was determined (C, N, S, P, K, Ca, Mg, Al, Mn, Fe) using standard methodology of the laboratory at the International Institute of Tropical Forestry [25]. Briefly, in acid digested samples, C, N, and S were determined using a LECO elemental analyzer, whereas P and cations were measured using Inductively coupled plasma atomic emission spectroscopy (ICP-AES) techniques.

In addition, the natural abundance of C (δ^{13}C) and N isotopes (δ^{15}N) in leaves was determined at the Stable Isotope Laboratory of the University of Miami, Coral Gables, using mass spectrometric techniques. The δ^{13}C values are used as indicators of long-term water use efficiency, because they are related to lower leaf conductance [26]. The leaf $\delta^{15}N$ values are related to the natural abundance of N isotopes in the substrate. More positive values point to higher rates of mineralization of organic N in the soil [27].

Element concentrations and C and N isotopic ratios of all species per site were submitted to a one-way analysis of variance and post-hoc tests for comparison of site averages. Inter-element correlations were estimated for the whole set of species to establish commonalities among sites. All statistical analyses were performed using the JMP statistical program.

Table 2. Species sampled at each site designated by name and municipality.

Guajataca, Isabela	El Tallonal, Arecibo	Cambalache, Barceloneta
Calophyllum antillanum	*Amyris elemifera*	*Coccoloba diversifolia*
Clusia rosea	*Coccoloba diversifolia*	*Crossopetalum rhacoma*
Coccoloba pubescens	*Exothea paniculata*	*Gymnanthes lucida*
Comocladia glabra	*Gymnanthes lucida*	*Krugiodendron ferreum*
Eugenia monticola	*Pimenta racemosa* var. *grisea*	*Sideroxylon salicifolium*
Lonchocarpus glaucifolius	*Prunus myrtifolia*	
Neolaugeria resinosa	*Tabebuia karsensis*	
Tetrazygia elaeagnoides		
Nevarez, Toa Baja	**Rio Lajas, Dorado**	**Hato Tejas, Bayamon**
Ardisia obovata	*Ardisia obovata*	*Coccoloba diversifolia*
Coccoloba diversifolia	*Calyptranthes pallens*	*Eugenia monticola*
Drypetes alba	*Coccoloba diversifolia*	*Ficus citrifolia*
Exothea paniculata	*Cojoba arborea.*	*Phyllanthus epiphyllanthu*
Garcinia portoricensis	*Exothea paniculata*	*Sideroxylon salicifolium*
Gaussia attenuata	*Krugiodendron ferreum*	
Guapira fragrans	*Licaria salicifolia*	
Licaria salicifolia	*Ottoschultzia rodoxylon*	
Ottoschultzia rodoxylon	*Sideroxylon salicifolium*	
Picramnia pentandra		
Guánica Ridge F., Guánica	**Guánica Dwarf F., Guánica**	
Amyris elemifera	*Amyris elemifera*	
Bourreria succulenta	*Bourreria virgata*	
Bucida buceras	*Canella winterana*	
Elaeodendrum xylocarpa	*Coccoloba diversifolia*	
Colubrina arborescens	*Colubrina arborescens*	
Erythalis fruticosa	*Crossopetalum rhacoma*	
Erythroxylum areolatum	*Erithalis fruticosa*	
Exostemma caribaeum	*Eugenia foetida*	
Gymnanthes lucida	*Ficus citrifolia*	
Pisonia albida	*Jacquinia arborea*	
Tabebuia heterophylla	*Jacquinia berteroi*	
Thouinia striata var. *portoricensis*	*Pisonia albida*	
	Quadrella cynophallophora	
	Reynosia uncinata	
	Sideroxylon salicifolium	
	Stenostomum acutatum	
	Strumpfia maritima	
	Tabebuia heterophylla	
	Thrinax morrisii	

Plant names after [24].

3. Results

3.1. Elemental Composition of the Substrate in Karst Areas in Puerto Rico

The analysis of the elemental composition of rocks from different karst types of the northern karst belt published by Monroe [12] indicate that nearly 83% is constituted by $CaCO_3$, and that they contain small but significant amounts of Mg, Al, and Fe. The elements P and particularly K are found at concentrations below 1 mmol/kg. Rock fragments from the sites selected for vegetation sampling showed important differences in the proportion of minor elements. The sequence of elemental concentrations in the northern karst site confirmed the sequence Ca >> Mg > Al > Fe, corresponding to the average values from Monroe [12] (Figure 2). The samples from the southern karst had a similar elemental distribution, but Na was present in the Guánica samples at higher concentrations than any other site. Phosphorus concentration was at the same level in all samples (≈1 mmol/kg). Potassium reached concentrations around 10 mmol/kg in the southern samples, whereas it was always well below the 1 mmol/kg level in the northern karst samples.

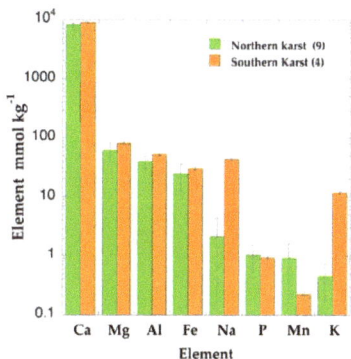

Figure 2. Average elemental composition of karst rocks from northern and southern sites in Puerto Rico.

The presence of non-calcareous sediments in cracks, holes, and crevices of the karstic substrate raised the question about their potential as a source of nutrients for the vegetation. They have been characterized in the Dominican Republic as lateritic in nature, with high concentrations of Al and Fe [1]. We do not have information on the elemental composition of those sediments in Puerto Rico. However, we calculated sediment concentrations from a nutrient inventory study in the northern karst belt [28]. Both A and C horizons indicate much higher concentrations of Fe, Mn, and Al, and lower concentrations of Ca and Mg compared to those measured in calcareous rocks (Figure 3). The higher concentrations of Ca and Mg of the A horizon reveal the influence of in situ weathering of karst rocks. It appears then that these sediments may be a source of nutrients for the vegetation, and due to their clayey texture, they have much higher water retention and cation exchange capacities than the calcareous substrates.

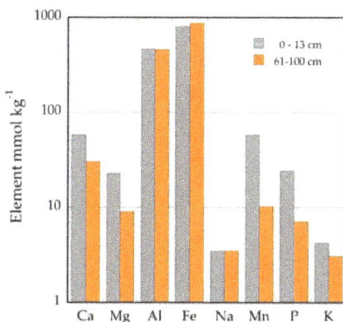

Figure 3. Elemental composition of clay soils on karst in El Tallonal, Arecibo, Puerto Rico. Values for the A horizon (0–13 cm) and the C horizon (61–100 cm) were generated from [28].

3.2. Leaf Elemental Composition

The concentration of the main organic matter elements did not differ significantly between sites. The coefficients of variation were small for C (<10%), intermediate for N and P (10 to 100%), and large for S (>100%) (Table 3). The concentration of metallic elements differed between sites for K, Ca, Fe, Mn, and Na (Table 4). The coefficients of variation were intermediate for K, Mg, Ca, and Fe, large for Na, and very large for Al and Mn (>250%). In the southern karst, average K concentration was higher, whereas for Ca and Mn concentration averages were lower than the overall mean. The average Na concentration in the Guánica Dwarf forest site was much larger than the overall mean, probably because this site is located near the coast line, under the influence of sea salt spray, and Na may be present in the rooting substrate, as indicated by the composition of the limestone from Guánica (Figure 2).

Table 3. Mean (±SE) concentrations of main elements of organic matter C in mol/kg, the rest in mmol/kg.

System	*n*	C	N	S	P
Guajataca	8	43.7 (1.2)	1247 (163)	60 (12)	18.7 (1.9)
Tallonal	7	43.7 (1.0)	1193 (100)	58 (11)	26.7 (3.0)
Cambalache	5	41.7 (1.1)	1315 (217)	102 (37)	22.8 (3.1)
Nevarez	10	41.7 (0.7)	1226 (179)	128 (34)	21.4 (1.8)
Río Lajas	10	43.2 (0.6)	1375 (179)	60 (8)	21.9 (3.1)
Hato Tejas	5	42.1 (1.2)	1144 (103)	77 (18)	23.1 (3.7)
Guánica Ridge	13	41.5 (0.7)	1361 (168)	75 (10)	25.4 (3.7)
Guánica Dwarf	26	42.3 (0.5)	991 (65)	94 (26)	20.5 (2.5)
Overall mean	84	42.4 (0.3)	1192 (51)	84 (10)	22.2 (1.1)
Coefficient of Variation (CV)		6	39	105	46
Analysis of Variance (ANOVA)		1.1 (ns)	1.3 (ns)	0.7 (ns)	0.6 (ns)

ns, not significant.

Table 4. Mean (± SE) concentrations of ash (%) and metallic elements in leaves per site in mmol/kg.

System	*n*	Ash	K	Mg	Ca	Al	Fe	Mn	Na
Guajataca	8	6.8 (1.3)	132 (20)	110 (17)	450 (95)	91.6 (86.7)	1.3 (0.4)	3.6 (1.7)	61 (21)
Tallonal	7	7.7 (1.4)	162 (29)	114 (17)	583 (145)	5.1 (1.1)	0.7 (0.1)	0.3 (0.1)	25 (8)
Cambalache	5	10.9 (1.6)	102 (32)	148 (22)	872 (178)	6.3 (1.0)	0.7 (0.1)	0.3 (0.1)	77 (19)
Nevarez	10	9.9 (1.0)	129 (20)	154 (36)	708 (75)	7.1 (0.5)	1.4 (0.1)	0.5 (0.1)	92 (21)
Río Lajas	10	9.8 (1.4)	231 (60)	100 (15)	574 (93)	5.6 (0.8)	1.2 (0.2)	0.4 (0.1)	83 (28)
Hato Tejas	5	9.4 (1.6)	206 (55)	127 (28)	733 (148)	7.1 (1.1)	1.4 (0.2)	0.3 (0.1)	90 (22)
Guánica Ridge	13	8.4 (0.7)	315 (41)	162 (36)	409 (45)	4.1 (0.4)	0.6 (0.1)	0.3 (0.1)	40 (12)
Guánica Dwarf	26	7.8 (0.5)	212 (22)	138 (17)	340 (40)	4.6 (0.4)	0.6 (0.0)	0.5 (0.1)	221 (34)
Overall mean	84	8.5 (0.4)	202 (14)	132 (9)	509 (33)	13.6 (8.4)	0.9 (0.1)	0.7 (0.2)	112 (14)
CV		37	65	65	59	560	62	250	115
ANOVA (F; P)		1.3 (ns)	3.3 < 0.01	0.6 (ns)	4.9 < 0.01	1.4 (ns)	6.3 < 0.01	4.4 < 0.01	5.4 < 0.01

ns, not significant.

Some species within the sampled set stand out, with concentrations of particular elements 2–3 times larger or smaller than the overall average (Table 5). These differences may be attributed to variations in element availability between sites, to differences in leaf age, or to actual physiological differences between species. Leaf age between samples of the same species at different sites cannot be discarded, as leaf sampling was conducted during several months; however, leaves collected were always fully expanded, healthy, and without visual signals of senescence or herbivory.

The design of the research was not oriented to differentiate species occurring at different sites; therefore, the data set is insufficient to conduct within species comparisons.

Elemental ratios between organic matter forming elements (C, N, S, P) do not differ significantly between sites (Table 6). However, K to Ca ratios were larger and the Ca to Mg ratios lower for the two sites on the southern karst in correspondence with the relative elemental concentration of those elements measured in the calcareous rocks. The other sites showed K to Ca ratios ranging from 0.14 to 0.47, revealing the high availability of Ca in the soil solution.

Table 5. Maximum and minimum concentrations of elements (mmol/kg).

Element	Species	Site	Max	Mean	Min	Species	Site
N	*Erythroxylum areolatum*	Grf	2731	1197	639	*Cassine xylocarpa*	Grf
	Guapira fragrans	N	2621		565	*Erithalis fruticosa* Grf, Gdf	
S	*Quadrella cynophallophora*	Gdf	707	83	30	*Coccoloba pubescens*	G
	Gaussia attenuata	N	367		19	*Prunus myrtifolia*	T
P	*Erythroxylum areolatum*	Grf	62	22	9	*Erithalis fruticosa* Gdf, Grf	
	Coccoloba diversifolia	Gdf	48		7	*Jacquinia berteroi*	Gdf
K	*Bourreria succulenta*	Grf	615	202	51	*Sideroxylon salicifolium*	C
	Canella winterana	Gdf	524		50	*Strumpfia maritima*	Gdf
Mg	*Pisonia albida*	Grf, Gdf	553	132	38	*Thrinax morrisii*	Gdf
	Guapira fragrans	N	415		22	*Licaria salicifolia* N, RL	
Ca	*Crossopetalum rhacoma*	C	1478	509	133	*Calophyllum brasiliense*	G
	Coccoloba diversifolia	T, HT, Gdf	1281		75	*Thrinax morrisii*	Gdf
Al	*Tetrazygia elaeagnoides*	G	699	14	1.7	*Tabebuia karsensis*	T
	Lonchocarpus lancifolius	G	13		1.4	*Calophyllum brasiliense*	G
Fe	*Tetrazygia elaeagnoides*	G	3.7	0.9	0.4	*Prunus myrtifolia*	T
	Krugiodendron ferreum	RL	2.7		0.3	*Clusia rosea*	G
Mn	*Tetrazygia elaeagnoides*	G	13.2	0.7	0.04	*Prunus myrtifolia*	T
	Clusia rosea	G	9.3		0.04	*Jacquinia berteroi*	Gdf

C, Cambalache; G, Guajataca; Gdf, Guánica dwarf forest; Grf, Guánica ridge forest; HT, Hato Tejas; N, Nevárez; RL, Río Lajas; T, Tallonal.

Table 6. Molar elemental ratios of leaves from trees on calcareous substrates.

Site	*n*	C/N	C/P	C/S	N/P	K/Ca	Ca/Mg
Guajataca	8	39	2507	900	67	0.40	4.08
Tallonal	7	39	1831	996	46	0.47	5.09
Cambalache	5	35	1992	654	58	0.14	6.43
Nevarez	10	39	2086	505	56	0.20	7.35
Río Lajas	9	39	2514	838	67	0.34	7.34
Hato Tejas	5	38	2006	717	53	0.31	6.10
Guánica Ridge forest	13	37	2082	639	56	0.91	3.29
Guánica Dwarf forest	25	47	2819	744	59	0.87	2.90
Overall mean	84	41	2328	732	58	0.59	4.63
ANOVA (F; P)		1.0 (ns)	1.0 (ns)	1.5 (ns)	1.3 (ns)	3.4 <0.01	4.2 <0.01

ns, not significant.

Inter-Element Correlations

The correlation between leaf N and P was positive and significant, as usually reported for forest foliage throughout the world [29] (Figure 4). As expected, the Ash% was inversely correlated with C concentration (R^2 adj = 0.57), but positively correlated with Ca (R^2 adj = 0.72), Mg (R^2 adj = 0.24), and Mn (R^2 adj = 0.08). Clearly Ca and Mg were the main constituents of ash, explaining more than 80% of the variance.

We found an unexpected highly significant correlation between Ca and Al concentrations for all sites (Figure 5). Only two species departed strongly from this relationship, *T. elaeagnoides* and *L. glaucifolius*, both from the Guajataca site. This relationship deserves special attention because of its implications regarding the mechanisms of nutrient uptake from karstic complex substrates. Considering the low concentrations of Al, Fe, and Mn in the calcareous substrate in karst areas, it is not surprising that these elements are only a minor component of leaf ash of trees growing there. If clayey sediment accumulated in cracks and crevices of the rocky substrates on *mogote* tops was a source of those metallic elements, we expected higher concentrations of them in tree foliage, but that was not the case in our data set, with the exception of *T. elaeagnoides*, identified as a metal accumulator (Table 5).

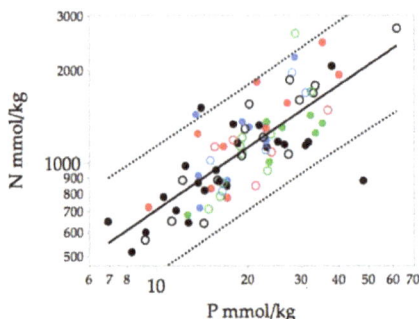

Figure 4. Correlation between N and P concentrations in adult leaves from different karst forest species. Guánica forests: dwarf, Black; ridge, empty black; Guajataca, solid blue; Cambalache, empty blue; Tallonal solid green; Nevarez, empty green; Río Lajas, solid red; Hato Tejas, empty red. Log (N mmol/kg) = 5.011 + 0.672 × Log (P); Fratio = 120; F < p < 0.001; R^2 adj = 0.592; n = 83.

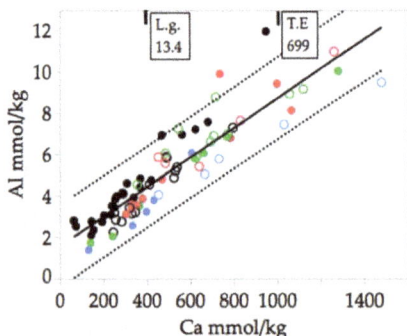

Figure 5. Linear correlation between the concentrations of Ca and Al in adult leaves of karst tree species. The arrows indicate the values of Al accumulators which were not included in the regression. Al (mmol/kg) = 1.55 + 0.007 × Ca (mmol/kg). R^2 adj = 0.825; F = 379, p < 0.0001, n = 81. Symbols are the same as in Figure 4.

3.3. Isotopic Ratios

The δ^{13}C values did not differ significantly in the northern karst forest locations (Figure 6). The only outlier was represented by *C. rosea*, a tree with crassulacean acid metabolism [30] commonly occurring in the northern karst belt forests. It was excluded from average calculations. As expected, the two sites on the southern karst had significantly higher δ^{13}C values, because of the drier conditions under which this vegetation grows. Average δ^{13}C of leaves from humid karst was −30.83 ± 1.85‰ (n = 52), and from dry karst was −27.37 ± 1.16‰, the difference was highly significant (Tukey Honest Significant Difference test p = 0.01).

The distribution of δ^{15}N values varied significantly among sites, showing a more complex pattern than that of the carbon isotopes (Figure 6). The sites on southern karst and the Nevarez site showed positive values, whereas the rest of the sites had negative δ^{15}N average values. Overall averages were positive in the dry karst, 3.49 ± 2.20‰ (*n* = 40), and slightly negative, −0.28 ± 1.82‰ (*n* = 51), in the moist karst forests. The degree of overlap was higher in the case of ^{15}N, revealing the more complex regulation of this factor under natural conditions.

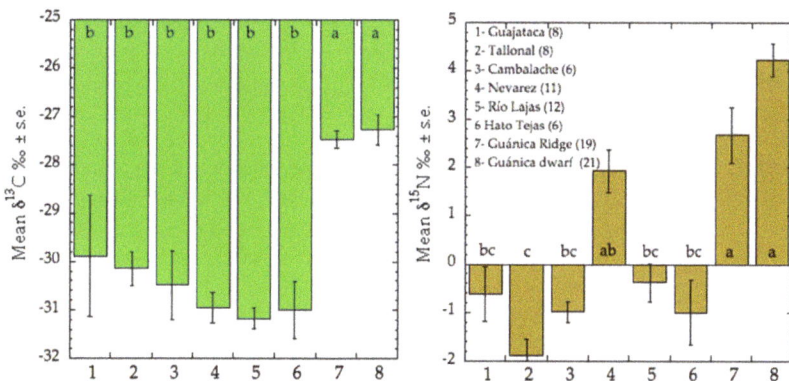

Figure 6. Average ± SE. values of N and C isotopes in adult leaves from different karst forests in Puerto Rico. Columns with the same letter at the base indicate statistical similarity (Tukey Honest Significant Difference, *p* = 0.01).

4. Discussion

4.1. Comparison between Humid and Dry Tropical Forests on Contrasting Soils

Comparison of the dry/moist karst leaf data with other assessments of leaf nutrient concentration in wet and seasonally dry tropical forests shows, in general, that karst forests in Puerto Rico are more limited by P than N, judging by their high N/P ratios, similar to those reported for P-poor forests in the Amazon basin (Table 7). A lower montane rain forest on volcanic soils in Puerto Rico (El Verde, Luquillo Experimental Forest) [31] has much higher concentrations of K and P, and lower N/P ratio than both the humid and dry karst leaves. Moreover, dry and humid karst leaves have quite similar average concentrations of Mg, P, and N, and large N/P ratios. Remarkably, dry karst leaves are richer in K, but poorer in Ca compared to humid karst leaves. Possible explanation for the high K may be the coastal location of the dry karst sites, whereas the higher Ca concentration in moist karst is probably associated with the prevailing humidity of the substrate, which leads to higher availability of soluble Ca.

Table 7. Average leaf nutrient concentrations in a range of humid and dry tropical forests (mmol/kg).

Site	*n*	N	P	K	Ca	Mg	N/P	Reference
Rainforests								
Oxisols	462	1371	26	-	-	-	58	[32]
Other	110	1307	37	-	-	-	38	[32]
French Guyana	45	1143	23	-	-	-	54	[33]
El Verde PR	41	1156	26	291	257	163	46	[31]
Seasonally dry forest								
Chamela, Mexico	19	1281	33	-	-	-	40	[34]
Goias, Brazil	13	1680	46	293	389	89	43	[35]
Karst region southwestern China								
	121	1224	47	210	517	125	26	[36]
Karst moist forest	44	1243	22	150	640	123	59	Present study
Karst dry forest	38	1123	21	248	360	141	58	Present study

The average values of C, N, and P from our data set are similar to those reported by Hättenschwiler et al. [33] for 45 tree species in a lowland wet neotropical forest. The range of variation of the species from karst forests (Table 3) is the same for C but much larger for N and P than that of the neotropical forest in French Guiana. The average N/P ratio of their data set amounts to 54, slightly lower than the average of our data set (58), but the C/N and C/P ratios are similar.

Absolute concentrations of N and P in karst forest species are also lower than those reported for many species from forests on P-poor oxisols in Amazonas by Townsend et al. [32], but the N/P ratio is quite similar, amounting to 58 on a molar basis. It may be concluded from these comparisons with moist and wet tropical forest sites that the karts forests from Puerto Rico are relatively limited by P.

Most seasonally dry forest soils in the neotropics are comparatively rich in Ca, but are as low in P as the karst forest reported here. Leaf P concentrations in seasonally dry forest in Chamela, Mexico [34], and a deciduous forest on calcareous outcrops in the cerrado of Goias in Brazil [35], are much higher than the values reported here (Table 7). Therefore, their average leaf N/P ratios are much lower than those measured in our karst forests.

A recent study from vegetation on karst in China slightly above 26° N reports surprisingly much higher P concentrations in leaves, and lower N/P values than those reported here for Puerto Rico [36]. Concentrations of the other elements are within the same range as those reported here (Table 7).

The influence of element availability in substrate is suggested by the fact that the species with higher K concentrations, and the mean of site values, correspond to the Guánica Ridge and Dwarf forests, in agreement with the results of rock analyses.

Physiological differences may be indicated for the case of *Tetrazygia elaeagnoides*, one of the few Melastomataceae tree species recorded for *mogote* tops in the northern karst belt [17,24]. This species is a strong Al accumulator, a characteristic common among the members of the Melastomataceae, but it is remarkable that it also showed the highest concentrations of Fe and Mn in the whole data set. In addition, the two legumes included in this data set, *P. arboreum* and *L. glaucifolius*, are among the species with highest concentrations of N.

Concentrations of heavy metals (Fe and Mn) in the Liu et al. study [36] were also about two times higher than in the present paper (2 vs. 0.7 mmol/kg). Data on Al in karst vegetation were not found in the available literature. Excluding the Al accumulators sampled in Guajataca site, concentrations reported here are below 10 mmol/kg (Table 4), much lower than values reported by Masunaga et al. [37] for non-accumulator trees in Sumatra rain forest (21.5).

4.2. The Ca/Al Relationship: Al Accumulators in Limestone Forests

In an extensive analysis of tree material in a tropical rain forest in West Sumatra, Masunaga et al. [37,38] were able to identify a wide variation in Al accumulation in tree foliage. They modified the criteria for defining Al-accumulator species using the 1 g/kg leaf concentration boundary (37 mmol/kg), introducing categories of accumulators between 1 and 3 g Al/kg (37–111 mmol/kg) and strong accumulators with >3 g Al/kg (>111 mmol/kg). For non-Al accumulators, Al was positively correlated with Ca ($r = 0.62$, $n = 469$) and Mg ($r = 0.23$). For strong accumulators, the correlation was not significant at the 1% level. For the non-accumulators, Masunaga et al. [37] indicate that the Al/Ca ratio averages 0.04 ± 0.02, whereas for our data set this ratio averaged 0.011 ± 0.005, due to the much higher Ca concentration in leaf tissues (509 vs. 367 mmol Ca/kg).

The elemental concentration of leaves from rainforests on Ca-rich soils in West Sumatra is characterized by a large range of variation of most metals, particularly Al, Fe, and Mn, and includes several metal accumulator species [38]. The rainforest leaves of the West Sumatra forest have quite similar concentrations of P, S, K, Ca, and Mg, but much higher concentrations of Al, Fe, and Mn than the karst forest species reported here. These comparisons show that in spite of huge differences in rainfall, the main factor regulating concentration of cations in leaves is the elemental availability from soil, with the exception of metal accumulators. Aluminum accumulators are numerous in the West Sumatra data set [37], whereas only two tree species were detected in the karst forests studied.

Why is there such a clear relationship between Ca and Al concentrations in non-Al accumulators, both in rain and moist and dry karst forests? We suggest that root activity in calcareous substrate includes the secretion of organic acids in the rhizosphere, leading to the dissolution of the carbonatic rocks, thereby releasing Ca and other metals present, as well as P. This is one of the specific mechanisms described to counteract low Fe availability in calcareous soils [14]. Organic acids may act as chelating agents facilitating Fe uptake, but also the uptake of other metallic cations present in the soil solution. In addition to the high Al-Ca correlation, our data set shows that Ca is linearly, but more weakly, related to the concentrations of Mg and Fe ($p = 0.05$), but no relationship was detected with Mn.

4.3. Isotopic Relationships

The significant higher values of $\delta^{13}C$ in both dry karst forest sites were expected. The restrictions in water supply determine long-term reductions in stomatal conductance leading to reduction in ^{13}C discrimination in the photosynthetic process [26].

Variation of N isotopic values under natural conditions can be associated with intrinsic biological characteristics of the species involved or with ecosystems nutrient cycling properties:

(a) Occurrence of mycorrhiza may decrease values compared to non-mycorrhizal plants [39,40].
(b) Symbiotic N_2 fixation (as in legumes-*Rhizobium* association) results in lower values because this process does not discriminate against the heavier N isotope [41].
(c) Conditions inhibiting denitrification processes (water saturated soils) lead to more negative values in vegetation, the opposite effect occurs in drier areas where denitrification operates freely [42,43].
(d) High levels of N availability and P limitation lead to higher values [40,44].

There are few studies on mycorrhizal associations in woody vegetation on calcareous substrates in the Caribbean. However, a recent report concluded that seasonal rainfall has a strong influence on the diversity of mycorrhizal fungi but not on infectivity [45]. Therefore, we may assume that mycorrhiza is not responsible for the differences observed. Furthermore, legumes represent only a small fraction of the species analyzed (2 out of 56). Finally, P deficiencies assessed through the N to P ratios were similar for moist and dry sites. The main factor determining lower leaf $\delta^{15}N$ values in the moist north karst species compared to the drier southern karst region seems to be the difference in rainfall. In *mogote* tops, however, humidity conditions may be highly heterogeneous due to low water retention capacity of calcareous substrates, and the random occurrence of rock cracks and crevices where water may be retained.

5. Conclusions

- Karst vegetation in Puerto Rico appears to be P limited based both on lower P concentrations and high N/P ratios in leaves. The latter are similar to those of rain forests on P poor soils.
- Source of cations originate both from calcareous rocks (Ca and Mg) and allochthonous sediments from volcaniclastic rocks (Al and Fe).
- Calcium is the predominant cation in leaf tissue and its concentration explains more than 70% of the variation in ash content.
- The leaf concentrations of K and Na are larger in the dry karst sites, probably because of their coastal location. Calcium was higher in moist karst sites, probably due the larger availability of soluble Ca in the substrate.
- The tree species of both moist and dry karst forest accumulate small amounts of Al, which is linearly correlated with the total accumulation of Ca in leaf tissue. This correlation may derive from the process of element solubilization in the rhizosphere driven by the secretion of organic acids.
- The $\delta^{13}C$ values of leaf tissue are significantly higher in the dry karst forest, revealing the limitations of water supply.

- The $\delta^{15}N$ values of leaf tissue were, with one exception, significantly lower in moist karst sites. The probable cause is that denitrification tends to predominate in drier environments.

Acknowledgments: Field work and chemical analyses were conducted using IITF facilities. Alberto Rodríguez (formerly at IITF) provided assistance during field work, and Marcos Caraballo (University of Puerto Rico) helped with the species identification. Edwin López Soto from IITF lab revised the methods section and the whole data set.

Author Contributions: E.M., E.C. and A.E.L. conceived the project; E.M. and E.C. conducted the field sampling; E.M. conducted the statistical analyses and wrote the paper with contributions from E.C. and A.E.L.

Conflicts of Interest: The authors declare no conflict of interest.

References

1. Goldich, S.S.; Bergquist, H.R. *Aluminous Lateritic Soil of the Sierra de Bahoruco Area Dominican Republic, W. I.*; Bulletin 953-C; United States Department of the Interior, Geological Survey, United States Government Printing Office: Washington, DC, USA, 1947.
2. Asprey, G.F.; Robbins, R.G. The vegetation of Jamaica. *Ecol. Monogr.* **1953**, *23*, 359–412. [CrossRef]
3. Lötschert, W. Die Übereinstimmung von geologischer Unterlage und Vegetation in der Sierra de los Organos (Westkuba). *Ber. Deutsch. Bot. Ges.* **1958**, *71*, 55.
4. Lugo, A.E.; Miranda Castro, L.; Vale, A.; López, T.M.; Hernández Prieto, E.; García Martinó, A.; Puente Rolón, A.R.; Tossas, A.G.; McFarlane, D.A.; Miller, T.; et al. *Puerto Rican Karst—A Vital Resource*; General Technical Report WO-65; USDA Forest Service: Washington, DC, USA, 2001.
5. Díaz del Olmo, F.; Cámara Artigas, R. *Karst Tropical de Colinas, Tipología y Evolución en el Pliocuaternario en República Dominicana*; Flor, G., Ed.; Actas de la XI Reunión Nacional de Cuaternario, XI Reunión Nacional de Cuaternario: Oviedo, Spain, 2003; pp. 123–129.
6. Day, M. Human interaction with Caribbean karst landscapes: Past, present and future. *Acta Carsologica* **2010**, *39*, 137–146. [CrossRef]
7. Kelly, D.L.; Tanner, E.V.J.; Kapos, V.; Dickinson, T.A.; Goodfriend, G.A.; Fairbairn, P. Jamaican Limestone Forests: Floristics, structure and environment of three examples along a rainfall gradient. *J. Trop. Ecol.* **1988**, *4*, 121–156. [CrossRef]
8. Borhidi, A. Dry coastal ecosystems of Cuba. In *Ecosystems of the World. Dry Coastal Ecosystems Africa, America, Asia and Oceania*; van der Maarel, E., Ed.; Elsevier: Amsterdam, The Netherlands, 1993; Volume 2, pp. 423–452.
9. Areces-Mallea, A.E.; Weakley, A.S.; Li, X.; Sayre, R.G.; Parrish, J.D.; Tipton, C.V.; Bouche, T. *A Guide to Caribbean Vegetation Types: Preliminary Classification System and Descriptions*; Panagopoulos, N., Ed.; The Nature Conservancy International Headquarters: Washington, DC, USA, 1999; p. 166.
10. Alemán-Gonzalez, W.B. *Karst Map of Puerto Rico*; US Geological Survey: Reston, VA, USA, 2010.
11. Briggs, R.P. The blanket sands of northern Puerto Rico. In Proceedings of the Transactions of the 3rd Caribbean Geology Conference, Kingston, Jamaica, 2–11 April 1962; Jamaica Geological Survey Publication: Kingston, Jamaica, 1966.; pp. 60–69.
12. Monroe, W.H. *The Karst Landforms of Puerto Rico*; U.S. Geological Survey Professional Paper 899; Government Printing Office: Washington, DC, USA, 1976; p. 69.
13. Mengel, K.; Kirkby, E.A. *Principles of Plant Nutrition*, 5th ed.; Kluwer Academic Publishers: Dordrecht, The Netherlands, 2001; p. 849.
14. Marschner, H. *Mineral Nutrition of Higher Plants*, 2nd ed.; Academic Press: Cambridge, MA, USA, 1995.
15. Kinzel, H. Influence of limestone, silicates and soil pH on vegetation. In *Physiological Plant Ecology III. Responses to Chemical and Biological Environment*; Lange, O.L., Nobel, P.S., Osmond, C.B., Ziegler, H., Eds.; Encyclopedia of Plant Physiology; Springer: Berlin, Germany, 1983; Volume 12, pp. 201–244.
16. Medina, E.; Cuevas, E.; Molina, S.; Lugo, A.E.; Ramos, O. Structural variability and species diversity of a dwarf Caribbean dry forest. *Caribb. J. Sci.* **2012**, *46*, 203–215. [CrossRef]
17. Alvarez Ruiz, M.; Acevedo Rodriguez, P.; Vazquez, M. Quantitative description of the structure and diversity of the vegetation in the Limestone Forest of Rio Abajo, Arecibo-Utuado, Puerto Rico. *Acta Cient.* **1997**, *11*, 21–66.

18. Acevedo-Rodríguez, P.; Axelrod, F.S. Annotated checklist for the tracheophytes of Río Abajo forest reserve, Puerto Rico. *Caribb. J. Sci.* **1999**, *35*, 265–285.

19. Murphy, P.G.; Lugo, A.E. Structure and biomass of a subtropical dry forest in Puerto Rico. *Biotropica* **1986**, *18*, 89–96. [CrossRef]

20. Monsegur-Rivera, O.A. Vascular Flora of the Guánica dry Forest, Puerto Rico. Master's Thesis, University of Puerto Rico, Mayagüez Campus, San Juan, Puerto Rico, 2009.

21. Chinea, J.D. The Forest Vegetation of the Limestone Hills of Northern Puerto Rico. Ph.D. Thesis, Cornell University, Ithaca, NY, USA, 1980.

22. Aukema, J.E.; Carlo, T.A.; Collazo, J.A. Landscape assessment of tree communities in the northern karst region of Puerto Rico. *Plant Ecol.* **2007**, *189*, 101–115. [CrossRef]

23. Molina Colon, S.; Lugo, A.E. Recovery of a subtropical dry forest after abandonment of different land uses. *Biotropica* **2006**, *38*, 354–364. [CrossRef]

24. Axelrod, F.S. *A Systematic Vademecum to the Vascular Plants of Puerto Rico*; Botanical Research Institute of Texas: Fort Worth, TX, USA, 2011.

25. Sánchez, M.J.; Lopez, E.; Lugo, A.E. *Chemical and Physical Analysis of Selected Plants and Soils from Puerto Rico (1981–2000)*; Gen. Tech. Rep. GTR-IITF-45; U.S. Department of Agriculture, Forest Service, International Institute of Tropical Forestry: Rio Piedras, PR, USA, 2015; p. 85.

26. Farquhar, G.D.; Hubick, K.T.; Condon, A.G.; Richards, R.A. Carbon isotope fractionation and plant water use efficiency. In *Stable Isotopes in Ecological Research*; Rundel, P.W., Ehleringer, J.R., Nagy, K.A., Eds.; Ecological Studies; Springer: New York, NY, USA, 1989; Volume 68, pp. 21–40.

27. Handley, L.L.; Raven, J.A. The use of natural abundance of nitrogen isotopes in plant physiology and ecology. *Plant Cell Environ.* **1992**, *15*, 965–985. [CrossRef]

28. Viera Martínez, C.A.; Abelleira Martínez, O.J.; Lugo, A.E. Estructura y química del suelo en un bosque de *Castilla elastica* en el carso del norte de Puerto Rico: Resultados de una calicata. *Acta Cient.* **2008**, *22*, 29–35.

29. McGroddy, M.E.; Daufresne, T.; Hedin, L.O. Scaling of C:N:P stoichiometry in forests worldwide: Implications of terrestrial Redfield-type ratios. *Ecology* **2004**, *85*, 2390–2401. [CrossRef]

30. Tinoco-Ojanguren, C.; Vázquez-Yánes, C. Especies CAM en la selva húmeda tropical de Los Tuxtlas, Veracruz. *Bol. Soc. Bot. Mex.* **1983**, *45*, 150–153.

31. Ovington, J.D.; Olson, J.S. Biomass and Chemical Content of El Verde Lower Montane Rain Forest Plants. In *A Tropical Rain Forest: A Study of Irradiation and Ecology at El Verde, Puerto Rico*; Odum, H.T., Pigeon, R.F., Eds.; National Technical Information Service: Springfield, VA, USA, 1970; Volume H-2, pp. H53–H75.

32. Townsend, A.R.; Cleveland, C.C.; Asner, G.P.; Bustamante, M.M.C. Controls over foliar N:P ratios in tropical rain forests. *Ecology* **2007**, *88*, 107–118. [CrossRef]

33. Hättenschwiler, S.; Aeschlimann, B.; Coûteaux, M.-M.; Roy, J.; Bonal, D. High variation in foliage and leaf litter chemistry among 45 tree species of a neotropical rainforest community. *New Phytol.* **2008**, *179*, 165–175.

34. Rentería, L.Y.; Jaramillo, V.J. Rainfall drives leaf traits and leaf nutrient resorption in a tropical dry forest in Mexico. *Oecologia* **2011**, *165*, 201–211. [CrossRef] [PubMed]

35. Rossatto, D.R.; Alvim Carvalho, F.; Haridasan, M. Soil and leaf nutrient content of tree species support deciduous forests on limestone outcrops as a eutrophic ecosystem. *Acta Bot. Bras.* **2015**, *29*, 231–238. [CrossRef]

36. Liu, C.; Liu, Y.; Guo, K.; Wang, S.; Yang, Y. Concentrations and resorption patterns of 13 nutrients in different plant functional types in the karst region of south-western China. *Ann. Bot.* **2014**, *113*, 873–885. [CrossRef] [PubMed]

37. Masunaga, T.; Kubota, D.; Hotta, M.; Wakatsuki, T. Mineral composition of leaves and bark in aluminum accumulators in a tropical rain forest in Indonesia. *Soil Sci. Plant Nutr.* **1998**, *44*, 347–358. [CrossRef]

38. Masunaga, T.; Kubota, D.; Hotta, M.; Wakatsuki, T. Nutritional characteristics of mineral elements in tree species of tropical rain forest, west Sumatra, Indonesia. *Soil Sci. Plant Nutr.* **1997**, *43*, 405–418. [CrossRef]

39. Högberg, P. [15]N natural abundance in soil-plant systems. *New Phytol.* **1997**, *137*, 179–203. [CrossRef]

40. Craine, J.M.; Elmore, A.J.; Aidar, M.P.M.; Bustamante, M.; Dawson, T.E.; Hobbie, E.A.; Kahmen, A.; Mack, M.C.; McLauchlan, K.K.; Michelsen, A.; et al. Global patterns of foliar nitrogen isotopes and their relationships with climate, mycorrhizal fungi, foliar nutrient concentrations, and nitrogen availability. *New Phytol.* **2009**, *183*, 980–992. [CrossRef] [PubMed]

41. Shearer, G.; Kohl, D.H. N$_2$-fixation in field settings: Estimations based on natural ^{15}N abundance. *Aust. J. Plant Physiol.* **1986**, *13*, 699–756. [CrossRef]

42. Martinelli, L.A.; Piccolo, M.C.; Townsend, A.R.; Vitousek, P.M.; Cuevas, E.; Mcdowell, W.; Robertson, G.P.; Santos, O.C.; Treseder, K. Nitrogen stable isotopic composition of leaves and soil: Tropical versus temperate forests. *Biogeochemistry* **1999**, *46*, 45–65. [CrossRef]

43. Medina, E.; Izaguirre, M.L. N$_2$-fixation in tropical American savannas evaluated by the natural abundance of ^{15}N in plant tissues and soil organic matter. *Trop. Ecol.* **2004**, *45*, 87–95.

44. Nardoto, G.B.; Ometto, J.P.H.B.; Ehleringer, J.R.; Higuchi, N.; Bustamante, M.M.; Martinelli, L.A. Understanding the influences of spatial patterns on N availability within the Brazilian Amazon forest. *Ecosystems* **2008**, *11*, 1234–1246. [CrossRef]

45. Guadarrama, P.; Castillo, S.; Ramos-Zapata, J.A.; Hernández-Cuevas, L.V.; Camargo-Ricalde, S.L. Arbuscular mycorrhizal fungal communities in changing environments: The effects of seasonality and anthropogenic disturbance in a seasonal dry forest. *Pedobiologia* **2014**, *57*, 87–95. [CrossRef]

![forests logo] *forests*

MDPI

Article

Land Use, Conservation, Forestry, and Agriculture in Puerto Rico

William A. Gould [1,*], Frank H. Wadsworth [1], Maya Quiñones [1], Stephen J. Fain [2] and Nora L. Álvarez-Berríos [1]

[1] United States Department of Agriculture, Forest Service, International Institute of Tropical Forestry, Jardín Botánico Sur, 1201 Ceiba St., Río Piedras, San Juan 00926, Puerto Rico; frankhwadsworth@gmail.com (F.H.W.); mquinones@fs.fed.us (M.Q.); nalvarezberrios@fs.fed.us (N.L.Á.-B.)

[2] Pinchot Institute for Conservation, Western Regional Office: 721 NW 9th Ave Ste 240, Portland, OR 97209, USA; sjfain@gmail.com

* Correspondence: wgould@fs.fed.us; Tel.: +1-787-764-7790; Fax: +1-787-766-6302

Received: 12 June 2017; Accepted: 3 July 2017; Published: 7 July 2017

Abstract: Global food security concerns emphasize the need for sustainable agriculture and local food production. In Puerto Rico, over 80 percent of food is imported, and local production levels have reached historical lows. Efforts to increase local food production are driven by government agencies, non-government organizations, farmers, and consumers. Integration of geographic information helps plan and balance the reinvention and invigoration of the agriculture sector while maintaining ecological services. We used simple criteria that included currently protected lands and the importance of slope and forest cover in protection from erosion to identify land well-suited for conservation, agriculture and forestry in Puerto Rico. Within these categories we assessed U.S. Department of Agriculture (USDA) farmland soils classification data, lands currently in agricultural production, current land cover, and current land use planning designations. We found that developed lands occupy 13 percent of Puerto Rico; lands well-suited for conservation that include protected areas, riparian buffers, lands surrounding reservoirs, wetlands, beaches, and salt flats, occupy 45 percent of Puerto Rico; potential working lands encompass 42 percent of Puerto Rico. These include lands well-suited for mechanized and non-mechanized agriculture, such as row and specialty crops, livestock, dairy, hay, pasture, and fruits, which occupy 23 percent of Puerto Rico; and areas suitable for forestry production, such as timber and non-timber products, agroforestry, and shade coffee, which occupy 19 percent of Puerto Rico.

Keywords: Caribbean; land use planning; tropical agriculture; tropical forests; geospatial analyses

1. Introduction

The question of how to best use land that provides food, forest products, water, and shelter, is as old as civilization. People have answered in a way that has allowed us to inhabit all corners of the earth, and thrive under a wide range of environments. Problem solved? Not exactly. As the population expands, technology advances, climate changes, and resource demands shift, questions persist as to how to sustain the flow of food, fiber, and ecosystem services. Globally, one of the most pressing challenges is population growth and the equitable distribution of resources. Recent projections by the United Nations Food and Agricultural Organization estimate global populations will reach 9.1 billion by 2050 (compared to 7.4 in 2017). Providing adequate nutrition to this many people will require an estimated 70 percent increase in food production. While the majority of this increase is projected to come from increased yields and cropping intensity on existing lands, agricultural land is expected to expand by 70 million hectares worldwide [1]. These projections highlight the need for effective land use planning, to balance local food production with other demands, such as urban development, water, biodiversity, forest products, and recreation.

The world is more interconnected than at any time in our history. Goods, services, energy, and information, flow at ever increasing rates. Connectivity and the pace of change challenges global, national, and local structures established to govern resources and implement land use decisions. Information about the distribution and state of resources is valuable in decision making processes. Planners and resource managers navigate complex landscapes of competing demands. Islands, like Puerto Rico, have limited land area and sharp boundaries between imported and local resources, such as food and water, which are dependent on climate and land use practices.

A key decision for any society, and particularly those on islands, is how to partition lands among the potentially competing uses of urbanization and residential use, conservation, forestry, and agriculture. In this paper, we develop a set of simple landscape characterizations to guide land use decisions toward lands most suitable for agriculture, forestry, and conservation. The central premise is that water, food, and forest products are valuable services generated from the land, along with recreation, conservation of biodiversity, energy production, and other services. Land use decisions can assess suitability, conflict, and compatibility, depending on the prevailing vision and needs of a society. These characterizations can help frame the discussion of what may be gained or lost in promoting a particular use in a given area.

Puerto Rico is one of over thirty island nations or territories in the Caribbean that share many similarities in climate, landscape features, flora, fauna, and agricultural crops. Since the 1960s, while there has been a greater-than-world average increase in agricultural productivity in Latin America and the Caribbean due to technological advances, the Caribbean islands have seen either little increase or a decrease in productivity [2]. With agricultural production in the region historically oriented toward crops produced for export, such as sugar cane, coffee, and tobacco, domestic food production has long been inadequate to satisfy domestic demand [3,4]. The decline in food production, and a subsequent increase in food imports, is contributing to reduced employment and increased impoverishment of rural communities in many countries [4,5]. Additionally, dependency on food imports makes regional food security vulnerable to fluctuations in global food prices, shortages and export blockades, transportation fuel prices, and the effects of climate change [3,5–7].

The history of forestry and agriculture in Puerto Rico is central to understanding the current matrix of forests, cities, protected areas and agricultural lands. Francisco Watlington [8] made some assessments of the agricultural carrying capacity of pre-Columbian Puerto Rico. Early accounts record a population of 600,000 indigenous inhabitants [9], not including women and children. Watlington makes the assumption of equal numbers of men and women, with on average four children, and arrives at an estimate of 3.6 million people—roughly equal to today's population. Cassava (*Manihot esculenta*) was a staple food for indigenous people. Watlington estimates this population could have required about 90,000 ha of cultivated land to support the indigenous population.

Political and economic forces within and outside of Puerto Rico have generally driven land use over the last century [10]. Forest cover declined steadily from the late 1800s to the mid 1900s as the population grew, and much of the island was converted to intensive agriculture. During the second half of the 20th century, Puerto Rico transitioned from an agrarian economy based primarily on sugar cane, to one based on industry and services [4,10,11]. The abandonment of agricultural land was followed by rapid forest recovery across the island. Puerto Rico's forest cover went from 6 percent in the 1950s [11], to 55 percent as of 2009 [12]. Loss of forest cover in the early part of the century led to the loss of a thriving timber industry, loss of traditional knowledge of using forest products, and greater importation of wood products. Agricultural abandonment in the second half of the 20th century, along with a boom in the industrial sector, led to a decrease in the relative economic importance of agriculture and an increase in food imports.

Puerto Rico currently imports over 80 percent of its food supply [13]. However, a new wave of initiatives is attempting to ensure food security by rebuilding a vital and ecologically conscious agrarian sector within the island. This includes developing new products and markets, and improving supply chains. The mission statement of the Puerto Rico Department of Agriculture is "food security through sustainable agriculture that is ecologically responsible." One of its primary objectives is promoting sustainable agricultural practices and the expansion of local food production [14]. Other efforts toward improving food security and increasing local food production are arising in the island led by municipal and federal governments, as well as private and non-governmental organizations. These efforts reflect a broader, global movement to transform food production and supply chains toward models that empower local farmers, reconnect urban populations with food sources, improve farming practices, respect local knowledge and ecological conditions, and build climate resilience [15,16].

Prime quality agricultural land is a limited resource in Puerto Rico, and is defined as having soils with the necessary qualities to produce high crop yields when properly managed [17]. Urbanization and land degradation, including loss of topsoil, reduced soil water holding capacity, and loss of soil carbon, can depress agricultural production by reducing the availability of highly productive land, decreasing the sustainability of agricultural systems, and encouraging the use of less productive marginal lands [18]. Urban development in Puerto Rico has grown steadily over the last five decades, even with the population declining since 2000. Recent growth has been characterized as urban sprawl, with construction on soils suitable for agriculture [19,20]. López et al. [19] found that 42 percent of urban areas constructed in Puerto Rico between the years 1977 and 1994, were built on potential agricultural land, and that urban growth in Puerto Rico tends to occur on prime farmland, making the preservation of remaining agricultural land important to assuring food security for future generations.

Trends towards greater movement of people and goods lead to questions of what people value and want to see in their local landscape, i.e., what combination of living and industrial use, food and fiber production, recreation, and conservation, is most sustainable and leads to the highest level of human well-being. Land use conflicts can be a result of conflicting visions and a lack of knowledge of land suitability to deliver services [21]. While clearer, shared information about services may not resolve conflicting visions, it may provide a sound platform for decision making. This paper addresses a need for shared common knowledge about the suitability of specific components of the landscape to deliver services related to agriculture, forestry, and conservation.

In this study, we characterized the lands of Puerto Rico into four categories: impervious surfaces (developed lands and roads) based on remote sensing analyses [22], lands best suited for conservation based on the current protected areas network and other conservation priorities [23], and two categories of working lands with potential for agriculture and forestry, and less prone to erosion, based on slope and land use [24,25]. Within these broad categories, we assessed the current land cover, current zoning classification under the Puerto Rico Land Use Plan, and Natural Resources Conservation Service (NRCS) farmland soil characterizations. These characterizations of suitability support the assumption that clean water and healthy soils provide ecosystem services and are important to a sustainable society. These services can be lost by conversion of open space to developed land, and by erosion of exposed slopes leading to sedimentation in reservoirs, estuaries, and coastal waters [26].

2. Materials and Methods

We developed a set of simple criteria to identify areas well-suited to mechanized agriculture, well-suited to non-mechanized agriculture on moderate to steep slopes, and areas suitable for forestry practices, including timber harvest potential, where greater forest cover has benefits in terms of soil conservation and water management. These are steeper slopes where timber production may be integrated with agroforestry, shade coffee, non-timber forest product uses, or other forms of sustainable activity that maintain a high degree of forest cover. Criteria were developed based on literature review, expert opinion, and geospatial data availability. The agriculture and forestry models used slope and land cover parameters to identify the land with the highest potential for these activities, while excluding areas with developed land and a high degree of conservation potential. We characterized areas with high agriculture and forestry potential by patch size, and summarized by municipality.

We also assessed lands identified with agricultural potential following the USDA Natural Resource Conservation Service (NRCS) soils and farmland classification. Finally, we developed a map of areas with highest potential for conservation, agriculture, and forestry, excluding currently developed lands. We assess the spatial distribution of current agricultural production, land cover, and land use planning objectives, in terms of our classification.

2.1. Lands Well-Suited for Agriculture

We identified land with agriculture potential within two slope ranges. One identified relatively flat land (under 10 percent slope), optimal for mechanized agriculture [27], while the other identified potential agricultural land with moderate slopes (10 to 20 percent). We calculated slope percentage using the 10 m pixel Digital Elevation Model (DEM) from the US Geological Survey (USGS) National Elevation Dataset (NED) for Puerto Rico [28]. We used the land cover of Puerto Rico for the year 2000 [22] with a spatial resolution of 15 m. We excluded wetlands, developed land surface, and natural barrens (i.e., fresh and ocean water, mudflats, riparian and other natural barrens, gravel and sandy beaches, and rocky cliffs) from lands with agricultural potential. We also excluded Puerto Rico's protected areas [23], and all the cays and small islands. We excluded riparian zones, identified as 50 m on each side of perennial streams and rivers, and reservoirs, to protect water bodies from erosion and contaminants [29,30]. The zones were created by delineating a buffer around the rivers using the National Hydrography Dataset [31], and reservoirs, using data from the Puerto Rico Department of Natural and Environmental Resources.

2.2. Lands Well-Suited for Forestry

We identified land with forestry (timber production) potential as having slopes from 20 to 50 percent using the 10 m NED derived slope dataset. We did not include protected areas [23], wetlands, developed land surface, natural barrens [22], riparian zone 50 m buffers, or watersheds that contain reservoirs, as areas suitable for timber production. These watersheds were excluded given their role in reducing sedimentation and protecting important water sources for Puerto Rico's reservoirs.

2.3. Farmland Soils

We identified soil classes of the U.S. Department of Agriculture (USDA) Natural Resources and Conservation Service (NRCS) farmland classification of soils within lands we identified as well-suited for conservation, agriculture and forestry. The farmland classification attribute identifies NRCS soil map units under the categories of prime farmland, farmland of statewide importance, conditional prime farmland (i.e., prime farmland if irrigated, prime farmland if drained, prime farmland if irrigated and reclaimed of excess salts and sodium), and farmland of statewide importance, if irrigated. The classification system is based on a combination of physical and chemical characteristics of soil desirable for producing food, feed, forage, fiber, and oilseed crops [17]. We quantified the NRCS farmland soils that were outside of those we classified as well-suited to agriculture, identifying according to the following characteristics: Protected areas, developed land, wetlands, natural barrens, slopes over 20 percent, and river and reservoir 50 m buffers.

2.4. Lands Well-Suited for Conservation, Agriculture, and Forestry Uses

We characterized all open space, or unbuilt lands in Puerto Rico, as well-suited to conservation, agriculture, and forestry uses. These exclude developed land, based on Gould et al. [22], and additional impervious road surfaces [32], and include lands well-suited for agriculture, including row crops, orchards, hay, pasture, and dairy; areas well-suited for forestry, including timber production, agroforestry, shade coffee, livestock grazing, and non-timber forest product uses; and areas well-suited for conservation, including the protected areas of Puerto Rico [23], wetlands, lands over 50 percent slope, and natural barrens (i.e., fresh and ocean water, mudflats, riparian and other natural barrens, gravel and sandy beaches, and rocky cliffs) based on Gould et al. [22], cays and small islands, and riparian zones of 50 m to each side of the rivers and reservoirs.

2.5. Current Agricultural Productivity

We assessed the relationship of lands we identified as well-suited to agriculture with the current spatial extent of areas under agricultural production. We used information from the Farm Service Agency (FSA) common land units. The common land unit (CLU) dataset consists of digitized farm tracts and field boundaries, and associated attribute data. These identify farm tracts enrolled in FSA programs and eligible for USDA support. Farm tracts are defined by FSA as sets of contiguous fields under single ownership. Common land units are used to administer USDA farm commodity support and conservation programs in a GIS environment. Not all land under production is enrolled in these programs so they underestimate land in production in that sense, and not all of tract area is in production so they overestimate land under production in that sense.

2.6. Current Land Cover and Zoning

We assessed the relationship of lands we identified as well-suited to agriculture, forestry, and conservation in terms of their current land cover based on Gould et al. [22], and in terms of lands identified in the Puerto Rico Land Use Plan [32] as either water, roads, specially protected rustic (rural) lands (for agriculture or in combination with agriculture), common rural lands, specially protected rustic lands (for conservation, not agriculture), lands with urban potential, or urban lands [32].

3. Results

3.1. Lands Well-Suited for Agriculture

Vicente-Chandler [33] identified 106,120 ha (262,228 acres), or 12 percent of Puerto Rico, as suitable for mechanized agriculture [27]. In this assessment, we identify 124,187 ha (306,873 acres), or 14 percent of Puerto Rico, as well-suited to mechanized agriculture, with slopes under 10 percent (Table 1, Figures 1 and 2). This land is mainly located in the coastal plains and interior valleys, with the largest patches located in the northwest and south of the island. The coastal and interior plains of Puerto Rico encompass a total of 240,000 ha (27 percent of all land). Of these, 142,292 ha (16 percent) are classified by the Puerto Rico Department of Agriculture as agricultural reserves, and 98,247 ha (11 percent of all land) are developed. Within the agricultural reserves, 21,774 ha (15 percent) are wetlands, 16,072 ha (11 percent) are currently forested, and 6015 ha (4 percent) are conservation protected areas. The difference between the areas identified in this analysis, and land within the agricultural reserves not identified above (25,589 ha), include riparian and reservoir buffers, saline mudflats, beaches, interior waters, and other barrens not suitable for agriculture. The municipalities with the largest amount of land with agriculture potential under 10 percent slopes include Arecibo, Salinas, Lajas, Santa Isabel, and Cabo Rojo, all with over 4000 hectares ($\approx$10,000 acres) (Appendix A, Figure A1, Table A1).

Table 1. Land classes well-suited to conservation, agriculture, and forestry, with developed land of Puerto Rico, excluding the protected islands of Mona, Monito, and Desecheo.

Land Class	Hectares	Percent
Well-suited to mechanized agriculture (<10% slope)	124,187	14
Well-suited to non-mechanized agriculture (10–20% slope)	84,574	9
Well-suited to forestry (20–50% slope)	169,125	19
Developed (built-up, artificial barrens and roads)	115,859	13
Well-suited to conservation	399,673	45
TOTAL	893,418	100

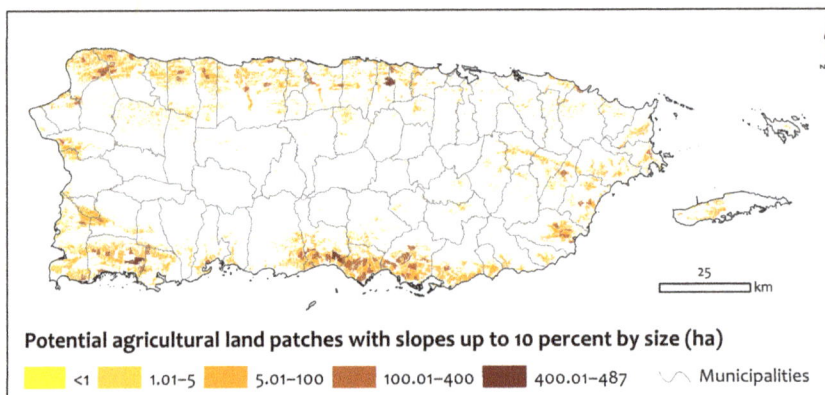

Figure 1. Map of lands well-suited to mechanized agriculture on relatively flat terrain with slopes up to 10 percent, classified by patch size.

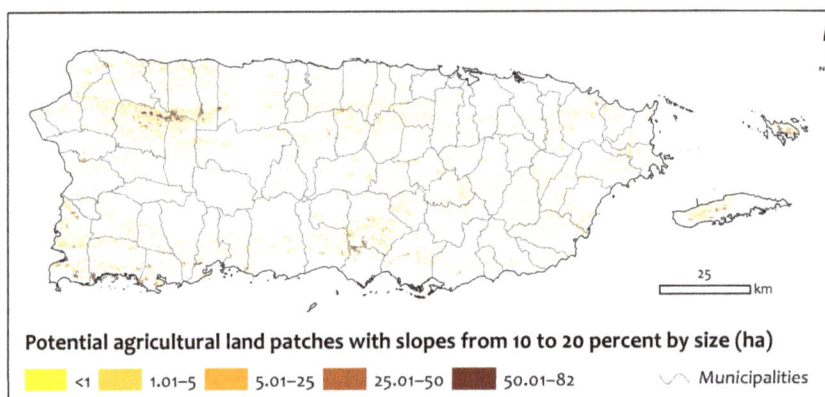

Figure 2. Map of lands well-suited to non-mechanized agriculture on slopes from 10 to 20 percent, classified by patch size.

We identified 84,574 ha (208,987 acres) of land—nine percent of Puerto Rico—as well-suited to non-mechanized agriculture on moderate (10–20 percent) slopes (Figure 2). The largest areas were located in the northern karst belt, and the southwest of the island, with noteworthy areas in Vieques and Culebra. The municipalities with the largest amount of land well-suited to non-mechanized agriculture on moderate slopes are San Sebastián, Arecibo, Cabo Rojo, and Coamo, all with over 4000 hectares (≈10,000 acres) (Appendix A, Figure A2, Table A1).

3.2. Lands Well-Suited to Forestry

A total of 169,125 ha (417,917 acres)—about 19 percent of Puerto Rico—were identified as well-suited for forestry production, while excluding watersheds that supply water to Puerto Rico's reservoirs (Figure 3). These lands are located across the hills and mountains in the main island of Puerto Rico, with large areas in the central mountains, northern karst hills, southern hills, and small patches in Vieques. The municipalities with the largest amount of land with forestry potential are Arecibo, Coamo, San Germán, Corozal, and Ciales, all with over 4000 hectares (≈10,000 acres) (Appendix A, Figure A3, Table A1).

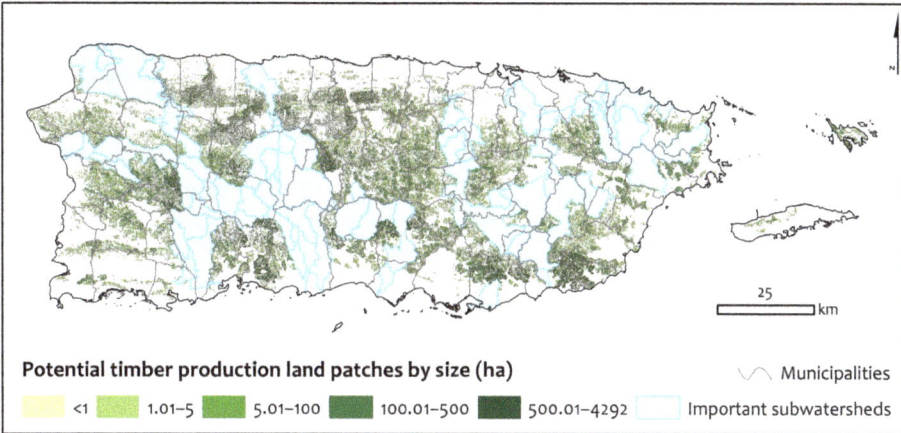

Figure 3. Map of lands well-suited to forestry, including agroforestry, such as shaded pastures for livestock, coffee, and the use of non-timber forest products including beekeeping and honey production. Watersheds surrounding reservoirs are excluded.

3.3. Farmland Classification of Soils

Classified farmland soils cover just under 26 percent of Puerto Rico. Following NRCS classifications, 9 percent of Puerto Rico's soils are classified as prime farmland (Figure 4); 11 percent are classified as farmland soils of statewide importance, and 6 percent as conditional farmland soils. This includes farmland soils of statewide importance, if irrigated (<1 percent), prime farmland if irrigated (4 percent), prime farmland if irrigated and reclaimed of excess salts and sodium (<1 percent), and prime farmland if drained (2 percent).

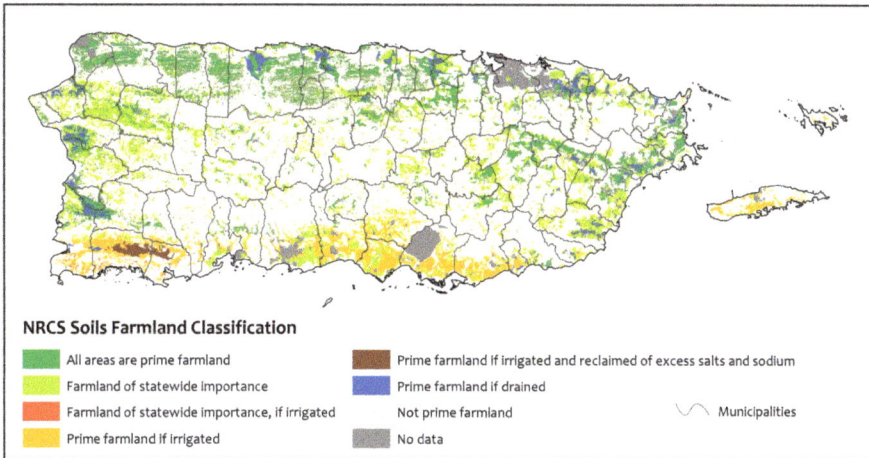

Figure 4. Map of Natural Resource Conservation Service (NRCS) soils farmland classification in Puerto Rico.

Sixty-nine percent of the lands we have identified as well-suited for mechanized agriculture are classified as farmland (42 percent) or conditional farmland (27 percent). Of the non-conditional farmland soils, 26 percent were classified as prime farmland soils and 16 percent as farmland soils of statewide importance, while 27 percent included soils classified as not prime farmland, and 4 percent had no data (Figure 5, Table A2). The numbers were very different for lands we characterized as well-suited for non-mechanized agriculture with slopes from 10 to 20 percent. More than half of the land resulting from this model contained soils classified as not prime farmland (60 percent) and only 14 percent and 18 percent were classified as prime farmland soils and farmland of statewide importance respectively; 6 percent were classified as conditional prime farmland and 2 percent had no data (Figure 5, Table A2). The prime farmland classification identifies soils with the best quality, dependable moisture supply, favorable temperature and growing season, acceptable acidity or alkalinity, not excessively erodible or saturated for long periods, an acceptable salt and sodium content, and few or no rocks, i.e., characteristics needed to economically produce sustained high yields of a wide variety of crops, including row crops, fruit trees, and forage, when properly managed and using modern farming techniques [34]. Prime farmland soils are considered a limited resource by the USDA. However, the classification does not imply that soils classified as not prime farmland cannot be cultivated successfully.

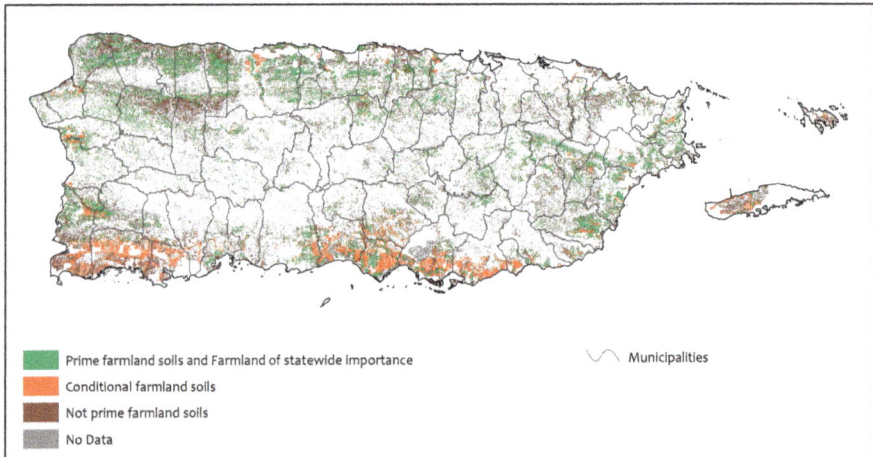

Figure 5. Map of lands well-suited to mechanized and non-mechanized agriculture, characterized by NRCS soils farmland classification of prime farmland soils and soils of statewide importance, conditional farmland soils (typically needing irrigation), not prime farmland soils, and areas of no soils information.

A total of 120,030 hectares of prime farmland soils (97,150 ha) and conditional prime farmland soils (22,879 ha) were located outside land modeled to be well-suited for mechanized and non-mechanized agriculture. These farmland soils were located across all the features excluded from lands well-suited to agriculture, but were mostly found in land with slopes over 20 percent, river and reservoir 50 m buffers, protected areas, developed areas, and wetlands (Figure 6).

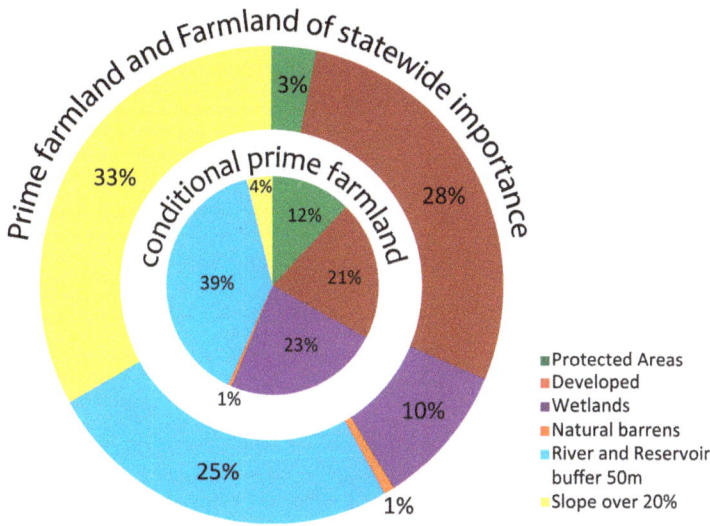

Figure 6. Distribution of NRCS farmland soils that occur outside of lands characterized as well-suited to agriculture in this study.

3.4. Characterizing Suitability for Conservation, Agriculture, and Forestry

We used the aforementioned classifications to characterize the Puerto Rican landscape by lands currently developed, and by lands well-suited to conservation, agriculture, and forestry (Figure 7). These characterizations are not exclusionary. Complimentary uses exist that cut across all categories. For example, conservation, urban forestry, and urban agriculture are all compatible to some extent with development; conservation of soil, water, and wildlife are compatible with agriculture and forestry practices; and forestry activities in particular, can be sustainable and complementary with protected areas and conservation. Lands designated as well-suited to forestry can contribute to the agricultural potential of Puerto Rico, using practices such as agroforestry, coffee production, apiculture, livestock grazing, agrotourism, and renewable non-timber forest products. Lands well-suited to conservation practices, and excluded from agriculture and forestry lands, have potential to sustainably produce forest products while maintaining conservation priorities. This includes sustainable forestry practices that retain forest cover in state and national forests and in riparian and reservoir buffers.

We find that current impervious surfaces (roads and developed lands) occupy 13 percent of Puerto Rico, Vieques and Culebra; lands well-suited for conservation—including protected areas riparian and reservoir buffers, subwatersheds surrounding reservoirs, wetlands, and barrens such as salt and mudflats, beaches, slopes greater than 50 percent and water bodies, occupy 45 percent of Puerto Rico; lands well-suited for mechanized and non-mechanized agriculture—including row crops, livestock and dairy, hay and pasture, fruits, and other specialty crops, occupy 23 percent of Puerto Rico; and areas suitable for forestry production—including timber, non-timber forest products, agroforestry, coffee, apiculture, livestock grazing, and agrotourism, occupy 19 percent of Puerto Rico.

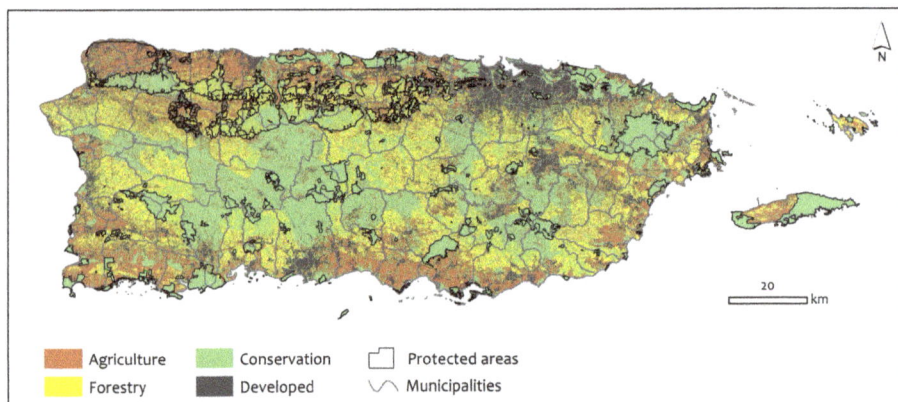

Figure 7. The Puerto Rican landscape characterized by lands well-suited to conservation, agriculture, forestry, and development indicating current developed lands (13 percent); lands well-suited for conservation—including protected natural areas and riparian and reservoir buffers, subwatersheds surrounding reservoirs, wetlands, and barrens such as salt and mudflats, beaches, and water bodies (45 percent); lands well-suited for mechanized and non-mechanized agriculture—including row crops, livestock and dairy, hay and pasture, fruits, and other specialty crops (23 percent); and areas suitable for forestry production—including timber, non-timber forest products, agroforestry, livestock, shade coffee, apiculture, and agrotourism (19 percent).

3.5. Lands in Agricultural Productivity

Various methods are used to assess land under cultivation or other agricultural use, and the location of those lands. Each assessment has strengths and limitations. The National Agricultural Statistical Survey (NASS) takes place every 5 years and captures information reported by farmers [35]. These assessments indicate 219,109 ha were under cultivation in 2007, and 229,900 ha in 2012, or 26 to 28 percent of Puerto Rico, respectively. The FSA assesses farm tract locations, and the extent of those tracts that are registered as active farms and eligible for conservation and farm assistance from the USDA. From 2007 to the present, 106,955 ha or 13 percent of Puerto Rican lands have been enrolled with FSA, and their lands are mapped as Common Land Units. This is about half of the farmland captured by the NASS. Of these areas, it is estimated that about 80 percent are in cultivation, and 20 percent in other uses, such as conservation buffers. Of the lands identified by FSA, 38 percent are located on those lands well-suited for agriculture, and 62 percent on those lands well-suited for forestry production, indicating many of the farms in the FSA program are on steeper slopes and within areas well-suited for shade coffee, agroforestry, timber, and non-timber forestry production. Only about 19 percent of those lands we have identified as well-suited for agricultural production have been enrolled in the FSA program in the last five years.

3.6. Current Land Cover and Land Use

Twenty-two percent of the lands characterized as well-suited for agriculture are currently forested, with 13 percent as woodland or shrubland, and 65 percent with grassland, row crops, or other agriculture (Table 2). Fifteen percent of the forested lands in Puerto Rico occur on lands well-suited to agriculture. Sixty-eight percent of the lands well-suited to forestry are currently forest or woodland. Seventy-four percent of the lands well-suited to conservation are currently forest or woodland. This includes all of the forested wetlands, both coastal mangroves, and montane cloud forests. Seventy-eight percent of lands well-suited to agriculture are currently non-forest or shrubland; most of these lands are either lands under agricultural production, pasture, abandoned agriculture, or otherwise managed non-forest lands.

The Puerto Rico Land Use Plan [32], approved by the Puerto Rico legislature, designates 28 percent of the island as specially protected rural lands for agriculture, or in combination with agriculture. Thirty-four percent of these occur on lands we designate as well-suited for agriculture, 19 percent on lands we designate as well-suited for forestry, and 46 percent on lands we designate as well-suited for conservation (Table 3). Additionally, the Land Use Plan designates 32 percent of Puerto Rico as specially protected rural lands for conservation, not agriculture. We designate 64 percent of these lands as well-suited for conservation, 21 percent as well-suited for forestry, and 14 percent as well-suited for agriculture. The Land Use Plan designates 14 percent of Puerto Rico as urban or potentially urban lands. We designate 54 percent of these lands as either well-suited for conservation, agriculture, or forestry. While the Land Use Plan and this study are in general agreement, important differences are that much of what the Land Use Plan designates as urban, or potentially urban, are currently open spaces that, in this study, are characterized as well-suited for forestry, conservation, or agricultural production. Additionally, many of the lands designated for agriculture in the Land Use Plan, are characterized as well-suited for conservation in this study (46 percent), and 34 percent of the lands designated for conservation in the Land Use Plan are identified as well-suitable for forestry or agriculture in this study.

Table 2. Relative amounts of land cover types based on Gould et al. [22] within areas well-suited for mechanized and non-mechanized agriculture, forestry, and conservation for Puerto Rico.

Land Cover Classification [20]	Mechanized Agriculture <10% Slope		Non-Mechanized Agriculture 10–20% Slope		Agriculture <20% Slope		Forestry 20–50% Slope		Conservation	
	Hectares	% of Total	Hectares	% of Total	Hectares	% of Total	Hectares	% of Total	Hectares	% of Total
Forest	17,686	14	27,598	33	45,285	22	85,918	51	210,182	53
Natural barren	0	0	0	0	0	0	0	0	1063	0
Grassland and agriculture	93,219	75	42,657	50	135,876	65	53,679	32	87,281	22
Woodland and shrubland	12,948	10	14,264	17	27,212	13	29,493	17	58,224	15
Forested wetlands	0	0	0	0	0	0	0	0	8482	2
Non-forested wetlands	0	0	0	0	0	0	0	0	26,586	7
Water	66	0	0	0	66	0	0	0	7660	2
TOTAL	123,919	100	84,519	100	208,438	100	169,090	100	399,478	100

Table 3. Relative amounts of land use categories based on the official Puerto Rico Land Use Plan [32] within areas well-suited for mechanized and non-mechanized agriculture, forestry, and conservation for Puerto Rico.

Land Use Plan [32]	Mechanized Agriculture <10% Slope		Non-Mechanized Agriculture 10–20% Slope		Combined Agriculture <20% Slope		Forestry 20–50% Slope		Conservation	
	Hectares	% of Total	Hectares	% of Total	Hectares	% of Total	Hectares	% of Total	Hectares	% of Total
Water	200	0	76	0	276	0	55	0	9478	2
Specially protected rustic lands for agriculture or in combination with agriculture	59,091	48	26,112	31	85,203	41	47,211	28	117,564	29
Common rustic lands	20,812	17	24,249	29	45,061	22	51,457	30	68,960	17
Specially protected rustic lands for conservation, not agriculture	18,830	15	22,370	26	41,200	20	59,190	35	184,102	46
Potential urban	1522	1	1141	1	2662	1	1759	1	2033	1
Urban, not necessarily built	23,590	19	10,608	13	34,197	16	9445	6	17,533	4
TOTAL	124,044	100	84,556	100	208,600	100	169,116	100	399,671	100

4. Discussion

Increasing the productivity of food and forest products and services from Puerto Rican lands can serve to increase economic stability, increase food security, improve the freshness and quality of food products, and reduce the risks associated with climate change and food insecurity [6,13]. Additionally, Puerto Rico is a tropical island with a long history of agricultural and ecological research, high capacity for using technological tools in planning and in agriculture, highly educated population, and rich in natural resources. Puerto Rico has the potential to be a leader in demonstrating how to increase food security while maintaining a balance of agriculture and forestry production, conservation, and urban, residential, and commercial uses of a finite landscape. Addressing this problem in Puerto Rico can be broadly useful, as many nations with less capacity are addressing similar problems as global populations grow, and food, water, and living space needs increase, and climate change adds uncertainty to the future.

We classified and mapped all lands in Puerto Rico at a fine spatial resolution as to whether they are built or unbuilt surfaces, and as to the suitability of currently open space for conservation agriculture, and forestry, based on slope, and criteria such as proximity to rivers, presence of wetlands, and protected status. These categories are far from mutually exclusive, nor are potential practices homogeneous in terms of sustainability, service delivery, or broader effects. The classification provides a basis for quantifying the extent of suitable areas, and for estimating the effects of land use choices within the context of the broader picture of what potential services the land provides. The classes broadly mirror current uses in the sense that lands well-suited to agriculture are primarily non-forested, and lands well-suited to forestry and conservation are forested. However, much of lands well-suited to agricultural production are pasture or abandoned pasture, and not intensively managed. Additionally, timber production is almost nonexistent on the lands suitable for that use. Finally, the majority of the lands well-suited to conservation are outside of protected areas or other conservation mechanisms [23].

In assessing the urban and residential component of the landscape, we used mapped developed land, or impervious surfaces derived from satellite remote sensing analyses from the year 2000 [22], and the current road network [32]. This is likely a conservative estimate of lands not suited for agriculture and forestry due to urban uses, as it excludes lawns, road right-of-ways, and golf courses, among other things. Notwithstanding, innovative approaches to urban agriculture, such as backyard conservation, roof top gardens, hydroponic production, or vertical agriculture [36–38], can lead to food production, even within this component. Permeable surfaces that have potential for conservation, agriculture, and forestry, but are typically managed for other uses, such as residential, recreational, or transportation corridors, are of interest as mechanisms to increase food security, habitat, and ecosystem services [39]. As such, we included these surfaces as components of the lands well suited to either conservation, agriculture, or forestry. The 13 percent of lands characterized as "developed" in this analysis represent the impervious infrastructure of Puerto Rico. There is continued pressure to revamp the construction sector and convert permeable lands to impervious surfaces, even in light of the declining population over recent decades [40]. Projected population levels for Puerto Rico are expected to continue to decline over the next decade, due to economic conditions and emigration [41]. Additionally, much of the existing infrastructure is underused as population declines have been greater in urban centers than suburban and rural areas [20,42], leaving room for potential redevelopment and modernization of current infrastructure without encroaching on permeable lands. Likely increases in impermeable surfaces include transportation corridors, and commercial and residential development. Future conversion from permeable to impermeable surfaces may relate to the reduced cost of building on existing open space vs. redeveloping urban space, desired ambiance of non-urban settings for development, proximity to existing development, or other reasons. This analysis serves as a basis for decisions as to what well-suited uses will be lost in conversion from open space to impermeable surfaces.

In assessing those lands most well-suited to conservation uses, we broadly defined those areas as including all formally protected lands [23,43,44], with the addition of marginally productive lands, such as salt flats, and small cays and islands, wetlands, lands susceptible to erosion, such as slopes over 50 percent, and lands that protect water resources, such as riparian buffers and catchments for the reservoirs which supply Puerto Rico with water. This is the largest component of the landscape in this study, and represents nearly half of the land area of Puerto Rico. Twenty percent of these lands are set aside as protected areas. Over 60 percent are characterized as well-suited to conservation due to their value in conserving water quantity and quality, such as riparian buffers or catchment protection, and over 30 percent are on steep slopes and unsuitable for agriculture and forestry. Much of the steep slopes, catchment and riparian areas co-occur within protected areas, increasing their conservation value. All of these lands have varying degrees of potential for forestry and agriculture production co-occurring with conservation uses and maintaining conservation value. Designating them as lands well-suited to conservation, indicates that they are also well-suited to providing ecosystem services to the larger society. Innovative practices on public and private lands that value ecosystem services or integrate conservation, forestry, and agricultural goals, can help increase the benefit to individual owners, while maintaining value, such as water services, which benefit the larger society [45].

We characterize nearly 20 percent of the island as land well-suited to forestry. We balance the interest in forestry productivity with interests in agricultural productivity and water conservation, as this use is intermediate between intensive agriculture and forest preservation. Low impact forest harvest methods provide forest cover, habitat, and watershed protection from erosion [46–48]. Puerto Rico, at one time, had a thriving timber industry, producing fuel, furniture, and building materials to meet all of its needs [49,50]. The decline of forest cover due to increasing agricultural activity, led to the loss of the timber industry and associated markets and supply chains. A reinvigoration of that industry has economic and social benefits, in that value added post-harvest wood product development can substantially increase the value of timber production, provide jobs, and serve as an intermediate land use option that maintains many essential ecosystem services, and protects soils from degradation [51–54]. We consider the lands well-suited to forestry in a broad context. This includes agroforestry, such as shaded pastures for livestock, shade coffee, and the use of non-timber forest products, including beekeeping and honey production as uses well-suited to this component of the landscape.

Finally, we characterize 208,761 ha, nearly one quarter of the island, as well-suited for mechanized and non-mechanized agriculture. Current estimates indicate about 28 percent of the island is farmland, but a much smaller proportion cultivated as cropland (50,000 ha), and a large portion as idle lands, rangeland, brush, or other farm uses (90,000 ha) [54]. Current practices are producing only about 15 percent of food needs for Puerto Rico. Vicente Chandler [33] describes in detail how better utilization of the landscape, i.e., improved multisectoral planning, matching crops with optimal soil and water availability, modernizing practices, and taking advantage of the diversity of soils and environments to develop diversified farming operations, can greatly increase productivity on the lands well-suited for agriculture. These estimates [32], along with the spatial analyses of this study, indicate the potential to increase agricultural and timber production in Puerto Rico. Lands well-suited for agriculture also have the potential to integrate conservation and forestry practices that can provide ecosystem services, including riparian buffers, woodlots, and agroforestry.

Characterizations of what lands are well-suited for agriculture, forestry and conservation indicate that forest cover, biodiversity and ecosystem services can be maintained while increasing agricultural productivity on flatter lands and lower slopes, and integrating agroforestry, shade coffee, low impact timber harvest, and non-timber forest product uses on steeper slopes. Given the relatively small size and mountainous terrain of Puerto Rico, innovation will be important to keep key watersheds and mountain slopes forested, and to increase sustainability and productivity on all working lands. Additionally, best practices in all agricultural operations will improve productivity per land unit area. These include improved crop varieties, improved water and nutrient management, and integrating value-added farming operations that include specialty crops, livestock, timber products, and agrotourism [13].

The interface of each of the four land uses assessed in this study have potential for integration and for conflict. Perceived conflict between urban development and agriculture are critical to address, as are perceived conflict among agriculture, forestry and conservation. The definition of prime farmland from the USDA NRCS states that prime farmland cannot be "urban or built-up land or water areas" [36]. Our findings, however, show that 15,254 hectares of soils classified as prime farmland (20 percent of all prime farmland soils), 12,217 hectares as farmland of statewide importance (12 percent of all farmland soils of statewide importance), and 4716 of conditional farmland (9 percent of all conditional farmland soils) overlap with developed (i.e., built-up) surfaces. As part of the farmland classification definition, the NRCS documented the recent land cover conversion trend from prime farmland to industrial and urban uses, which puts pressure on marginal lands to be used for agriculture, although these are generally more erodible, prone to drought, and less productive [34]. Our results indicate that as of 2000, Puerto Rico had lost about 14 percent (32,186 ha) of this important and limited resource to development. Other conflicts include those between lands set aside for conservation, working forested lands, and non-forested agricultural lands. The difference between forested and non-forested lands probably has the most striking effect on ecosystem properties, including water and nutrient cycling, and biodiversity. The timber industry has been virtually non-existent in Puerto Rico for several decades. Innovative practitioners are reviving interest, and developing markets for local timber and value-added wood products. Increasing timber production has the capacity to greatly increase the economic productivity of working lands given the relatively high growth rates and highly valued tropical wood species found in Puerto Rico [50,55]. Low impact and selective timber harvest have the potential to minimize conflicts between working lands and conservation lands, and between forestry and agricultural uses. By developing markets for local timber, non-timber forest products, and value-added wood products, land owners and managers of both conservation and agricultural lands can take advantage of these markets for controlled harvests that maintain the ecological or agricultural services of a forest tract. For example, occasional timber from woodlots or farm buffers can be a source of income to farmers, as can thinning of plantations and secondary forest on conservation lands, or even salvage harvest of timber from urban lands.

These results highlight the agricultural and forest product potential that is currently relatively untapped. High unemployment rates, issues of food security, and the rising cost of importing agricultural products, are just a few key examples of the issues that are pressing Puerto Rico toward a revitalization of its working lands sector [13]. If this revitalization is to be experienced in a sustainable way that works to protect ecosystem services such as water quality and biodiversity, comprehensive planning efforts will be a great benefit. Planning should be an "all sector" activity, as market forces and economics, government regulation and incentives, as well as public knowledge and perception all shape land use decisions. The areas of suitability we have identified are not mutually exclusive, but require coordination between landowners, communities, and government entities. In the absence of planning and intervention, short-term economic needs may overcome longer-term concerns over soil degradation and the erosion of key watersheds. In recent decades, 14 percent of the island's prime agricultural land has been converted to urban use through development, with arguably more being restricted by non-agricultural uses, such as residential (lawns) or recreational (golf courses) uses. In addition to centralized government planning, regional and local efforts can be important planning tools to balance land use interests. Many communities in temperate regions have developed, or are seeking to establish, land trusts at the municipality level to help alleviate developmental pressure, by offering landowners economic options in the form of easements. This model has proven successful in preserving timber and agriculture lands in many regions throughout North America, most notably in New England. In the tropical island landscape of Puerto Rico, regional efforts such as the 'Bosque Modelo' (Model Forest) project, may provide useful prototypes for integrated planning, community involvement, and co-environmental and economic benefits.

Finally, while there is a great deal of information on landscape characteristics that can help in land use planning, there is much less spatially explicit information about how people are using their working lands, and what farming and forestry practices, for them, are sustainable ecologically and economically. Additionally, there is a lack of information on the road blocks, incentives, and motivations that favor one land use or farming decision, over another, for working lands. A well-known road block is that generally, innovative uses and products may lack both available technical support and available markets—so implementing an innovative practice may prove economically unfeasible without parallel innovation in support capacity, marketing, and supply chains.

5. Conclusions

Puerto Rico, like many tropical landscapes, particularly islands, is rich in landscape and ecological diversity. This characteristic provides many opportunities and options for working lands. Hundreds of crop and tree species will grow in its frost-free, highly productive, tropical climate. However, the island's complex and diverse landscape make planning for sustainability and productivity on working lands a challenge. Farm and forestry planning methods and practices that may prove economically sustainable in temperate zones, or in regions with large expanses of land under single ownership, often do not work in tropical islands. Puerto Rico has over a century of excellent research in tropical agricultural and forestry practices, and this research has been exported successfully around the world. To fully realize the potential of its working lands, managers, advisors, farmers, and foresters benefit from diverse and innovative techniques and programs that connect a new generation with the right combination of scientific and traditional knowledge, incentive programs, global and local markets, and the technological support necessary to convert planning into productive and sustainable farm and forest activities.

Acknowledgments: All research at the International Institute of Tropical Forestry is done in collaboration with the University of Puerto Rico. We thank Grizelle González, Ariel Lugo, and three anonymous reviewers for reviews of this manuscript. Thanks to Jessica Castro, Sandra Soto, and Eva Holupchinski for help with spatial analyses, and Winston Martínez for providing information from the Farm Service Agency. Thanks to Mario Rodríguez, NRCS, for advice on the slope ranges and crops per range.

Author Contributions: William A. Gould conceived and designed the analysis, presenting the initial findings at the 75th anniversary of the USDA Forest Service International Institute of Tropical Forestry in 2014; Frank H. Wadsworth provided expertise on developing the criteria associated with suitability of forestry and agriculture, Maya Quiñones did the geospatial analyses and prepared an initial draft of the manuscript, Stephen J. Fain provided writing and context for the introduction, Nora L. Álvarez-Berríos provided information and analyses on the Farm Service Agency Common Land Unit attributes. All authors contributed substantially in review and editing.

Conflicts of Interest: The authors declare no conflict of interest.

Appendix A

Distributions of land well-suite for mechanized agricultural (slopes less than 10 percent), non-mechanized agriculture (slopes 10–20 percent), and forestry—including timber, non-timber products, agroforestry, and shade coffee by municipality.

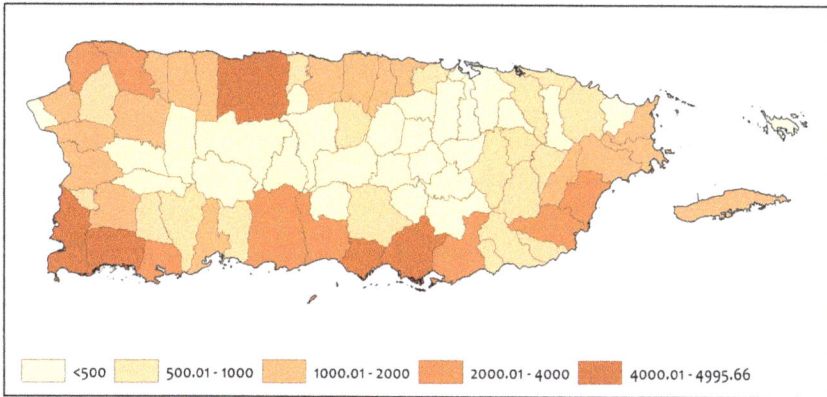

Figure A1. Map of Puerto Rico municipalities colored by the amount of land in hectares well-suited to mechanized agriculture with under 10 percent slope.

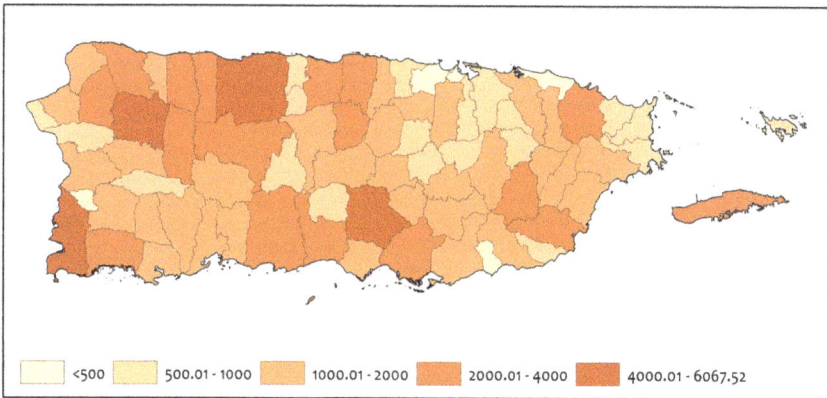

Figure A2. Map of Puerto Rico municipalities colored by the amount of land in hectares well-suited to non-mechanized agriculture on moderate slopes.

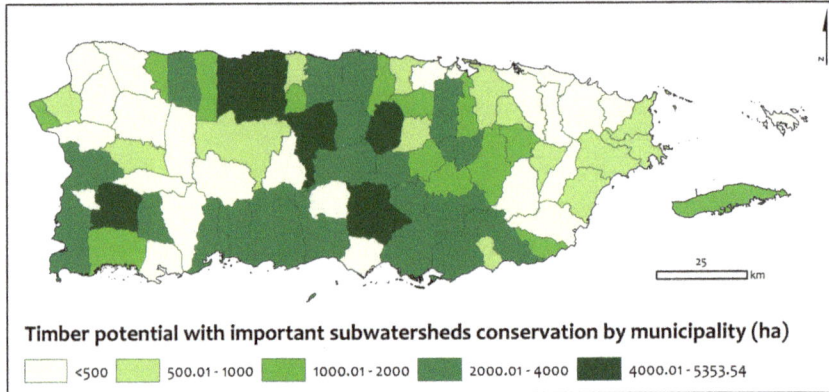

Figure A3. Map of Puerto Rico municipalities colored by the amount of land in hectares well-suited to forestry production Excluding key subwatersheds and areas of extreme rainfall important to water conservation.

Table A1. Area in hectares of lands well-suited for conservation, agriculture and forestry by municipality.

Municipality	Lands Well-Suited (ha)			
	Land Well-Suited for Mechanized Agriculture: <10 Percent Slope	Land Well-Suited for Non-Mechanized Agriculture: 10 to 20 Percent Slope	Land Well-Suited for Forestry	Lands Well-Suited for Conservation
Añasco	1653	672	1041	5874
Adjuntas	244	1037	1086	14,302
Aguada	1223	747	2334	2261
Aguadilla	3880	1031	416	1698
Aguas Buenas	216	762	3363	2801
Aibonito	470	923	2486	3306
Arecibo	6930	3161	6205	12,964
Arroyo	965	206	905	1268
Barceloneta	1245	394	588	1739
Barranquitas	249	867	3941	3004
Bayamón	704	743	1942	2757
Cabo Rojo	6021	2749	2361	5394
Caguas	1063	1136	1126	7935
Camuy	3100	2024	2982	2741
Canóvanas	692	778	793	5044
Carolina	1040	752	2255	3846
Cataño	97	1	0	520
Cayey	559	922	2578	7870
Ceiba	1255	708	1072	3557
Ciales	473	1328	5885	8979
Cidra	652	1245	1706	4535
Coamo	1756	2880	5076	9313
Comerío	163	574	1193	4942
Corozal	487	1358	5031	3235
Culebra	313	575	1108	895
Dorado	1629	396	713	1815
Fajardo	1223	550	606	3895
Florida	484	509	1261	1335
Guanica	2719	874	835	4363
Guayama	3462	1006	3416	7270

Table A1. *Cont.*

Municipality	Land Well-Suited for Mechanized Agriculture: <10 Percent Slope	Land Well-Suited for Non-Mechanized Agriculture: 10 to 20 Percent Slope	Land Well-Suited for Forestry	Lands Well-Suited for Conservation
		Lands Well-Suited (ha)		
Guayanilla	1363	1022	3159	4565
Guaynabo	409	547	1870	1487
Gurabo	1151	522	1789	2500
Hatillo	3263	1882	2373	1833
Hormigueros	1016	188	473	622
Humacao	2535	1129	2845	2825
Isabela	4302	1461	10	6867
Jayuya	250	729	603	9404
Juana Díaz	4426	1397	3300	4745
Juncos	1305	752	1086	2610
Lajas	5693	1586	1853	5249
Lares	1188	2466	2928	8373
Las Marías	248	994	4036	6338
Las Piedras	1554	945	2069	3079
Loíza	770	33	7	3765
Luquillo	687	601	1471	3215
Manatí	2212	1235	2807	4050
Maricao	108	541	3024	5532
Maunabo	662	403	2266	1706
Mayagüez	1384	1177	5062	9611
Moca	1400	2315	3517	4461
Morovis	1100	1407	3251	3403
Naguabo	2065	1030	1790	7585
Naranjito	153	550	961	4731
Orocovis	245	1218	6479	7793
Patillas	836	675	2410	7687
Peñuelas	884	850	3757	4871
Ponce	3245	1709	2700	16,826
Quebradillas	1711	871	1075	1530
Río Grande	1393	1473	387	10,847
Rincón	349	402	1405	834
Sabana Grande	989	804	2331	4117
Salinas	5675	1164	2290	7379
San German	1908	1368	4882	4575
San Juan	675	515	486	2596
San Lorenzo	1073	1651	2639	7303
San Sebastian	2593	4329	3571	6256
Santa Isabel	4889	606	341	1996
Toa Alta	890	1150	1066	2328
Toa Baja	1012	162	312	2352
Trujillo Alto	408	625	1103	1861
Utuado	928	2538	4326	20,804
Vega Alta	1596	758	1718	2011
Vega Baja	2496	1156	2397	4077
Vieques	1904	1278	1009	8496
Villalba	241	755	219	7746
Yabucoa	3147	1553	5051	3413
Yauco	831	1137	313	13,961
Total	124,131	84,564	169,120	399,672

Table A2. Area in hectares of the farm classification of the soils and the agriculture potential models.

NRCS Soils Farm Classification	Area (ha)			
	Land Well-Suited for Mechanized Agriculture: (Slope <10 Percent)	Land Well-Suited for Non-Mechanized Agriculture: 10 to 20 Percent Slope	Land Better Suited for Forestry, Conservation, or Developed Land Uses	Total
All areas are prime farmland	30,513	9153	37,656	77,323
Farmland of statewide importance	20,817	14,805	63,180	98,803
Farmland of statewide importance, if irrigated	167	99	174	440
Not prime farmland	40,185	56,449	568,113	664,746
Prime farmland if drained	6318	312	8152	14,783
Prime farmland if irrigated	19,850	2616	18,883	41,350
Prime farmland if irrigated and reclaimed of excess salts and sodium	1378	17	1921	3316
Prime farmland if protected from flooding or not frequently flooded during the growing season			30	30
no data	4837	1104	22,539	28,481
Total	124,065	84,555	720,648	929,270

References

1. Food and Agriculture Organization of the United Nations (FAO). Global Agriculture Towards 2050: How to feed the world in 2050. In Proceedings of the High Level Expert Forum, Rome, Italy, 12–13 October 2009; pp. 1–35.
2. Ludena, C.E. *Agricultural Productivity Growth, Efficiency Change And Technical Progress in Latin America and the Caribbean*; Research Dept. II. Title. III. Working Paper Series; Inter-American Development Bank: Washington, DC, USA, 2010; p. 186.
3. Walters, L.M.; Jones, K.G. Caribbean Food Import Demand: Influence of the Changing Dynamics of the Caribbean Economy. In Proceedings of the Southern Agricultural Economics Association Annual Meetings, Birmingham, AL, USA, 4–7 February 2012; p. 19.
4. Carro-Figueroa, V. Agricultural Decline and Food Import Dependency in Puerto Rico: A Historical Perspective on the Outcomes of Postwar Farm and Food Policies. *Caribb. Stud.* **2002**, *30*, 77–107.
5. Kendall, P.; Petracco, M. The Current State and Future of Caribbean Agriculture. *J. Sustain. Agric.* **2009**, *33*, 780–797. [CrossRef]
6. Comas, M. Vulnerabilidad De Las Cadenas De Suministros, El Cambio Climático Y El Desarrollo De Estrategias De Adaptación: El Caso De Las Cadenas De Suministros De Alimento De Puerto Rico. Ph.D. Thesis, University of Puerto Rico, Mayaguez, Puerto Rico, 2009.
7. Galtier, F.; Vindel, B. *Managing Food Price Instability in Developing Countries: A Critical Analysis of Strategies and Instruments*; Agence Française de Développement: Paris, France, 2013.
8. Watlington, F. Cassava and carrying capacity in aboriginal Puerto Rico: Revisiting the Taino downfall at conquest. *Southeast. Geogr.* **2009**, *49*, 394–403. [CrossRef]
9. Vázquez de Espinosa, A. *Compendio Y Descripción De Las Indias Occidentales*; Atlas: Madrid, Spain, 1948.
10. Dietz, J.L. *Economic History of Puerto Rico: Institutional Change and Capitalist Development*; Princeton University Press: Princeton, NJ, USA, 1986.
11. Birdsey, R.A.; Weaver, P.L. World Forestry: Puerto Rico's Timberland. *J. For.* **1983**, *81*, 671–699.
12. Brandeis, T.J.; Turner, J.A. *Puerto Rico's Forests, 2009*; USDA Forest Service Southern Research Station: Asheville, NC, USA, 2013.

13. Gould, W.A.; Fain, S.J.; Pares, I.K.; McGinley, K.; Perry, A.; Steele, R. *Caribbean Regional Climate Sub Hub Assessment of Climate Change Vulnerability and Adaptation and Mitigation Strategies*; United States Department of Agriculture: San Juan, Puerto Rico, 2015.

14. Asamblea Legislativa de Puerto Rico. Ley del Plan Estratégico Integral Agrícola de Puerto Rico. Ley Núm. In *131 de 6 de agosto de 2014 (P. de la C. 1284)*; Commonwealth of Puerto Rico: San Juan, Puerto Rico, 2014.

15. Altieri, M.A.; Nicholls, C.I.; Henao, A.; Lana, M.A. *Agroecology and the Design of Climate Change-Resilient Farming Systems*; Springer Verlag/EDP Sciences/INRA: Berlin, Germany, 2015; Volume 35.

16. Godfray, C.H.J.; Garnett, T. Food security and sustainable intensification Food security and sustainable intensification. *Philos. Trans. R. Soc.* **2014**. [CrossRef] [PubMed]

17. Natural Resources Conservation Service (NRCS). Identification of Important Farmland. Available online: https://prod.nrcs.usda.gov/Internet/FSE_DOCUMENTS/nrcs142p2_010970.pdf (accessed on 11 February 2015).

18. Hulme, T.; Grosskopf, T.; Hindle, J. Agricultural land classification. *Agfact AC* **2002**, *25*, 15.

19. López, T.M.; Aide, T.M.; Thomlinson, J.R. Urban expansion and the loss of prime agricultural lands in Puerto Rico. *Ambio* **2001**, *30*, 49–54. [CrossRef] [PubMed]

20. Martinuzzi, S.; Gould, W.A.; Ramos González, O.M. Land development, land use, and urban sprawl in Puerto Rico integrating remote sensing and population census data. *Landsc. Urban Plan.* **2007**, *79*, 288–297. [CrossRef]

21. Valle Junior, R.F.; Varandas, S.G.P.; Sanches Fernandes, L.F.; Pachecp, F.A.L. Environmntal land use conflicts: A threat to soil conservation. *Land Use Policy* **2014**, *41*, 172–185. [CrossRef]

22. Gould, W.A.; Alarcón, C.; Fevold, B.; Jiménez, M.E.; Martinuzzi, S.; Potts, G.; Quiñones, M.; Solórzano, M. *The Puerto Rico Gap Analysis Project*; U.S. Department of Agriculture Forest Service; International Institute of Tropical Forestry: Río Piedras, Puerto Rico, 2008.

23. Quiñones, M.; Gould, W.A.; Castro-Prieto, J.; Martinuzzi, S. Spatial Analysis of Puerto Rico's Terrestrial Protected Areas. *US For. Serv. Res. Map Ser.* **2011**, *66*. [CrossRef]

24. Gellis, A.C.; Webb, R.M.T.; McIntyre, S.C.; Wolfe, W.J. Land-use effects on erosion, sediment yields, and reservoir sedimentation: A case study in the Lago Loiza basin, Puerto Rico. *Phys. Geogr.* **2006**, *27*, 39–69. [CrossRef]

25. Gellis, A.C. Factors influencing storm-generated suspended-sediment concentrations and loads in four basins of contrasting land use, humid-tropical Puerto Rico. *CATENA* **2013**, *104*, 39–57, ISSN 0341-8162. [CrossRef]

26. Larsen, M.C.; Webb, R.M.T. Potential Effects of Runoff, Fluvial Sediment, and Nutrient Discharges on the Coral Reefs of Puerto Rico. *J. Coast. Res.* **2009**, *25*, 189–208. [CrossRef]

27. Jasinski, E.; Morton, D.; DeFries, R.; Shimabukuro, Y.; Anderson, L.; Hansen, M. Physical landscape correlates of the expansion of mechanized agriculture in Mato Grosso, Brazil. *Earth Interact.* **2005**, *9*, 1–18. [CrossRef]

28. Gesch, D.; Oimoen, M.; Greenlee, S.; Nelson, C.; Steuck, M.; Tyler, D. The National Elevation Dataset. *Photogramm. Eng. Remote Sens.* **2002**, *68*, 5–32.

29. Mayer, P.M.; Reynolds, S.K.; McCutchen, M.D.; Canfield, T.J. Meta-Analysis of Nitrogen Removal in Riparian Buffers. *J. Environ. Qual.* **2007**, *36*, 1172–1180. [CrossRef] [PubMed]

30. Bentrup, G. Conservation Buffers: Design Guidelines for Buffers, Corridors, and Greenways. Literature Cited; In *Conservation Buffers: Design Guidelines for Buffers, Corridors, and Greenways*; USDA Forest Service Southern Research Station: Asheville, NC, USA, 2008.

31. U.S. Geological Survey. National Hydrography Geodatabase: The National Map Viewer available on the World Wide Web. 2013. Available online: http://viewer.nationalmap.gov/viewer/nhd.html?p=nhd (accessed on 1 May 2017).

32. Junta de Planificación. *Plan De Uso De Terrenos, Guías De Ordenación Del Territorio*; Puerto Rico Planning Board: San Juan, Puerto Rico, 2015; p. 220.

33. Vicente-Chandler, J. *Una Nueva Agricultura Para Puerto Rico: Año 2000*; Agricultural Experimental Station: Río Piedras, Puerto Rico, 2000.

34. Natural Resources Conservation Service (NRCS). Prime & Other Important Farmlands Definitions. Available online: http://www.nrcs.usda.gov/wps/portal/nrcs/detail/pr/soils/?cid=nrcs141p2_037285 (accessed on 25 February 2015).

35. *National Agricultural Statistics Service. 2012 Census of Agriculture*; USDA: Washington, DC, USA, 2017.

36. Torreggiani, D.; Dall'Ara, E.; Tassinari, P. The urban nature of agriculture: Bidirectional trends between city and countryside. *Cities* **2012**, *29*, 412–416. [CrossRef]

37. Sanyé-Mengual, E.; Anguelovski, I.; Oliver-Solà, J.; Montero, J.I.; Rieradevall, J. Resolving differing stakeholder perceptions of urban rooftop farming in Mediterranean cities: Promoting food production as a driver for innovative forms of urban agriculture. *Agric. Hum. Values* **2016**, *33*, 101–120. [CrossRef]

38. Stolhandske, S.; Evans, T.L. On the Bleeding Edge of Farming the City: An Ethnographic Study of Small-scale Commercial Urban Farming in Vancouver. *J. Agric. Food Syst. Community Dev.* **2017**, *7*, 1–21. [CrossRef]

39. Robertson, G.; Mason, A. (Eds.) *Assessing the Sustainability of Agricultural and Urban Forests in the United States*; FS-1067; USDA Forest Service: Washington, DC, USA, 2016.

40. Castro-Prieto, J.; Martinuzzi, S.; Radeloff, V.C.; Helmers, D.P.; Quiñones, M.; Gould, W.A. Declining human population but increasing residential development around protected areas in Puerto Rico. *Biol. Conserv.* **2017**, *209*, 473–481. [CrossRef]

41. United States Census Bureau Population and Housing Units Estimate. Available online: https://www.census.gov/popest/ (accessed on 1 November 2016).

42. Ngaini, Z.; Wahi, R.; Halimatulzahara, D.; Mohd Yusoff, N.A.N. Chemically modified sago waste for oil absorption. *Pertanika J. Sci. Technol.* **2014**, *22*, 153–162.

43. Caribbean Landscape Conservation Cooperative (CLCC). Puerto Rico Protected Areas Database [version of December 2016]. In *GIS Data*; USDA Forest Service; International Institute of Tropical Forestry: Río Piedras, Puerto Rico, 2016.

44. Castro-Prieto, J.; Quiñones, M.; Gould, W.A. Characterization of the Network of Protected Areas in Puerto Rico. *Carib. Nat.* **2016**, *29*, 1–16.

45. Nelson, E.; Polasky, S.; Lewis, D.J.; Plantinga, A.J.; Lonsdorf, E.; White, D.; Bael, D.; Lawler, J.J. Efficiency of incentives to jointly increase carbon sequestration and species conservation on a landscape. *Proc. Natl. Acad. Sci. USA* **2008**, *105*, 9471–9476. [CrossRef] [PubMed]

46. Putz, F.E.; Sist, P.; Fredericksen, T.; Dykstra, D. Reduced-impact logging: Challenges and opportunities. *For. Ecol. Manag.* **2008**, *256*, 1427–1433. [CrossRef]

47. Corlett, R.T. Applied Ecology of Tropical Forests. *Trop. For. Handb.* **2016**, 511–518. [CrossRef]

48. Sasaki, N.; Asner, G.P.; Pan, Y.; Knorr, W.; Durst, P.B.; Ma, H.O.; Abe, I.; Lowe, A.J.; Koh, L.P.; Putz, F.E. Sustainable Management of Tropical Forests Can Reduce Carbon Emissions and Stabilize Timber Production. *Front. Environ. Sci.* **2016**. [CrossRef]

49. Cristóbal, C.D. *Panorama Histórico Forestal De Puerto Rico*; La Editorial, UPR: San Juan, Puerto Rico, 2000.

50. Forero-Montaña, J. Potential of Subtropical Secondary Forests for Sustainable Forestry in Puerto Rico. Doctoral Dissertation, University of Puerto Rico, Río Piedras, Puerto Rico, 2015.

51. Wadsworth, F.H. Conserva a Puerto Rico con bosques maderables. *Acta Cient.* **2009**, *23*, 73–80.

52. Wadsworth, F.H.; Bryan, B.; Figueroa-Colón, J. Cutover tropical forest productivity potential merits assessment, Puerto Rico. *Bois et Forets des Tropiques* **2010**, *305*, 33–41.

53. Ratnasingam, J.; Liat, L.C.; Ramasamy, G.; Mohamed, S.; Senin, A.L. Attributes of Sawn Timber Important for the Manufacturers of Value-Added Wood Products in Malaysia. *BioResources* **2016**, *11*, 8297–8306. [CrossRef]

54. Gilani, H.R.; Kozak, R.A.; Innes, J.L. The state of innovation in the British Columbia value-added wood products sector: The example of chain of custody certification. *Can. J. For. Res.* **2016**, *46*, 1067–1075. [CrossRef]

55. Wadsworth, F.H. *Forest Production for Tropical America*; USDA Forest Service: Río Piedras, Puerto Rico, 2000. Available online: http://ageconsearch.umn.edu/bitstream/119724/2/119724.pdf (accessed on 7 July 2017).

forests

MDPI

Commentary

A Forest Service Vision during the Anthropocene

Michael T. Rains [1,2,†]

[1]   The Northern Research Station, United States Department of Agriculture (USDA) Forest Service,
      Newtown Square, PA 19073, USA; mtrains7@verizon.net
[2]   The Forest Products Laboratory, USDA Forest Service, Madison, WI 53726, USA
[†]   The USDA Forest Service (the "Forest Service") (now retired) for the Keynote Talk at the Institute's 75th
      Anniversary Symposium to be held at the Inés María Mendoza Park located on the grounds of the
      Luis Muñoz Marín Foundation in San Juan, on 21 May 2014.

Academic Editors: Grizelle González and Ariel E. Lugo
Received: 21 January 2017; Accepted: 17 March 2017; Published: 22 March 2017

Abstract: During the history of the Forest Service, human activity has been the dominant influence on climate and the environment; the time being called the *Anthropocene*. As we look ahead and strive to continue our mission of sustaining the health, diversity, and productivity of the Nation's forests and grasslands to meet our current and future needs, we must be more flexible to focus our actions to better meet the contemporary conservation challenges now and ahead. During this era of intense human activity, a changing climate; development and loss of open space; resource consumption; destructive invasive species; and diversity in core beliefs and values will test our task relevant maturity—ability and willingness to meet the growing demands for services. The Forest Service is now on a transformative campaign to improve our abilities and meet these challenges, including forest resiliency through restorative actions. There are several things we must do to ensure we are brilliantly competitive to address the contemporary conservation needs along a complex rural to urban land gradient, now and ahead. The intent of this paper is to present one person's view of what this "campaign of our campaign" should include.

Keywords: Anthropocene; Forest Service; vision; contemporary conservation

1. Introduction

My good friend, Dr. Ariel Lugo, Director of the International Institute of Tropical Forestry asked me to give a presentation about the Forest Service vision during the *Anthropocene*. My first response was, " ... what is the *Anthropocene*?" This was quickly followed by a driving question. That is, " ... why me?" I will tend to the definition in a moment. However, the second question deserves some attention: " ... why me?"

Over the years, Dr. Lugo and I have been together in countless events. I have come to know him as a voice of authority, careful to speak, seeming always to seek just the right time to capture a point. I, on the other hand, have not been so wise. I tend to wear my heart on my sleeve and speak—oftentimes when I should probably be listening. Still, Dr. Lugo accepts me and we share common ground on many points and positions. So, " ... why me?", honestly, I am not sure. Probably because I have simply been with the Forest Service for a long time—about 45 years now—and that has indeed shaped some perspective, and, in the end I agreed, because he asked.

However, to be clear, there are many others who can speak to this subject—a Forest Service Vision—equally well or better. Additionally, the views presented in this paper may not be corporate. That is, not shared by all. Never-the-less, I am happy to share my thoughts about the Forest Service that I love and how the agency can be most effective in "caring for the land and serving people, *where they live*" during the *Anthropocene*.

Anthropocene

According to Webster's Dictionary, the word *Anthropocene* (An·thro·po·cene) means, "the period of time during which human activities have had an environmental impact on the Earth regarded as constituting a distinct geological age". So, when I talk about a Forest Service vision during the *Anthropocene*, it means to me " . . . a vision for the agency at our beginning; now; and, for a very long time into the future."

2. Discussion

2.1. The Forest Service (In the Beginning)

Following the Civil War, a dominant culture in acquiring land; money; material things; and, exploitation " . . . were the spirit of the times, with little regard for the ethics of conservation or the needs of the future [1]." Concerns by influential visionaries such as George Marsh, Wesley Powell, Bernard Fernow, and John Muir—and others—helped surface a call to action. Accordingly, in 1876, Congress created the office of Special Agent in the Department of Agriculture to assess the quality and conditions of forests in the United States. Franklin B. Hough was appointed the head of the office. In 1881, the office was expanded into the newly formed Division of Forestry. The Forest Reserve Act of 1891 authorized withdrawing land from the public domain as "forest reserves", managed by the Department of the Interior. In 1901, the Division of Forestry was renamed the Bureau of Forestry. The Transfer Act of 1905 transferred the management of "forest reserves" from the General Land Office of the Interior Department to the Bureau of Forestry, Department of Agriculture. This Bureau of Forestry became known as the United States Forest Service, with Gifford Pinchot the first Chief Forester under the Administration of President Theodore Roosevelt. The forest reserves became known as the National Forests.

The culture of America that shaped the beginning of the Forest Service is both different and in some ways the same as the culture today; perhaps now the differences are more acute due to the intensity of human activity. With populations across planet Earth continuing to rise, behaviors will continue to cause exploitative impacts, even if intentions are noble. So, as Dr. Lugo asks, " . . . should the Forest Service cruise on with our mission with the conservation paradigms that we inherited from Pinchot and Leopold, or is there a need for another leap or evolution in our relationship with forestlands?" The answer to this driving question will of course depend on our corporate view of the future; the challenges we face; and the accuracy of forecasts on the demands for our services.

2.2. Gifford Pinchot: First Chief of the Forest Service, 1905–1910

The Chief of the new Forest Service had a strong hand in guiding the fledgling organization toward the utilitarian philosophy of the "greatest good for the greatest number in the long run". Gifford Pinchot is generally regarded as the founder of American conservation because of his great and unrelenting concern for the protection of the American forests. Significant legislation that has shaped the Forest Service, from the beginning until now, is shown below in Box 1.

Box 1. Significant Legislation Affecting the Forest Service.

Significant federal legislation affecting the Forest Service includes the: Weeks Act of 1911; Multiple Use—Sustained Yield Act of 1960, P.L. 86–517; Wilderness Act, P.L. 88–577; National Forest Management Act, P.L. 94–588; National Environmental Policy Act, P.L. 91–190; Cooperative Forestry Assistance Act, P.L. 95–313; and, Forest and Rangelands Renewable Resources Planning Act, P.L. 95–307.

2.3.Thomas Tidwell: 17th Chief of the Forest Service, 2009-Present

Tom Tidwell, our current Chief, has spent about 38 years in the Forest Service. Under his leadership, the Forest Service—through an "all-lands" approach—is helping to restore healthy, resilient forest and grassland ecosystems along a complex urban to rural land gradient. Chief Tidwell is helping to deploy a "transformational campaign" so the Forest Service can be more competitive in addressing the contemporary conservation challenges now and ahead.

2.3. A City Kid Joins the Forest Service

I was born in East Los Angeles, spending most of my early youth in California. We were poor, my parents being classic depression-era Americans. Eventually, we moved to Sacramento and then to a very rural area called Sly Park, just east of Placerville—another small town in northern California between Sacramento and Lake Tahoe. It was there where I became associated with the Forest Service.

My mother, looking for a job, was offered a secretary position on the El Dorado National Forest. Her supervisor—the Forest Supervisor—was a grand gentleman named Douglas Leisz. Mr. Leisz would be instrumental in getting me my first job in forestry, as a fire fighter for the California Division of Forestry (CDF) at Mt. Danaher Fire Station, Camino, CA (just north of Placerville). I lived at the CDF Barracks and received a monthly salary of $255.15 for the summer before my senior year in high school.

Following high school, I would get selected from a permanent Civil Service roster and became a General Schedule (GS)-2 Biological Aid. I carried stakes for a "P-Line" crew laying out "cut and fill" points for logging roads on the El Dorado National Forest. Perhaps not glamorous, but it was a full-time, permanent job. That is all I wanted; a real, full-time job. As far as I was concerned, I was on my way.

I had no intention of going to college. I did not think I could afford it. Besides, I had a full-time job with the Forest Service and if I played my cards right, maybe someday I would be a "Survey Party Team Leader". At a GS-7 level, I would be set.

Just before my summer ended on the "P-Line" crew, my supervisor Mr. James Floyd pulled me aside. He told me that if I wanted to go to college, the Forest Service would offer me "educational leave without pay". Then, when next summer would come, I would be hired back and that time would count toward my career.

Honestly, I was not thinking about a career, really. I just wanted to get to that GS-7 level as quickly as possible. The idea seemed sound, but there was one significant problem. I still did not have enough money to pay my way to college—unless maybe I went to a Community College. In this case, the savings from my GS-2 appointment just might be enough. So, that is just what I did for two years before transferring to a small forestry school near a town called Arcata, California—Humboldt State University. The Community College was called Sierra Junior College near Rocklin, CA. To this day, I owe most everything to that school. It allowed me to get started and, just like Jim Floyd said to me earlier, right before each subsequent summer began, I received a letter telling me where to report for work with the Forest Service. Like Forrest Gump™ said, " ... just one less thing to worry about". That was very nice.

2.4. A Junior Forester

I can clearly recall the day, about 45 years ago, when I walked into the Supervisor's Office (SO) on the El Dorado National Forest in Placerville, California. Before the "SO" was moved to a new location at Forni Road, I had often visited the old building perched on top of the hill overlooking old "Hangtown".

I was a freshly minted "JF"—Junior Forester—out of Humboldt State University. Today, we might take exception to being labeled with the letters "JF", but I was proud to be able to have the initials, while still looking forward to the end of the first year and being called a "Forester" for the Lake Valley Ranger District. While not a "Survey Party Team Leader", I was a GS-7 and now a "forester" for the Forest Service. What could be better?

After a brief introduction, my Forest Supervisor Irwin Bosworth directed me to "get to work" at my new position. The ride to Meyers—the District Office headquarters just a few miles from South Lake Tahoe—was exciting. I had taken the ride many times before, but never as a "JF" for the Forest

Service. I recall as if it were yesterday, the admonishment by the District Ranger when I walked into his office. He said, " … listen up young fella. If you want to make it to the short-go around here, you will do whatever it takes". I quickly said, " … yes sir", wondering what he meant by the "short-go". Later, I would find out the phrase was a rodeo term meaning the final go-around or the finals of a competition. I got the message. The expectations were very clear.

My mother, now a personnel clerk for the El Dorado National Forest, always told me I would like the Forest Service. She was right.

Over the last four decades and then some, I have watched and participated with the agency in almost a continual transformational campaign, striving to stay contemporary in addressing conservation issues. It has been a magnificent ride for me. The era of human domination has tested the Forest Service and will continue to do so as we move deeper into the 21st Century.

When asked about the agency, I always provided the three brief statements: " … I like being employed. It is an honor to work for the Department of Agriculture. And, I work for the greatest organization in the world, the Forest Service."

2.5. The Forest Service Mission

Since the beginning of the Forest Service, our mission has been remarkably clear—conservation of our forest and rangeland resources for most of the people for the long haul. Taken directly from our website, the mission statement of the Forest Service is " … to sustain the health, diversity, and productivity of the Nation's forests and grasslands to meet the needs of present and future generations." The mission statement is characterized by the motto: "caring for the land and serving people." Over the years, we have strengthened the mission statement with new words, but generally our mission has been remarkably stable and clear.

> … To sustain the health, diversity, and productivity of the Nation's forests and grasslands to meet the needs of present and future generations.
>
> —Mission Statement of the USDA Forest Service

What has changed is the way we carry out the mission to better address the contemporary conservation challenges that have evolved. Our scope has expanded along the rural to urban land gradient. We take more of an "all-lands" approach than before. We are trying to be more inclusive as we strive to attract a more representative workforce, and, to be fair, our ability—actually, our *flexibility*—has been questioned. There are times when we seem to be just a bit stodgy in both our *influence* and *deployment* strategies. Often times this gets cast as being confused about the mission. For me, this is not accurate. When we become confused, it is typically over implementation tactics and being too cautious, not program direction.

To help deploy our mission, we have a clear vision; guiding principles; and, several current points of focus. For example, our vision calls for us to be the recognized leader in land conservation and public service. We have a "Shared Intention Statement" for inclusivity. We desire a workforce that is representative of those we serve and excels in helping the Forest Service meet its contemporary conservation challenges.

> … To create a culture of inclusion that awakens and strengthens all people's connections to the land.
>
> —Shared Intention Statement for Diversity and Inclusion, Forest Service, 27 February 2013

Our mission has foundational guiding principles that include a science-based, ecological approach to stewardship across all lands along a complex rural to urban land gradient. We fully understand the power of partnerships; we cannot accomplish our mission alone.

When the "forest reserves" were first set aside, much of the land had been abused. In fact, for the eastern part of the country some called many of these landscapes " … the lands nobody wanted".

After decades of management, protection, and wise use, these lands have now become productive, healthy " ... jewels of envy". This could not have happened without the Forest Service.

However, contemporary conservation issues continue to emerge and our ability and willingness to be optimally adaptive is challenged. In recent years, the issue of Urban Natural Resources Stewardship, for example, has surfaced as a dominant need for a stronger Forest Service role. Since the agency still has a dominant rural culture, I think it would be safe to say that our flexibility to emphasize the urban portion of the rural to urban land gradient is, well, not so flexible.

2.6. New Forestry

Between 1989 and 2001, just twelve years, the Forest Service changed dramatically in the way it carried out its mission. In 1989, a concept of "new perspectives" or "new forestry" was launched following a critical meeting—it was deemed " ... the walk in the woods"—with our 12th Chief, F. Dale Robertson and Senator David Pryor of Arkansas regarding the Ouachita National Forest. Simply put, many thought we were cutting too much wood on this and the other National Forests—clear-cutting. At the time, the Forest Service was harvesting about 11 billion board feet annually, through "traditional forestry". In 1992, "New Perspectives" was launched—an approach that looked at "ecosystem management and sustainability" and placed timber management in line with other forest uses.

> ... New Perspectives was about institutional change in the Forest Service. Through on-the-ground demonstrations, problem-focused research, and constituent engagement, New Perspectives was designed to stimulate imitative and innovation.
>
> —Pinchot Institute for Conservation, Volume 11, No. 1, 1994

Also during this time, we were experiencing an unprecedented rate of "catastrophic wildfires", leading way to a report entitled, "Managing the Impacts of Wildfires on Communities and the Environment"—the National Fire Plan [2]. A critical feature of the National Fire Plan was "hazardous fuels reduction". A cornerstone to a successful hazardous fuels reduction program was the expansion and new development of high value markets from this low value wood. We thought then (and now) that by creating cost-effective ways to enable enough hazardous fuels to be removed from America's forests, wildfires would remain smaller and begin again to be a tool for improved forest health as opposed to destructive behemoths that destroy lives, communities, and landscapes.

By most standards, the results of the National Fire Plan have not materialized as planned; fires and suppression costs are higher than ever before. Part of the problem is, indeed, a changing climate. When the original report was drafted, climate change was not considered as much as it should have been. Thus, long-term, severe weather patterns have made much of America's forests vulnerable to disturbances with longer, more intense fire seasons. Furthermore, the continued expansion of the "Wildland-Urban Interface", whereby development and fire prone forests come face to face, make protecting lives and property from wildfires a very dangerous and expensive proposition.

In 2001, the fire budget represented about 22 percent of the total Forest Service budget, up from 16 percent just a few years earlier. It is now about one-half of the total budget and increasing. More and more funds are being diverted from other uses to fight fires. Fire management in this century has replaced timber management of the 1980s as the dominate focus of the agency.

In the late 1990s, the General Accounting Office (GAO) concluded that "the most extensive and serious problem related to the health of forests in the interior West is the over-accumulation of vegetation, which has caused an increasing number of large, intense, uncontrollable, and catastrophically destructive wildfires". In developing the National Fire Plan in 2001, about $850 million annually was thought to be required to more effectively address the issue of hazardous fuels removal. More recently (2013), the GAO concluded it would take about $69 billion over a 16-year period—$4.3 billion each year. Relying on taxpayer dollars, the Forest Service has only managed an average of about $300 million annually for hazardous fuels treatment.

The cost estimations for reducing hazardous fuels vary. What does not vary is the fact that fire suppression costs are increasing and the impacts are more severe. If we want a future where wildfires are not destructive behemoths, we must create new large-scale markets for forest biomass uses. In terms of the future, this has to be a "dominating common thread" of the Forest Service mission, lest we become the USDA *Fire* Service. Accelerated forest restoration could be the answer if we concentrate on high value, high volume markets from low value wood, while keeping an eye on the ultimate "brass ring": healthy, sustainable trees, forests, and forest ecosystems that are more resilient to disturbances that are, in part, caused by intense human activity.

2.7. A Paradox Exists

Even with the understanding of some shortfalls, by most standards, the Forest Service is a premier conservation agency. The Forest Service holds important keys to sustaining our planet Earth (clean air and water, conserving natural resources). Yet, according to a broad range of authors, " ... our work (protecting the environment) generally does not directly challenge major economic or material concerns". We seem to lack relevancy in many minds to be truly competitive. While I do not agree with this, many who "decide" do agree. How could keeping our air clean, for example, not be completely relevant, I ask rhetorically? In simple terms, our work holds a key to America's economic and social vibrancy. Yet, much of what we do and who we are is not viewed as mainstream and essential.

The issue, it seems to me, is a profound lack of understanding by the general population about our environment, its condition, and what we as humans do to harm or help its condition. The situation (lack of awareness) may be more acute now than at any time I can recall. Author Jay Gould says, " ... you do not fight for what you do not love". I think it is even more basic. That is, you do not fight for what you do not know. We could alter the paradox in our next 100 years (actually by the next decade) by reaching out more and improving the nation's environmental literacy. In other words, emphasize inclusion *and* education—the two gems that have surfaced repeatedly during most of my career in government and especially during our current transformative campaign.

Thus, when we think of a Forest Service future, we must include an aggressive component to help create an informed citizenry about our natural resources and the impact these resources have on our lives and how *we* affect the health and sustainability of these natural resources. As Pinchot concluded in Breaking New Ground, " ... natural resources must be about us from our infancy or we cannot live at all" [3].

2.8. Conservation Along the Urban Land Gradient

The Forest Service has a direct and indirect role on about 80 percent of our nation's forests: 885 million acres, including 138 million acres of urban forests where most Americans live.

As "Chief Forester for America's Forests", the Chief of the Forest Service has a conservation and restoration responsibility for a complex rural-to-urban land gradient to help ensure that forested landscapes, including those in urban areas, are healthy, sustainable, and provide the required green infrastructure that effectively links environmental health with community resiliency and stability. Today, 83 percent of our population lives in cities and towns. Fully one-fourth of the nation's counties are urbanized. How federal, state, and local governments and a wide range of other partnerships band together to ensure the proper care of America's urban natural resources is a fundamental part of improving people's lives. The slogan that illustrates the mission of the Forest Service is: "caring for the land and serving people". As we face new conservation demands along the entire rural-to-urban land gradient, it may be more fitting now to think of this slogan as "caring for the land and serving people *where they live*". As we look ahead, caring for America's urban natural resources—Urban Natural Resources Stewardship (UNRS)—must be a signature piece of our program direction. In simple terms, the Forest Service needs to be more attentive to the urban side of the rural to urban land gradient. The recent Forest Service Chiefs, perhaps especially, Tom Tidwell and Abigail "Gail" Kimbell, have become strong advocates of UNRS.

2.9. The Mission Areas

> ... The mission area designations may prove to be the demise of the Forest Service ability
> to effectively address the contemporary conservation issues of the 21st Century.
>
> —Michael T. Rains

The Forest Service has a comprehensive stewardship role—in collaboration with others—for the management, protection, and use on all forest and rangelands. We recognize that this role extends along a complex rural-to-urban land gradient, yet we struggle at times to efficiently fulfill this role. Earlier, I used the term stodgy. If this is true—that the Agency is a bit stodgy—then why? I think, in part, it is due to the Mission Areas designations; that is, the National Forest System; Research and Development; State and Private Forestry. Fundamentally, I think we may get in our own way. This has prompted me to conclude that "the Mission Area designations may prove to be the demise of the Forest Service ability to effectively address contemporary conservation issues of the 21st Century". I contend that the Mission Areas block our way of being Corporate; being one cohesive organization with a common purpose. In my view, the designations of Mission Area perpetuate "turf guarding". At one time, the designations were helpful. Now, I am not so sure.

For example, the Forest Service is now establishing a network of urban field stations to bring forest stewardship capacity closer to where people live. From the iconic Baltimore Ecosystem Study, to a research work unit just outside of Chicago, to a laboratory at Ft. Totten (Queens, New York City), to the new Philadelphia Field Station, and other areas including San Juan, Los Angeles, and Seattle, we are bringing science-based information to city governments and other practitioners so they can effectively balance the health and sustainability of their urban forests with community needs. When city leaders, for example, see these field stations in action, they think only (and correctly so) of one Forest Service—not State and Private Forestry; Research and Development; or, the National Forest System.

Yet, at times when we try to be most creative in our deployment of this work, it is not uncommon for someone within the agency to pop up and say, " ... hey, that's our job". Sometimes the stance from a particular Mission Area is so aggressive that assistance is halted altogether, apparently forgetting that our role is ultimately public service.

I have seen this happen from time to time over the years, but much more often in the past decade. Oftentimes, this issue gets embroiled in a "federal role" question. For example, should the core business of the Forest Service be limited to the management of the National Forests? This was discussed aggressively in 1995, and again in a less formal way in 2011, but I suspect the notion is always just below the surface.

There is little doubt that we are a "National Forest System-Centric" organization. Yet, our role is clearly much broader. However, when constraints surface—oftentimes around budgets—it is not atypical (albeit somewhat counterintuitive) for the agency to group itself into the traditional Mission Areas vs. a stronger corporate stance.

There is one clear point, however. At the end of the day, most people know us only as the "Forest Service", if they know us at all. Thus, if we want to be more competitive in addressing the contemporary conservation challenges ahead, like the challenges associated with Urban Natural Resources Stewardship and accelerated forest restoration, for example, we must improve our ability to act as one Forest Service vs. a series of independent, inconsistent units.

Perhaps one more example of how the Mission Area designations may stand in the way of corporate behavior: The Forest Inventory and Analysis (FIA) program. Currently the FIA program is viewed as a Research and Development program for the States. Actually, it is the "forest census" program for America's Chief Forester—the Forest Service Chief. Accordingly, there is perhaps no other program that is more corporate in nature in terms of utility for forest managers than the FIA program. Yet, because of our more narrowly defined view of FIA, this long-term forest census and its ultimate promise is never fully realized.

There are more examples, of course, about corporate behavior, or the lack thereof. The primary point is that in order to more optimally carry out the mission of the Forest Service, a cohesive, comprehensive approach works better than a series of solid, well-intended independent actions.

2.10. A More Optimal Organization

Not too long ago, I was informally asked in my role as Station Director about a more *optimal* organizational structure for the Forest Service in the Northeast and Midwest. I immediately recited my concern about the Mission Area designations and the lack of conservation decision-making flexibility and came up with an option: A "Regional Administrator for Forest and Rangeland Conservation". Candidly, I was thinking of the model from the Environmental Protection Agency that I become familiar with while involved in the Urban Waters Federal Partnership. The organization might look like the following (Figure 1) using the twenty-state area of the northeast and Midwest as the example:

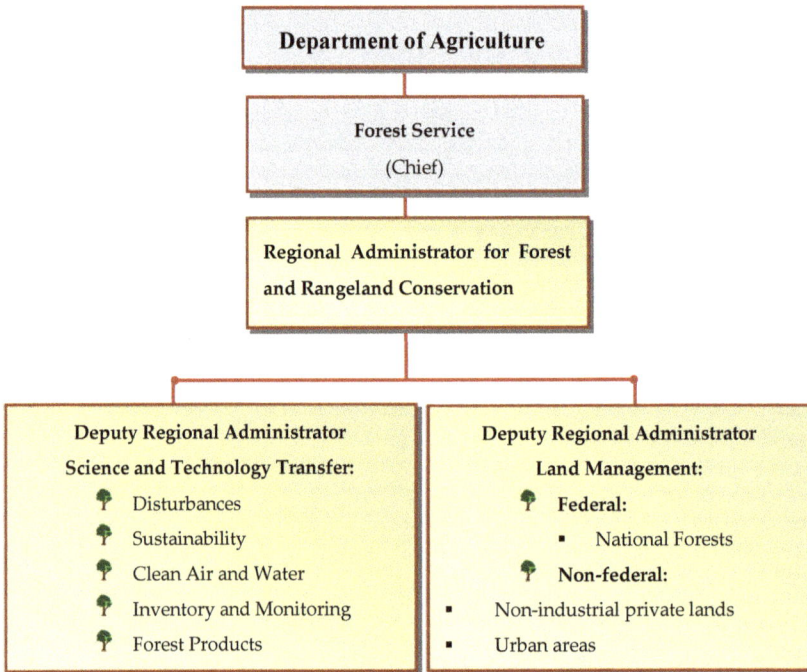

Figure 1. Example Forest Service Field Organization.

The benefits of such an organization seem apparent, exciting, and full of promise. The dominant feature is a more corporate Forest Service that will emerge bolstered by a consistent, powerful voice of one overall leader vs. separate administrative units. Program direction immediately becomes more cohesive, consistent, and comprehensive. You do not need to close down units. However, combining some functions does become easier and the promise of immediate and future savings while sustaining and improving services becomes very real. Leadership development would need to be enhanced and chronic concerns such as those from the Federal Employees Viewpoint Survey would need addressing to make the above configuration most effective. No doubt, the current Forest Service organization, largely unchanged since the early part of the 20th Century, is outdated. In order to be competitive for the next century, adjustments need to be made.

Actually, if we think about it, we may not be too far from this organizational configuration right now. In the west, for example, the National Forest System and State and Private Forestry are under one "Regional Administrator"—the Regional Forester. In the northeast and Midwest, all the science activities are under one leader. With some modest adjustments, a "Regional Administrator for Conservation" could be pilot tested to ensure effective program delivery. It seems sensible that the current "Regional Forester" configuration would become the overall Administrator. Perhaps someday, maybe taking advantage of a vacancy, two Regions could be administered by one executive (read, a "Regional Administrator"). Again, you do not have to close offices; there is no need. The promise of success is quite high. Public service should be better. Savings will be real. One, stronger Forest Service will emerge. The promise of this could be fun to envision.

2.11. Adjusting Is Nothing New

Earlier, I stated that during my time with the Forest Service, I have watched and participated with the agency in almost a continual transformational campaign. To be clear, adjusting is nothing new. In the early 1990s, for example, we had "Reinvention". In 2002 we were creating "efficiency plans" within the notion of "workforce restructuring". Over the last two years or so the Forest Service has been involved in "Cultural Transformation". My key point is that the Forest Service is always adjusting. That is good. Now, however, I think we may need something much more transformative to enable us to go from "good to great". We have a population makeup like never before and their demands for services are equally diverse. We have a changing climate and forest species are changing and moving. While our mission can be the same, the way we address the mission needs to be transformative. Perhaps the "Regional Administrator" might be one of those "transformative" actions. Emphasizing Urban Natural Resources Management could be another. Reinventing our approach to working with people who "decide" (vs. people who "play") could be another. Let me explain the latter.

2.12. Working More with Those That Decide

Early in my Forest Service career, I took a class from Dr. Paul Hersey, who along with Dr. Ken Blanchard developed the theory of "situational leadership" [4]. During that time, Dr. Hersey talked about the effectiveness of working more with people who decide your fate as opposed to those who do not. He used the phrase " . . . people who 'decide' vs. people who 'play'". It is my opinion, especially during more recent times, that the Forest Service has tended to work more with those that "play". Why? Because it is easier; more comfortable. I have to be careful here, because people can become easily offended by the phrase and words.

The word "decide" tends to be linked with "influence" or "leadership". The word "play" can be linked with "follow" or "deploy". We all like to be with people that are like us, and we tend to "talk to each other"—we play. More difficult work comes with influencing and working with critics who can make or break us with their support, or lack thereof. Constantly working or being associated with those that do not "decide" still takes time and the gains are only marginal. People who "decide" can also be your friends. So, this is not a proposition of ignoring your friends (of course, not); this is a proposition about balance—maximizing your time wisely by making real, significant differences.

To be more effective in the 21st Century, the Forest Service needs to embrace the notion of working more with people who can shape its future.

2.13. Our Core Beliefs and Values

Within the overall framework of "Cultural Transformation", the Forest Service embarked on several "Field Leadership Forums" as we develop a new, more robust "Community of Leaders". Part of this effort is attempting to address "mission clarity", albeit I personally believe our mission is quite clear. Our problem seems to be the inability—as one leader so aptly said—" . . . to hit the refresh button". This is very consistent with my beliefs. We know what to do, but sometimes our tentative or cautious behavior makes us appear to be stodgy.

Perhaps we need to keep our set of core beliefs and values in front of all the employees so we can remain more contemporary. For example, an expression of our relationship with the land, communities we serve, and the people we employ is illustrated below in Figure 2.

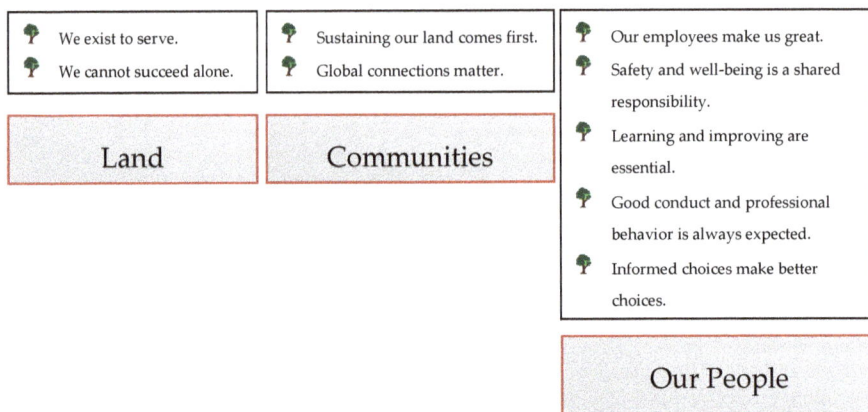

Figure 2. Forest Service Core Beliefs and Values.

2.14. The Chief's Prerogative

Even with core values and beliefs and a clear mission statement, the horizon can become hazy when a Chief exercises his or her prerogative to augment the current program direction to achieve the basic mission. To be clear, the Chief's right to place his or her imprimatur on the basic agency mission is altogether fitting. We simply need to recognize that the mission is not changing. What is changing is the work focus that the Chief deems appropriate to help advance the mission. Simply put: the Chief's prerogative does not change the mission.

For example, the 15th Chief, Dale Bosworth, declared that the Forest Service under his leadership should concentrate on areas that significantly threaten achieving long-term outcomes. Chief Bosworth termed these the "four threats".

1. Uncontrolled fires.
2. Destructive invasive species.
3. Irresponsible use of the National Forests (with an immediate emphasis of controlling the use of off-highway vehicles).
4. Loss of open space.

Chief Tom Tidwell targeted "five focus points for the future".

1. Enhanced safety.
2. Creating a culture of inclusion.
3. Forest restoration.
4. Fire management.
5. Community engagement.

Additionally, some may recall the "natural resources agenda" from our 14th Chief, Mike Dombeck and his focus:

1. Watershed Health and Restoration.
2. Development of a Long-Term Forest Roads Policy.

3. Sustainable Forest Management.
4. Recreation.

The key point is that our mission is timeless and all the Forest Service Chiefs strive to target work that they believe will enable the agency to stay more contemporary; sometimes they succeed, sometimes they do not.

2.15. A Common Thread

As we look ahead, perhaps it is prudent to look back. The three Chiefs of the Forest Service mentioned above all mentioned forest restoration in their prerogatives. Chief Bosworth linked his forest restoration concern primarily to "uncontrolled fires".

Today, we grow about twice as much wood as we use from America's forests. Our forests are getting over-crowded with hazardous fuels. Hazardous fuels lead to catastrophic fires. We have seen the devastating impacts of these fires again this summer with lives lost, homes destroyed, and millions of acres blackened. The cost of fighting these catastrophic wildfires is enormous—sometimes exceeding $1 million per hour. We approached $3 billion in federal fire suppression costs for the 2013 fire season alone; expenditures in 2014 are projected to be as high, perhaps higher. Finding high value, high-volume economically-viable uses for forest biomass from hazardous fuels reduction and forest restoration activities has been identified by Forest Service land managers as one of the most important barriers that must be overcome, as well as the need for a diverse array of strategies for promoting the use of woody biomass [5].

Wood-based nanotechnology, for example, offers a market-based solution to this wide-spread catastrophic fire loss. Wood-based nanomaterials, when used as an additive to a wide-range of commercial products (car bodies, concrete, laptops, body armor, containers, etc.), makes these products lighter and stronger. It is estimated that a strong, well-established program in wood-based nanotechnology could create high value markets from low value wood (hazardous fuels) that could help reasonably restore 8-11 million forested-acres annually, although a "moderate to high" rate of forest restoration up to 19 million acres annually from all ownerships could be attained, thereby reducing future fire suppression costs by as much as 12–15 percent; informal discussions suggest this figure could be as high as 23 percent. In some measured ways, woody biomass for energy also offers some higher value market opportunities. The initial work in Green Building Construction is another example that could help create a market-based incentive to remove "crummy, rotten wood" into higher value market economic streams.

Enhanced skills in wood-based market expansion and development are needed within the agency to make the "formula for success" (low value wood being processed and sold for higher value products that create new jobs and enhance the economy) a reality.

In 1979, the Forest Service began a major initiative in the "improved utilization of wood". This included advancing some of the items described above, especially "wood for energy" and creating new markets and expanding others. The emphasis did not last.

In the mid-1990s the agency advanced "rural development through forestry"—another way to address the conundrum of reducing low value wood from our forests. Except for a small provision whereby limited funds from hazardous fuels treatment are directed to grants for "woody biomass for energy", this effort has also fizzled.

It seems we cannot make this type of essential work—reducing low value wood from our forests at a pace that makes a real difference to accelerate forest restoration—a "campaign of our campaign". Simply put, we tend to be impatient and perhaps give up too easily. Nevertheless, we need again to make accelerated forest restoration through the reduction of low value wood a "campaign of our campaign" for the future. New efforts from the old "improved utilization of wood" program would be so relevant during the *Anthropocene* and provide strength to a common thread. In simple terms: we need to emphasize a wide-range of "biomass uses".

2.16. Gifford's Maxims

The first "Chief Forester for America's Forests", Gifford Pinchot, continues today to be a mythical figure. Chief Pinchot instinctively understood the role of the Forest Service and landscape scale conservation. He developed maxims on how a forester should behave. Today, these maxims, summarized in my own words below (Figure 3), are basic for all our employees and should be part of our contemporary thinking. As we look ahead, holding onto these types of core values and beliefs will be fundamental to our continued success as we go from "good to great".

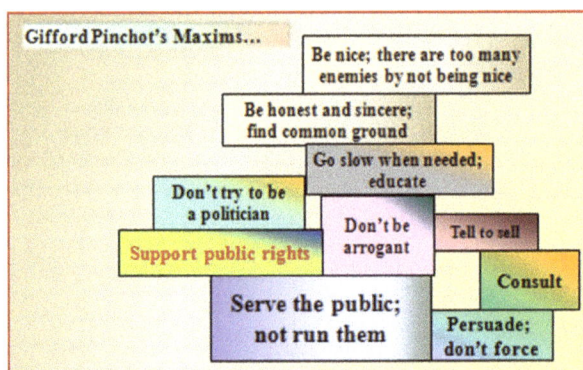

Figure 3. Gifford's Maxims (paraphrased).

2.17. The Next 100 Years

... Look to the vision and follow the mission. We know what to do. We feel pretty good. Sometimes we just need to know how we look. The Chief Forester for America's Forests can be a great mirror.

—Michael T. Rains

It is 2014 [6] at the time of writing this paper, and the Forest Service has existed for 109 years (1905—2014). The agency is recognized as a leader in the world of conservation, but we may not be that recognizable, overall. We have a very stable mission that still can be contemporary if we can be just a little less stodgy in our tactics and think more in terms of connected work along complex rural to urban land gradients. Our footprint is and should continue to be planet Earth. So, what do we need to do during the Anthropocene to be brilliantly competitive in addressing the contemporary conservation challenges now and ahead? I would like to focus on the following:

■ **Stay the course with both our current mission and statement.** We do have a clear, timeless mission and mission statement. We should leave it alone: To sustain the health, diversity, and productivity of the Nation's forests and grasslands to meet the needs of present and future generations. What we do need to do, however, is become less concerned about hitting the "refresh button" to remain contemporary. Part of this "refreshing" is the ability and willingness to more easily adapt to the "Chief's Prerogatives", whereby the Chief places his or her imprimatur on the basic agency mission to augment, not to change, the mission direction. We have a tendency of late to be too cautious, creating a somewhat stodgy appearance, if not an actual reputation. We need to be able to "hit the refresh button" without being so tentative.

■ **Adjust the slogan.** We do love our current slogan of "caring for the land and serving people". However, as the role of urban stewardship continues to grow, I enjoy adding "where they live" to the end of the existing slogan: " ... caring for the land and serving people, *where they live*". I doubt we will make this change because we are such traditionalists, but the adjustment does react to a more contemporary time. It is fun to think about and it would inform a wider range of people about the Forest Service intentions to help improve their lives.

■ **Create a culture of inclusion.** Inclusion is a belief system shared by all within the agency whereby all employees feel welcome and valued, and their contributions are fully utilized to advance the mission of the organization. The agency is working hard to "create a culture of inclusion", and this is just the right thing to do. Advancing our Shared Intention Statement (" … to create a culture of inclusion that awakens and strengthens all people's connections to the land") will be fundamental to our success. An inclusive culture will become a magnet for a more representative workforce that will enhance our abilities within a wide-range of diverse landscapes.

■ **Be more responsive to the complex rural to urban land gradient.** In the early part of our history, we were "rural oriented". When the Forest Service was created in 1905, only 13 cities worldwide had populations of one million people or more. Eighty years later, 230 cities had one million plus populations. In the new millennium, it is projected there will be over 400 cities with a population of one million people and 26 mega-cities with populations of over 10 million. Looking nationally, our population was about 50 percent urban in 1920; today 83 percent of our people live in cities and towns. Simply put, this is the first century in our history that the majority of humans live in urban areas. This fact is particularly significant, where the demand for natural resources and green space is high. If we take better care of what we have across all landscapes, the benefits from our natural resources will extend to everyone across a broad spectrum of physical, social, and economic conditions. This approach embodies the notion of "All Lands, All People" and represents an important venue to create an informed citizenry about natural resources. Accordingly, in order for the Forest Service to be more mainstream, working more on the urban side of the "rural to urban land gradient" in an enhanced, cohesive way will be important.

…Urban trees are the hardest working trees in America.

—Tom Tidwell, Chief, USDA Forest Service

■ **Adjust the Mission Area structure.** A more cohesive, consistent, and comprehensive Forest Service would be better able to meet 21st century challenges. Currently, we may not be too far from achieving this. With some measured adjustments, we could make significant strides in creating an organizational configuration that better enables a "one Forest Service" to emerge. Oftentimes, the Mission Areas block our way of being Corporate; that is, being one cohesive organization with a common purpose. Creating a field organization with a "Regional Administrator for Forest and Rangeland Conservation" would enable program direction to immediately become more cohesive, consistent, and comprehensive—let me call this the 3C Model. You do not need to close down units with the 3C Model. However, combining appropriate activities does become much easier and the promise of immediate and future savings while sustaining and improving services becomes very real.

■ **Create resilient forests through restorative actions.** Recent Chiefs of the Forest Service have all mentioned forest restoration in their prerogatives. With the current rate of growth and impacts such as a changing climate, our forests along the rural to urban land gradient are getting distressed and less healthy. This creates conditions prone to disturbances like catastrophic wildfires. By creating high value, high-volume uses, we can create cost-effective ways to enable enough hazardous fuels to be removed from America's forests so that wildfires remain smaller and begin again to be a tool for improved forest health as opposed to destructive behemoths that destroy lives, communities, and landscapes. Restoring fire to the landscape is essential. We need to make creating more resilient forests through a wide range of restorative actions—like the reduction of low value wood—a "campaign of our campaign" for the future. To make the "formula for success" (low value wood being processed and sold for higher value products that create new jobs and enhance the economy) a reality will require enhanced skills in wood-based market expansion and development and targeted resources in science-based technology development and transfer. Of course, restorative actions include more than improving the condition of fire prone areas. We have to view our work in terms of the entire ecosystem so entire landscapes become more productive and resilient to disturbances so that the linkage between environmental health and community stability is assured.

> ... Through restorative actions, we will help create sustainable, productive, and resilient forests so the linkage between environmental health and community stability can be more fully realized.
>
> —Michael T. Rains

■ *Influence* **more, play less.** To be more effective in the 21st Century, the Forest Service needs to work more with those who can shape our future; " ... people who 'decide' vs. people who 'play'". It is my opinion, especially during more recent times, that the Forest Service has tended to work more with those that "play". This is a proposition about balance—maximizing time wisely by making real, significant differences.

Conflicts of Interest: The author declares no conflict of interest.

Appendix A. The Mission, Vision, and Guiding Principles of the USDA Forest Service

The Mission Statement. " ... To sustain the health, diversity, and productivity of the Nation's forests and grasslands to meet the needs of present and future generations."

The Mission Slogan. " ... Caring for the land and serving people."

The Mission Includes. The mission of the USDA Forest Service includes:

- Advocating a conservation ethic.
- Listening to people and being responsive.
- Embracing the multiple-use concept.
- Assisting states to help them in the stewardship of non-federal forestlands.
- Assisting cities and towns to improve their natural resources.
- Providing international assistance and technical exchanges.
- Strengthening local economic conditions.
- Developing and using good science.
- Helping those in need.

Vision. The USDA Forest Service will strive to be:

- Recognized worldwide as a conservation leader.
- Multicultural and diverse.
- Efficient and productive.

Guiding Principles. To realize its mission and vision, the USDA Forest Service is guided by the following principles:

1. Use ecological approaches to land stewardship.
2. Use the best science available in helping make decisions.
3. Be good neighbors; respect private property rights.
4. Strive for quality and excellence, always.
5. Build partnerships.
6. Collaborate.
7. Build trust and share.
8. Value a representative organization.

9. Maintain high professional and ethical standards.
10. Be responsible and accountable.
11. Accept conflict; deal with it professionally.

References

1. Williams, G.W. *The USDA Forest Service—The First Century*; USDA Forest Service: Washington, DC, USA, 2005.
2. The Departments of Agriculture and Interior. *Managing the Impacts of Wildfires on Communities and the Environment* (known as the *National Fire Plan*). Unpublished work. 2001.
3. Pinchot, G. *Breaking New Ground*; Island Press: Washington, DC, USA, 1947.
4. Hersey, P.; Blanchard, K.H. *Management of Organizational Behavior—Utilizing Human Resources*; Pearson Prentice Hall: Upper Saddle River, NJ, USA, 1969.
5. Sundstrom, S.; Nielsen-Pincus, M.; Moseley, C.; McCaffery, S. Woody Biomass Use Trends, Barriers, and Strategies: Perspectives of US Forest Service Mangers. *J. For.* **2012**, *110*, 16–24. [CrossRef]
6. Rains, M. A Forest Service Vision during the Anthropocene. In Proceedings of the Institute's 75th Anniversary Symposium, the Inés María Mendoza Park, San Juan, Puerto Rico, 21 May 2014.

![forests logo] *forests*

MDPI

Editorial

Concluding Remarks: Moving Forward on Scientific Knowledge and Management Approaches to Tropical Forests in the Anthropocene Epoch

Grizelle González * and Ariel E. Lugo

United States Department of Agriculture, Forest Service, International Institute of Tropical Forestry, Jardín Botánico Sur, 1201 Ceiba St.-Río Piedras, PR 00926, USA
* Correspondence: grizelle.gonzalez@usda.gov; Tel.: +1-787-764-7800

Received: 2 July 2019; Accepted: 8 July 2019; Published: 10 July 2019

Abstract: The United States Department of Agriculture Forest Service International Institute of Tropical Forestry (the Institute) celebrates its 75th Anniversary with the publication of this Special Issue of *Forests*. This Issue is based on presentations delivered in a symposium held in San Juan, Puerto Rico in 2014. It augments a quarter century of scientific knowledge and capitalizes on a unique set of synergies chartered by a strategy based on shared stewardship, innovative transdisciplinary collaborations, and breakthroughs in science and technology. The manuscripts contained here present advancements in our approach to the development of policies for effective governance and stewardship, long-term focus for the understanding of ecosystem processes and functions, novelties given attention to cross-boundary collaborative approaches to science, and proposed alternative institutional visions in the Anthropocene. As the Institute continues to collaboratively explore new frontiers in science, we recognize advances in forestry, atmospheric sciences, modeling, hydrology, plant physiology, and microbial ecology as core to the understanding of tropical forests in the Anthropocene.

Keywords: conservation; American tropics; long-term ecological research; tropical forest management; Anthropocene; Puerto Rico

1. Scientific Knowledge and Management Approaches to Tropical Forests in the Anthropocene

The United States Department of Agriculture (USDA) Forest Service International Institute of Tropical Forestry (the Institute) has a long history of research. At its inception during the 1930s and 1940s, the Institute completed important silvicultural studies with rigorous controls to provide the basis for tropical forest production. In the 1950s, the ecology of natural, unmanaged forests was added to the research portfolio, while the 1960s marked a period of focus on endangered species being included in the research and development program. Biomass and climate change research began in the 1980s, and watershed and biogeochemical studies were in full swing by 1990. Landscape ecologists using remote sensing techniques and the study of the biology and ecology of soils were added components to the research unit in the 2000s. By the 2010s, the Institute had maintained almost all original lines of research [1], yet had expanded studies to encompass human and ecological systems in an effort of continued application of our research to science, society, and management. In the years ahead, we foresee that the Institute will continue providing society with long-term context, information synthesis, theory development, and deep knowledge of place while we continue working in multidisciplinary and collaborative teams with a focus on the American tropics and Caribbean region, much like the goals established for the United States by the National Science Foundation Long Term Ecological Research Network Program (LTER).

Looking back, one of the main goals of the manuscripts published in the volume that commemorated the 50th anniversary of the Institute [2] was to show the relevance of research to tropical forest management and propose that tropical forests could recover after human intervention, provided they were given the opportunity [3]. In this special issue, we augment a quarter century of scientific knowledge and are poised to capitalize on a unique set of synergies chartered by a strategy based on shared stewardship, innovative transdisciplinary collaborations, and breakthroughs in science and technology [4]. Moreover, the environment under which tropical forests function has dramatically changed over the past 25 years with the advent of the Anthropocene Epoch—the age of significant human impact on Earth's geology and ecosystems. The manuscripts contained in this special issue present advancements in our approach to the development of policies for effective governance and stewardship [5–8]; long-term focus for the understanding of ecosystem processes and functions [9–13], novelties given attention to cross-boundary collaborative approaches to science [14–17], and proposed alternative institutional visions [18] considering the Anthropocene Epoch.

2. Key Considerations Posed for Future Studies as Identified in This Special Issue Are:

- As human activities increasingly influence systems and processes at multiple scales, society may be more likely to see extraordinary and surprising events, making it difficult to predict the future with the level of precision and accuracy needed for broad-scale management prescriptions [5].
- Collaborative relationships with stakeholders, productive ties to the scientific community, and political support for adaptiveness and flexibility are critical elements in managing for the future resilience and sustainability of tropical forests [5].
- Deforestation in the dry tropics, with its artisanal basis for forest utilization, is likely to produce a more fragmented forest than the industrial-scale deforestation in the humid tropics [6].
- Urban knowledge systems can create and help transition to a sustainable and resilient future [7].
- Large-scale conservation partnerships are teams at their core, meaning that relevant land and sea managers are empowered to be part of a team that work toward a shared vision, understand the stressors on the system based on past observations and future projections, and see the opportunities for increased coordination to advance a conservation agenda [8].
- Organisms in the litter and soil of tropical forests in Puerto Rico independently and synergistically influence the rates of decomposition and availability of nutrients to tree roots [9].
- Novel ecosystems are expected to adapt to Anthropocene conditions and continue to function as carbon sinks in a new world order where the speed of ecological processes is accelerated [10].
- The effects from hurricane disturbances maintain Puerto Rico's forests in a constant state of structural and compositional change in response to the intensity and the cumulative effects of these events [11].
- Conservation efforts of migrant birds are most likely to be effective if based on research that uses a full annual cycle approach for the identification of factors that limit their population growth [12].
- Large-scale manipulative experiments in the Luquillo Experimental Forest have greatly enhanced our understanding of tropical forest function under different disturbance regimes and informed the development of management strategies [13].
- Novel dry forests contribute to the conservation of native plant species on highly degraded lands [14].
- Introduced and native trees can have different resource–investment strategies in tropical novel forests [15].
- Karst vegetation in Puerto Rico appears to be phosphorus limited [16].
- Potential working lands encompass 42% of Puerto Rico, these include lands well suited for mechanized and non-mechanized agriculture as well as for forestry production [17].

- High unemployment rates, issues of food security, and the rising cost of importing agricultural products are issues that are pressing Puerto Rico toward a revitalization of its working lands sector [17].
- The USDA Forest Service mission has foundational guiding principles that include a science-based, ecological approach to stewardship across all lands along a complex rural to urban land gradient [18].

3. Future Research Questions or Directions as Identified in This Special Issue Include:

- How can tropical ecosystems persist in human-modified landscapes, and which management strategies will be most effective at maintaining their structures and functions at different spatial and temporal scales? [5]
- There is an increase in the scientific understanding of tropical forests as complex social-ecological systems; yet variability in the dynamics across systems and related processes are expected to increase in the context of the Anthropocene. Are changes in policy or practice required for dealing with the Anthropocene? [5]
- How can landowners and managers strike the right balance of sustainably managing tropical forests for multiple uses? Furthermore, how do these management considerations, extreme events like fire, drought, or hurricanes interact with climate change and ultimately affect greenhouse gas emissions? [6]
- What are the institutional arrangements and stakeholder engagement processes most useful in the development of knowledge co-production systems to further advance urban sustainability issues? Are there social and institutional conditions that are more conducive to knowledge co-production efforts? [7]
- How can we re-think leadership in collaborative settings and on relational governance, cooperative teamwork procedures, and communications to ensure the long-term success of landscape conservation partnerships? [8]
- How does environmental variation affect the dynamics of different soil microbial and faunal assemblages? How does the variation in the composition of such organismal assemblages control the long-term sustainability and management of ecosystems that are subject to global change? [9]
- How do complexities of land use, cover, and climate change affect the carbon balance of whole tropical landscapes? [10]
- How can the dynamics in forest structure and composition during succession relate to the recovery and resilience of different components of the forest ecosystem after hurricane disturbance? [11]
- How would adaptation to environmental changes, genetic variation, strength, and spatial patterning of selection influence conservation efforts on migrant birds in the Caribbean Basin? [12]
- Will the forests in the Luquillo Experimental Forest continue to persist or will we see a significant shift in its size and composition as the world's climate and disturbance regimes continue to change? What future experiments should be conducted in the Luquillo Experimental Forest if we are to continue providing critical information for the development and refinement of forest management strategies? [13]
- What are the implications of novelty to the ecology, restoration, and conservation efforts of Puerto Rico's dry forests? [14]
- Do introduced and native species in novel forests in Puerto Rico differ in the efficient use of resources? Do they occupy distinct or overlapping positions in the leaf economic spectrum? [15]
- How different are the denitrification rates in karst forests under dry to sub-humid climatic conditions in the Caribbean Basin? [16]

- How can scientific and traditional knowledge, incentive programs, global and local markets, and technology be used to convert planning into productive and sustainable farm and forest activities in Puerto Rico? [17]
- What does the USDA Forest Service need to do during the Anthropocene to be competitive in addressing the contemporary conservation challenges now and ahead? [18]

4. Conclusions

The International Institute of Tropical Forestry exemplifies the United States Department of Agriculture Forest Service's mission of working cross-jurisdictions, building science capacity, and interpretation of the contemporary issues relevant to society and resource conservation. In addition, it has a long track record of doing so in cooperation with partners and stakeholders, while acting as co-conveners and co-facilitators for collaborative learning and decision making. As we continue to collaboratively explore new frontiers in science, we recognize the potential of forestry, atmospheric sciences, modeling, hydrology, plant physiology, and microbial ecology as core to the understanding of tropical forests in the Anthropocene.

Author Contributions: Conceptualization, G.G. and A.E.L.; Writing-Original Draft Preparation, G.G.; Writing-Review & Editing, G.G. and A.E.L.

Funding: This research was supported by the USDA Forest Service and conducted in collaboration with the University of Puerto Rico (UPR). Grants (DEB-0620910, DEB-0218039, DEB-0080538, DEB-9705814, DEB-1239764, DEB-1831952) were from the National Science Foundation to the Institute of Tropical Ecosystem Studies, UPR and the International Institute of Tropical Forestry as part of the Long-Term Ecological Research Program in the Luquillo Experimental Forest. The Luquillo Critical Zone Observatory (EAR-1331841) provided additional support for G. González.

Acknowledgments: We thank W.A. Gould for reviewing an earlier version of the manuscript.

Conflicts of Interest: The authors declare no conflict of interest.

References

1. Lugo, A.E.; González, G. Introduction to the Special Issues on Tropical Forests: Management and Ecology in the Anthropocene. *Forests* **2019**, *10*, 48. [CrossRef]
2. Lugo, A.E.; Lowe, C. *Tropical Forests: Management and Ecology*; Springer: New York, NY, USA, 1995.
3. Lugo, A.E. Tropical Forests: Their future and our future. In *Tropical Forests: Management and Ecology*; Lugo, A.E., Lowe, C., Eds.; Springer: New York, NY, USA, 1995.
4. United States, United States Department of Agriculture, Forest Service. *Toward Shared Stewardship Across Landscapes: An Outcome-Based Investment Strategy*; FS-1118; United States Department of Agriculture: Washington, DC, USA, 2018; 28p.
5. McGinley, K.A. Adapting tropical forest policy and practice in the context of the Anthropocene: Opportunities and challenges for the El Yunque National Forest in Puerto Rico. *Forests* **2017**, *8*, 259. [CrossRef]
6. Rudel, T.K. The dynamics of deforestation in the wet and dry tropics: A comparison with policy implications. *Forests* **2017**, *8*, 108. [CrossRef]
7. Muñoz-Erickson, T.; Miller, C.A.; Miller, T.R. How cities think: Knowledge co-production for urban sustainability and resilience. *Forests* **2017**, *8*, 203. [CrossRef]
8. Jacobs, K.R. Teams at their core: Implementing an "All LANDS approach to conservation" requires focusing on relationships, teamwork process, and communications. *Forests* **2017**, *8*, 246. [CrossRef]
9. González, G.; Lodge, D.J. Soil biology research across latitude, elevation and disturbance gradients: A review of forest studies from Puerto Rico during the past 25 years. *Forests* **2017**, *8*, 178. [CrossRef]
10. Brown, S.; Lugo, A.E. Trailblazing the Carbon Cycle of Tropical Forests from Puerto Rico. *Forests* **2017**, *8*, 101. [CrossRef]
11. Heartsill-Scalley, T. Insights on forest structure and composition from long-term research in the Luquillo Mountains. *Forests* **2017**, *8*, 204. [CrossRef]
12. Wunderle, J.M., Jr.; Arendt, W.J. The plight of migrant birds wintering in the Caribbean: Rainfall effects in the annual cycle. *Forests* **2017**, *8*, 115. [CrossRef]

13. Wood, T.E.; González, G.; Silver, W.L.; Reed, S.C.; Cavaleri, M.A. On the shoulders of giants: Continuing a legacy of large-scale ecosystem manipulation experiments in Puerto Rico. *Forests* **2019**, *10*, 210. [CrossRef]

14. Lugo, A.E.; Erickson, H.E. Novelty and its ecological implications to dry forest functioning and conservation. *Forests* **2017**, *8*, 161. [CrossRef]

15. Fonseca da Silva, J.; Medina, E.; Lugo, A.E. Traits and resource use of co-occuring introduced and native trees in a tropical novel forest. *Forests* **2017**, *8*, 339. [CrossRef]

16. Medina, E.; Cuevas, E.; Lugo, A.E. Substrate chemistry and rainfall regime regulate elemental composition of tree leaves in karst forests. *Forests* **2017**, *8*, 182. [CrossRef]

17. Gould, W.A.; Wadsworth, F.H.; Quiñones, M.; Fain, S.J.; Álvarez, N.L. Land use, conservation, forestry, and agriculture in Puerto Rico. *Forests* **2017**, *8*, 242. [CrossRef]

18. Rains, M.T. A Forest Service vision during the Anthropocene. *Forests* **2017**, *8*, 94. [CrossRef]

MDPI

St. Alban-Anlage 66

4052 Basel

Switzerland

Tel. +41 61 683 77 34

Fax +41 61 302 89 18

www.mdpi.com

Forests Editorial Office

E-mail: forests@mdpi.com

www.mdpi.com/journal/forests

www.ingramcontent.com/pod-product-compliance
Lightning Source LLC
Chambersburg PA
CBHW051730210326
41597CB00032B/5670